Amazon DynamoDB – The Definitive Guide

Explore enterprise-ready, serverless NoSQL with predictable, scalable performance

Aman Dhingra

Mike Mackay

Amazon DynamoDB – The Definitive Guide

Group Product Manager: Apeksha Shetty
Publishing Product Manager: Nilesh Kowadkar
Book Project Manager: Hemangi Lotlikar
Senior Editor: Tazeen Shaikh
Technical Editor: Seemanjay Ameriya
Copy Editor: Safis Editing
Proofreader: Tazeen Shaikh
Indexer: Hemangini Bari
Production Designer: Jyoti Kadam
DevRel Marketing Coordinator: Nivedita Singh

First published: August 2024

Production reference: 1160824

Published by Packt Publishing Ltd.
Grosvenor House
11 St Paul's Square
Birmingham
B3 1RB, UK.

ISBN 978-1-80324-689-5

www.packtpub.com

Every word in this book was written with Mike in mind. I am honored to contribute to Mike's legacy as a co-author. This work is dedicated to Oliver and the entire Mackay family.

Many thanks to my DynamoDB colleagues at AWS—engineering, product, and Worldwide Specialists alike. Some are friends first and colleagues later. I am grateful to those who have contributed to this remarkable technology. My fascination with distributed systems fuels my passion for each new DynamoDB project, and I am thrilled to contribute to its evolution!

Finally, a huge thank you to my family. Balancing the day job with authoring this book left little time for family and friends. I am deeply grateful to Divya, Dad, Mum, Cheets, and Janki for their support.

Aman Dhingra

Contributors

About the author

Aman Dhingra is a Senior DynamoDB specialist solutions architect at AWS, where he assists organizations in leveraging AWS to its fullest potential. He focuses on designing cost-efficient solutions and maximizing the agility and elasticity offered by the cloud and AWS. With a specialization in Amazon DynamoDB and NoSQL databases, Aman is also well-versed in AWS' big data suite of technologies, many of which are tightly integrated with open source projects. Aman is Dublin, Ireland-based.

You can connect with the author on LinkedIn at amdhing.

Mike Mackay was a senior NoSQL specialist solutions architect at AWS, known as one of the early AWS architects in Europe specializing in NoSQL technologies. Before AWS, he was technical director at Digital Annexe / Your Favourite Story and head of development for Outside Line.

Mike's freelance work includes collaborations with Warner Brothers (UK), Heineken Music, and Alicia Keys. He has also written PHP tutorials for .net magazine and Linux Format. His projects with Outside Line featured prominent clients such as The Hoosiers, Oasis, and Paul McCartney.

Outside of work, Mike loved to chat about F1 and indulge in DJ. Mike was London/Essex-based.

Check out Mike's LinkedIn at mikemackay82.

About the reviewer

Saumil Hukerikar is a distinguished software engineer with over 12 years of experience. Holding a Master's degree in computer science, Saumil specializes in backend distributed systems and database technologies, with a particular interest in NoSQL databases. As a former DynamoDB engineer, he played a pivotal role in developing its Global Admission Control system. This expertise ensures his reviews are both insightful and analytical. Beyond his professional achievements, Saumil is dedicated to giving back to the tech community. He actively participates in book reviews, serves as a judge in local and international competitions, and mentors the next generation of software engineers, sharing his knowledge and passion for technology.

Brandon Tarr is a seasoned IT professional with a diverse background encompassing infrastructure, software development, cloud technologies, and multi-SaaS product solutions. Currently, he specializes in architecting collaborative and SaaS solutions, with a deep focus on platforms like Microsoft 365, Atlassian, and Miro, and a particular emphasis on Slack Enterprise Grid. Brandon excels in translating business needs into technical solutions, driving digital transformation initiatives, and crafting strategic roadmaps. He is passionate about fostering collaboration and innovation within organizations.

I'd like to thank my wife, best friend, and love of my life, Samantha. Also, my twin girls, who mean the world to me. Lastly, I'm grateful to everyone who has given me the opportunity to work with them. It's always an honor.

Table of Contents

Part 2: Core Data Modeling

4

5

6

Read Consistency, Operations, and Transactions   107

7

Vertical Partitioning   127

8

Secondary Indexes   143

Part 3: Table Management and Internal Architecture

9

Capacity Modes and Table Classes   171

10

Request Routers, Storage Nodes, and Other Core Components   193

Part 4: Advanced Data Management and Caching

11

12

13

Global Tables 269

14

DynamoDB Accelerator (DAX) and Caching with DynamoDB 291

Part 5: Analytical Use Cases and Migrations

15

16

Preface

Welcome to *Amazon DynamoDB – The Definitive Guide*. This book was crafted by Mike Mackay and Aman Dhingra, specialist solutions architects at AWS who have closely worked with DynamoDB, witnessing its evolution and understanding its transformative impact on modern applications. Our aim is to equip you with the skills and knowledge to harness the full potential of DynamoDB.

In these pages, you'll find comprehensive guidance to leverage DynamoDB's capabilities for predictable performance at scale. We will help you build robust, highly available, and fault-tolerant systems with minimal ongoing management. Through a deep dive into NoSQL concepts, design patterns, DynamoDB features, and analytical implementations, you'll gain the expertise to create high-performing applications. Additionally, we will guide you through designing and executing migrations to DynamoDB, ensuring a smooth transition and optimal fit for your use case.

The AWS developer guide on DynamoDB is a comprehensive resource filled with intentionally non-opinionated information. In this book, we do not attempt to rewrite that material. Instead, we use the official documentation to complement the patterns and solutions presented, drawing from our extensive experience working with DynamoDB users.

Our passion for DynamoDB drives us to share this knowledge with you. By the end of this book, you will be well-versed in building scalable, performant systems backed by DynamoDB. Whether you're new to NoSQL or looking to deepen your understanding, we hope this guide will inspire and empower you to unlock new possibilities with DynamoDB.

Who this book is for

This book targets software architects designing scalable systems, developers optimizing performance and leveraging DynamoDB features, and engineering managers guiding teams in decision-making. Data engineers will learn to integrate DynamoDB into workflows and manage large-scale operations, while product owners will understand DynamoDB's capabilities for innovation. Traditional database administrators will find insights on transitioning to NoSQL and how DynamoDB complements RDBMS solutions.

A basic understanding of software engineering, Python, and cloud computing will be helpful. Hands-on AWS or DynamoDB experience is beneficial but not required

What this book covers

Chapter 1, Amazon DynamoDB in Action, explores DynamoDB's market position, evolution, real-world use cases, and suitability for various workloads. It prepares readers with the background and tools needed to start using DynamoDB effectively.

Chapter 2, The AWS Management Console and SDKs, builds on *Chapter 1* by introducing working with DynamoDB via the AWS Management Console, exploring layout and options without programming. It covers AWS SDKs for API interaction, running code with AWS Lambda, and using DynamoDB Local for offline development.

Chapter 3, NoSQL Workbench for DynamoDB, covers the cross-platform tool for designing and managing NoSQL databases. It includes setup, navigating the interface, and modifying data models, while reserving data modeling concepts for future chapters.

Chapter 4, Simple Key-Value, is the first chapter focusing on data modeling. It introduces the key-value concept in DynamoDB, detailing how data is stored and retrieved efficiently. The chapter covers key-value mechanics, DynamoDB data types, and limitations, laying the groundwork for advanced NoSQL design.

Chapter 5, Moving from a Relational Mindset, bridges DynamoDB data modeling with concepts from **Relational Database Management Systems (RDBMS)**. It contrasts RDBMS normalization with NoSQL denormalization, emphasizing strategies such as duplication for optimized compute and efficient data retrieval. The chapter explores when to use single versus multiple DynamoDB tables and discusses partitioning strategies and handling **Large Objects (LOBs)** in NoSQL.

Chapter 6, Read Consistency, Operations, and Transactions, explores DynamoDB's read consistency models, including eventually and strongly consistent reads. It covers read operations, details ACID-compliant transactions, and provides guidance on balancing performance with cost.

Chapter 7, Vertical Partitioning, explores this advanced data modeling technique through practical examples. It covers constructing read and write operations, understanding item collections, and evaluating the benefits and limitations of vertical partitioning, helping you implement it effectively for specific access patterns. This is likely the most interesting data modeling chapter in this book!

Chapter 8, Secondary Indexes, explores how to use secondary indexes in DynamoDB to optimize data retrieval and organization. It covers creating and modeling local and global secondary indexes, handling diverse access patterns, and best practices for cost-effective indexing.

Chapter 9, Capacity Modes and Table Classes, covers DynamoDB's two capacity modes: provisioned and on-demand. Learn to calculate and provision capacity, use auto-scaling, optimize costs with capacity reservations, and select the right table class for your needs in this chapter.

Chapter 10, Request Routers, Storage Nodes, and Other Core Components, explores DynamoDB's fundamental architecture. Understand how core components such as request routers and storage nodes work together for efficient, reliable database performance and secure data storage in this chapter.

Chapter 11, Backup, Restore, and More, covers DynamoDB's backup and restore features, including on-demand backups and **Point-in-Time Recovery** (**PITR**). It also explores data export/import options to and from S3, along with best practices for optimizing and securing backups.

Chapter 12, Streams and TTL, explores **Change Data Capture** (**CDC**) methods with DynamoDB Streams and Kinesis Data Streams, detailing stream records, consumers such as AWS Lambda and KCL, and cost implications. It also covers DynamoDB **Time to Live** (**TTL**) for managing DynamoDB data storage automatically.

Chapter 13, Global Tables, explores the features and benefits of DynamoDB global tables, including data replication, conflict resolution, and use cases for building highly available, fault-tolerant applications. It provides guidance on operating, managing, and troubleshooting global tables effectively.

Chapter 14, DynamoDB Accelerator (DAX) and Caching with DynamoDB, covers caching strategies and introduces **DynamoDB Accelerator** (**DAX**), which enhances performance by up to 10x. It includes setup, benefits, challenges, and comparisons with other caching alternatives.

Chapter 15, Enhanced Analytical Patterns, explores proven analytical patterns for DynamoDB data, including bulk processing and optimizations. It covers the need for analytics, best practices, and integrating services to build an effective data strategy.

Chapter 16, Migrations, provides guidance on transitioning to DynamoDB, covering signs that indicate migration needs, various migration strategies, and common challenges. Learn to identify when migration is needed and how to execute it effectively with practical advice and real-world examples.

To get the most out of this book

We recommend reading *AWS DynamoDB Developer Guide* alongside this book to get a comprehensive view of the specifics. Given that technology evolves, some limits or feature supports may have been updated since the book was published. Note also that console screenshots may differ from the latest version, but features and functionalities remain consistent. For the latest details, always refer to the AWS documentation. This book complements the documentation by offering insights and best practices drawn from more than a decade of combined DynamoDB and NoSQL experience.

While trying out the code examples can enrich your understanding, it is not mandatory. Note that using AWS features may incur costs, typically between $2 and $5 if resources are utilized and deleted within the same hour, with costs potentially increasing if resources remain active longer.

If you are using the digital version of this book, we advise you to type the code yourself or access the code from the book's GitHub repository (a link is available in the next section). Doing so will help you avoid any potential errors related to the copying and pasting of code.

Download the example code files

You can download the example code files for this book from GitHub at https://github.com/PacktPublishing/Amazon-DynamoDB---The-Definitive-Guide. If there's an update to the code, it will be updated in the GitHub repository.

We also have other code bundles from our rich catalog of books and videos available at https://github.com/PacktPublishing/. Check them out!

Conventions used

There are a number of text conventions used throughout this book.

Code in text: Indicates code words in text, database table names, folder names, filenames, file extensions, pathnames, dummy URLs, user input, and Twitter handles. Here is an example: "You must store binary data encoded in base64 when sending it to DynamoDB."

A block of code is set as follows:

```
import boto3

dynamodb = boto3.resource('dynamodb', region_name='eu-west-2')
table = dynamodb.Table('DefinitiveGuide01')
response = table.scan()
items = response['Items']


print(items)
```

Any command-line input or output is written as follows:

```
aws dynamodb get-item \
--table-name Employee \
--key '{"LoginAlias": {"S": "janed"}}' \
--endpoint https://dynamodb.eu-west-2.amazonaws.com \
--region eu-west-2
```

Bold: Indicates a new term, an important word, or words that you see onscreen. For instance, words in menus or dialog boxes appear in **bold**. Here is an example: "As you click on **Create replica**, you will see a colored balloon at the top of the AWS console that acknowledges that your request to create a replica has been received and work has begun by the DynamoDB service to copy the table data, followed by setting up online replication both ways."

> **Tips or important notes**
> Appear like this.

Get in touch

Feedback from our readers is always welcome.

General feedback: If you have questions about any aspect of this book, email us at customercare@packtpub.com and mention the book title in the subject of your message.

Errata: Although we have taken every care to ensure the accuracy of our content, mistakes do happen. If you have found a mistake in this book, we would be grateful if you would report this to us. Please visit www.packtpub.com/support/errata and fill in the form.

Piracy: If you come across any illegal copies of our works in any form on the internet, we would be grateful if you would provide us with the location address or website name. Please contact us at copyright@packtpub.com with a link to the material.

If you are interested in becoming an author: If there is a topic that you have expertise in and you are interested in either writing or contributing to a book, please visit authors.packtpub.com.

Share Your Thoughts

Once you've read *Amazon DynamoDB – The Definitive Guide*, we'd love to hear your thoughts! Scan the QR code below to go straight to the Amazon review page for this book and share your feedback.

https://packt.link/r/1-803-24689-8

Your review is important to us and the tech community and will help us make sure we're delivering excellent quality content.

Download a free PDF copy of this book

Thanks for purchasing this book!

Do you like to read on the go but are unable to carry your print books everywhere?

Is your eBook purchase not compatible with the device of your choice?

Don't worry, now with every Packt book you get a DRM-free PDF version of that book at no cost.

Read anywhere, any place, on any device. Search, copy, and paste code from your favorite technical books directly into your application.

The perks don't stop there, you can get exclusive access to discounts, newsletters, and great free content in your inbox daily

Follow these simple steps to get the benefits:

1. Scan the QR code or visit the link below

```
https://packt.link/free-ebook/978-1-80324-689-5
```

2. Submit your proof of purchase
3. That's it! We'll send your free PDF and other benefits to your email directly

Part 1: Introduction and Setup

This part introduces Amazon DynamoDB, highlighting its core features and real-world applications. It guides you through the AWS Management Console and SDKs, which are essential tools for interacting with DynamoDB. Additionally, the NoSQL Workbench is introduced, offering a user-friendly interface for modeling, visualizing, and querying data. These chapters lay a solid foundation for understanding and using DynamoDB effectively.

Part 1 has the following chapters:

- *Chapter 1, Amazon DynamoDB in Action*
- *Chapter 2, The AWS Management Console and SDKs*
- *Chapter 3, NoSQL Workbench for DynamoDB*

1

Amazon DynamoDB in Action

Amazon DynamoDB is a fully-managed NoSQL database service from **Amazon Web Services** (**AWS**). It offers single-digit millisecond performance and virtually unlimited data storage, and it can easily be tuned to support any throughput required for your workload, at any scale. Since its launch in January 2012, DynamoDB has come a remarkably long way and consistently evolves to support additional features, enabling more businesses and enterprises to leverage the scale of this NoSQL offering.

Welcome to the first chapter of *Amazon DynamoDB – The Definitive Guide*, where Mike and I, Aman, dive into all things NoSQL, focusing on DynamoDB. We'll explore its backstory, how it can supercharge your applications, and how it frees you from the hassles of managing clusters and instances. Learning NoSQL and DynamoDB isn't too tough, but our aim with this book is to share everything we've learned to make your journey smoother. That's why we've titled it *The Definitive Guide* – it's your go-to resource for mastering DynamoDB.

The book is for those of you who are in roles such as software architects, developers, engineering managers, data engineers, product owners, or traditional database administrators. The book itself will provide you with all the skills and knowledge needed to make the most of DynamoDB for your application. This means benefiting from predictable performance at scale, while also being highly available, fault-tolerant, and requiring minimal ongoing management. I could continue listing the service's benefits and why you should choose it for your applications, but it might be more valuable if you read through examples and success stories of other customers later in this chapter. Before we get started, let's learn about Mike's relationship with DynamoDB, in his own words.

My (Mike) journey into the world of DynamoDB has been a tale of two paths. The first path was to overall enlightenment, which was not as simple as I thought it might be – and that is what in part has led to this book. I wanted to share my knowledge, the way I approached data modeling, and how I had to unlearn some core concepts from the relational database world that then gave me that *lightbulb* moment when working with DynamoDB.

The second path was knowing where DynamoDB fits in my toolset, how to take advantage of the astonishing number of features that come baked into the service, and lastly, knowing when *not* to try and make workloads fit when there are more practical technologies out there. While DynamoDB can work for the majority of your database workloads, we need to be realistic and understand what it is not suited for, and that is where being armed with the right knowledge lets us make better, informed decisions about our database choices.

In this chapter, we will dive into DynamoDB's position in today's database market and trace its evolution in time. Subsequently, we will explore real-world use cases showcasing the impressive scale and volume regularly achieved by DynamoDB, demonstrating its battle-hardened capabilities beyond routine customer traffic.

We will then evaluate which workloads DynamoDB is suited for and how it can easily become one of the most used tools in your set of database technologies once you qualify DynamoDB as the right fit for the use case. There are many additional database technologies that do their own job extremely well, so knowing when to offload duties to the right technology is key in our decision making.

By the end of this chapter, you will have enough background knowledge and plenty of excitement to start setting up a working environment and looking at some useful tools we can leverage in preparation for getting hands-on with DynamoDB.

In this chapter, we will cover the following topics:

- NoSQL and DynamoDB in the current database market
- DynamoDB case studies
- Workloads not suited for DynamoDB

Reviewing the role of NoSQL and DynamoDB in the current database market

Let's take a look at how DynamoDB came to be – where it sits in the current NoSQL market, and how NoSQL compares to relational databases. Here, we cover a couple of core differences, both functional and technical, and learn of the very beginnings of the DynamoDB service.

My IT career started more than 20 years ago, in the early 2000s, when relational databases were the default choice of any application that required a data persistence layer. I had built up a fairly solid knowledge of leveraging MySQL in my builds and applications, as was the de facto for many similar developers and open source enthusiasts at the time. Like many before (and after) me, I had been at the mercy of maxing out CPU and connection limits of single-instance database servers, having headaches over performance, trying to manage uptime and reliability, and, of course, ensuring everything stays as secure as possible.

I remember sitting in the keynote session at the AWS London Summit at ExCeL, an exhibition center. An AWS customer was on stage talking about their use of DynamoDB. They were in the ad-space sector and bidding on real-time advertising impressions. Speed was crucial to their success (as is the case for many other industry verticals), and their database technology of choice was DynamoDB. I recall a slide where they discussed that their peak **request per second (RPS)** rate was frequently hitting 1 million RPS and that DynamoDB was easily handling this (and more when having to burst above that rate). I sat there thinking, "*If there's a database technology that lets me do 1M RPS without breaking a sweat or spending hours (if not days) having to configure it for that level, I want to be a part of it.*" It was from there that I became engrossed in DynamoDB, and quite frankly, I have not looked back since.

Comparing RDBMS and NoSQL – a high-level overview

Over the past few years, NoSQL database technologies have gained more and more in popularity and the increasing ease of use and low cost have attributed to this rise. With that, the **relational database management system** (**RDBMS**) is certainly not going anywhere anytime soon. There is, and always will be, a market for both NoSQL and RDBMSs. A key differentiator between the two is that NoSQL tables are typically designed for specific, well-known, and defined access patterns and **online transaction processing** (**OLTP**) traffic, whereas an RDBMS allows for ad-hoc queries and/or **online analytical processing** (**OLAP**) traffic.

Let's take a moment to step back. Relational databases have a rich history spanning many decades, excelling in their functionality. Whether open source or commercial, RDBMSs are widely used globally, with the choice of vendor often hinging on the application's complexity. While there are numerous distinctions between relational database offerings and NoSQL alternatives, we'll focus on one aspect here. In relational database design, a fundamental principle is normalization—a practice that involves organizing data and establishing relationships between tables to enhance data flexibility, ensure protection, and eliminate redundant data. This normalization is expressed in various normal forms (e.g., **First Normal Form** (**1NF**) to **Fifth Normal Form** (**5NF**)), typically achieving **Third Normal Form** (**3NF**) in most cases. In contrast, NoSQL embraces data duplication to facilitate storage and presentation in a denormalized and read-optimized fashion.

So why is normalization so crucial to relational databases? If we look back over the course of the relational history, we can note that storage cost and disk size/availability were significant limiting factors when using a relational database technology. Reducing the duplication of data in the database results in more storage available on the attached disk. More work was offloaded to the CPU to join data together via relationships and help with the overall integrity of the data. With today's technological advances, storage costs are fundamentally cheaper, and storage space is almost limitless so we can take advantage of these changes to help optimize the load on the CPU (which is attributed to the most expensive part of the database layer) to retrieve our data. While these advances haven't predominantly changed the way relational databases work, this is a core value proposition of most NoSQL technologies.

Additionally, one of the primary issues that NoSQL database technologies aim to address is that of extreme scale. What does **extreme scale** mean in this case? Over time and with the evolution of cloud services and the global nature of the internet, traffic to applications continues to grow exponentially, as do the requests to any underlying databases. While maintaining capacity and availability at this scale, databases need to remain highly performant and deliver results in ever-demanding low-latency environments – milliseconds and microseconds are considered normal, with the latter being often sought after. This can be achieved by **horizontally scaling** the data (and therefore the requests) across multiple storage devices, or nodes—a task that some relational databases can do but aren't necessarily performant at, at scale. To scale a relational database, you would typically **scale vertically** by upgrading the hardware on the server itself, although this has its limits.

Including high-level differences between RDBMS and NoSQL databases, the following table can be formulated:

Characteristic	RDBMS	NoSQL
Data Organization	Encourage normalization	Encourage de-normalization
Scalability	Typically scale vertically; has limitations	Typically scale horizontally; can be extremely scalable
Schema Support	Strict schema, enforced on read	Typically schema-less, flexible
Integrity Constraints	Enforced through relationships and constraints	Depends on the database; not enforced for DynamoDB
Storage/Compute Optimized	Optimized for storage	Optimized for compute
Read Consistency Support	Strong consistency by design; eventual consistency on read replicas	Depends on the database; strong and eventual supported by DynamoDB
Atomicity, Consistency, Isolation, and Durability (ACID)	Most RDBMSs are designed for ACID	Depends on the database; limited ACID transactions supported for DynamoDB
Query Language	Typically SQL	Depends on the database; proprietary for DynamoDB
Examples	MySQL, Oracle, and Postgres	DynamoDB, MongoDB, Neo4j, and Redis

Table 1.1 – RDBMS and NoSQL high-level comparison

It is important to realize that NoSQL is a bigger umbrella of different database technologies that differ from each other in terms of the data structures they support, their query languages, and the nature of the stored data itself. NoSQL broadly covers the following categories:

- Key-value
- Document
- Wide column
- Graph

DynamoDB is of the key-value kind, as we have established already, but other categories may support vastly different data models and have their own query languages.

Without digging into all of those, let us next understand how DynamoDB came about.

How and where does DynamoDB fit into all of this?

DynamoDB was simply born out of frustration and a need to solve problems that were directly impacting Amazon's customers. Toward the end of 2004, the Amazon.com retail platform suffered from several outages, attributed to database scalability, availability, and performance challenges with their relational database (1).

A group of databases and distributed systems experts at Amazon wanted something better. A database that could support their business needs for the long term that could scale as they needed, perform consistently, and offer the high availability that the retail platform demanded. A review of their current database usage revealed that ~70% of the operations they performed were of the key-value kind (2), and they were often not using the relational capabilities offered by their current database. Each of these operations only dealt with a single row at any given time.

Moreover, about 20% of the operations returned multiple rows but all from the same table (2). Not utilizing core querying capabilities of relational database systems such as expensive runtime JOINs in the early 2000s was a testament to the engineering efforts at Amazon. Armed with this insight, a team assembled internally and designed a horizontally scalable distributed database that banked on the simple key-value access requirements of typically smaller-sized rows (less than 1 MB), known back then as Dynamo.

The early internal results of Dynamo were extremely promising. Over a year of running and operating their platform backed by this new database system, they were able to handle huge peaks including holiday seasons much more effortlessly. This included a shopping cart service that served tens of millions of requests and resulted in about three million checkouts in a single day. This also involved managing the session state of hundreds of thousands of concurrent active users shopping for themselves and their nearest and dearest ones. Most of this was using the simple key-value access pattern.

This success led the team to go on to write Amazon's Dynamo whitepaper (3), which was shared at the 2007 ACM **Symposium on Operating Systems Principles (SOSP)** conference so industry colleagues and peers could benefit from it. This paper, in turn, helped spark the category and development of distributed database technologies, commonly known as NoSQL. It was so influential that it inspired several NoSQL database technologies, most notably Apache Cassandra (4).

As with many services developed internally by Amazon, and with AWS being a continually growing and evolving business, the team soon realized that AWS customers might find Dynamo as useful and supportive as they had. Therefore, they furthered the design requirements to ensure it was easy to manage and operate, which is a key requirement for the mass adoption of almost any technology. In January 2012, the team launched Amazon DynamoDB (5), the cloud-based, fully-managed NoSQL database offering designed from launch to support extreme scale.

In terms of the general key-value database landscape, where does DynamoDB sit? According to the current (at the time of writing) *DB-Engines Ranking* (6A) list, Amazon's DynamoDB is the second most popular key-value database behind Redis (7). Over the years, it has steadily and consistently increased in use and popularity and held its place in the chart firmly (6B).

The rest, as they say, is history.

> **Important note**
>
> We've only covered a very small fraction of NoSQL's history in a nutshell. For a greater explanation of the rise and popularity of the NoSQL movement, I recommend reading *Getting Started with NoSQL* by Gaurav Vaish (8).

We've learned about the advantages that NoSQL and DynamoDB offer, but what does that mean in terms of usage, and how are people utilizing DynamoDB's features? In the next section, we'll examine some high-profile customer use cases that show just how powerful DynamoDB can be.

DynamoDB case studies

DynamoDB is currently used by more than a million customers globally (9). These range from small applications performing a few requests here and there, to enterprise-grade systems that continually require millions of requests per day, and all these and more depend on DynamoDB to power mission-critical services.

There is an incredibly high chance that you have indirectly been a DynamoDB user if you have spent any time on the internet. From retail to media to motorsports and hospitality, DynamoDB powers a sizable number of online applications. One of the best ways to find out about the incredible workloads and systems that not only DynamoDB but also AWS helps to power is to hear it through the voice of their customers.

What's key to think about overall is that while these case studies shine a light on some of the large-scale DynamoDB solutions in use today, they may not completely represent what you or I want or need to use DynamoDB for right now. Importantly, the takeaway here is that you and I and every developer, start-up, or enterprise have the same power at our fingertips with DynamoDB as the following case studies. Every feature these customers have access to within the service, you do too, and that's where the power of not only DynamoDB but also AWS comes in.

The upcoming case studies illustrate how DynamoDB users adopted the service either to overcome specific challenges or to leverage their trust in the platform based on prior positive experiences with the technology.

Amazon retail

As we briefly covered in the previous section, Amazon.com, the retail platform, heavily relied on its existing relational databases but was struggling to scale its system to keep pace with the rapid growth and demand of the platform. The Amazon Herd team decided to move over to DynamoDB (10).

Migrating from a relational database to a NoSQL offering takes time, planning, and understanding of the workload that is going to be migrated. The internal teams (as we've highlighted) were able to identify that ~70% of the current database queries were that of the key-value type (the lookup was performed by a single primary key with which a single row was returned), which aligns well with NoSQL.

Once fully implemented, one of the many benefits the Herd team was able to make use of was being able to reduce their planning and the time required to scale the system for large events by 90%. This allowed the team to spend more time in other areas, such as creating new value-adding features for the Amazon retail platform.

On the back of this success, DynamoDB powers many of the mission-critical systems used within Amazon (including multiple high-traffic Amazon properties and systems). To further exhibit just how powerful and instrumental the move to DynamoDB was, and over the period of the multiple-day Amazon Prime Day event in 2023, in an official post by AWS (11), the following is stated:

> *"DynamoDB powers multiple high-traffic Amazon properties and systems including Alexa, the Amazon.com sites, and all Amazon fulfilment centers. Over the course of Prime Day, these sources made trillions of calls to the DynamoDB API. DynamoDB maintained high availability while delivering single-digit millisecond responses and peaking at 126 million requests per second."*
>
> *– Jeff Barr, Chief Evangelist for Amazon Web Services*

Let's just think about that for a second – 126 million requests per second?! For any service to be able to deliver this level of throughput is an incredible feat. What's more, this was delivered on top of all other AWS customers' workloads across the globe, without breaking a sweat. Whatever you can throw at DynamoDB, the service can handle it in its stride. To this day (and every year when I read the latest Amazon Prime Day stats), I'm in awe of the tremendous power that DynamoDB offers. This is one of the many reasons why so many customers choose DynamoDB to power their workloads.

What is Amazon Herd?

Amazon Herd is an internal orchestration system that helps to enable over 1,300 workflows within the Amazon retail system. Herd is used to manage services ranging from order processing and certain fulfillment center activities as well as powering parts that help Amazon Alexa to function in the cloud.

Disney+

In April 2021, the Walt Disney Company outlined in a press release (12) how they are using AWS to support their global expansion of Disney+, one of the largest online streaming video services in the world. Disney+ scales part of its feature set globally on DynamoDB.

If, like me, you have used Disney+ and have added a video to your watchlist, or perhaps you have started watching any of their video content, paused it, and have come back later, or even watched the rest of it on a different device, these features have data that persisted under the hood with DynamoDB. Globally, these events (and more) amount to billions of customer actions each day.

Disney+ takes advantage of AWS' global footprint to deploy its services in multiple regions (13), and with DynamoDB, they are making use of a feature called Global Tables (14) (see *Chapter 13, Global Tables*). In a nutshell, Global Tables provides a fully-managed solution for multi-region, multi-active tables within DynamoDB. You simply enable the feature and decide which regions you want to be "active" in, and DynamoDB takes care of regional provisioning, any data back-filling, and then all ongoing replication for you. It's incredibly powerful and requires no maintenance from you. It's just one of the mechanisms in place that we can refer to when thinking about DynamoDB's "click-button" scaling.

With Global Tables in place, many of the Disney+ APIs are configured to read and write content within the same region where the data is located, reducing overall latencies to and from the database. Global Tables also allows their services to failover geographically from one entire region to another, from both a technical failure perspective as well as being able to perform maintenance in a given region without incurring user downtime or affecting the availability and usability of the overall Disney+ service.

I highly recommend watching an AWS re:Invent video titled *How Disney+ scales globally on Amazon DynamoDB* (15). Next, let us review the characteristics of workloads that may not be suited for DynamoDB.

Workloads not suited for DynamoDB

Let's quickly cover a couple of areas that aren't as well supported with DynamoDB, such as search and analytics. While not technically impossible to implement, often, the requirements are strong enough to warrant a dedicated technology to which to offload these tasks. Just because we are using a NoSQL database, it doesn't mean that we should shoehorn every workload into it. As I called out in the opening paragraph, it's better to know when *not* to use a technology than forcing it to work and ending up with an overly and unnecessarily complex data model because of it.

While DynamoDB does have a very basic ASCII search operator available, this is only available on a sort key attribute or by using filtering on additional attributes. It can only perform left-to-right or exact string matching and does not offer any level of fuzzy matching or in-depth or regular expression matching. Unless your string matching operates at this level, you will need to utilize a dedicated technology for this. An approach often seen, and well documented, is to support search queries with Amazon OpenSearch Service (16, 17A, 17B), or similar.

Unlike some other NoSQL technologies, such as MongoDB (18), DynamoDB does not have any kind of aggregation framework built into it. This means you cannot easily perform analytical queries across results or datasets within your table. If you need to generate sums of data (for dashboarding, for example), this must be done within your application and the results written back into DynamoDB for later retrieval. Although an advanced design pattern using DynamoDB Streams and AWS Lambda could be leveraged to perform some kinds of aggregations, on a holistic level, the database engine itself does not support native operators for aggregation queries.

There is a misconception with NoSQL that once you have created your model, you cannot change it or work with newer and evolving access patterns – that's untrue. It's always worth pointing out that with many production applications, it is common for these access patterns to change organically over time. There could be several reasons for this: a change in business model, or a new team or service is introduced to the company that has an alternative access pattern that needs fulfilling. With DynamoDB, you are not strictly "stuck" with your original data model design. The DynamoDB team has got you covered with support for additional indices on your table by using what's called a **Global Secondary Index (GSI)** (19).

We will cover data modeling and design in a lot more depth over the coming chapters, giving plenty of opportunity to explore how we build data models and accommodate changes with GSIs – don't worry if this does not make sense right now, just know that you can support additional access patterns on the same table, without necessarily having to rebuild and restructure the entire table itself.

Summary

In this chapter, we learned about the beginnings of the service from Dynamo through to DynamoDB, and how some of the biggest companies in the world are leveraging DynamoDB to power their mission-critical workloads. We covered where DynamoDB sits within the NoSQL market, and briefly recapped how NoSQL technologies fit within the overall database landscape.

We read about some incredible examples of how well DynamoDB performs and operates at scale, and I encourage you to explore further case studies available on the DynamoDB product page (20).

We touched on the fact that the workloads best suited to DynamoDB are those that have well-known and defined access patterns and workloads that are predominantly, although not strictly limited to, OLTP. Search and analytical workloads typically aren't well supported, so knowing when to offload this need is important.

In the next chapter, we will explore how we interact with DynamoDB in order for us to prepare a working environment (either locally or remotely in the cloud) so we can start data modeling and working with some of the incredible features that DynamoDB has to offer.

References

1. SOSP announcement – `https://www.allthingsdistributed.com/2007/10/amazons_dynamo.html`

2. *A Decade of DynamoDB* article – `https://www.allthingsdistributed.com/2017/10/a-decade-of-dynamo.html`

3. *Dynamo: Amazon's Highly Available Key-value Store* – `https://www.allthingsdistributed.com/files/amazon-dynamo-sosp2007.pdf`

4. Apache Cassandra – `https://cassandra.apache.org`

5. Launch of Amazon DynamoDB – `https://www.allthingsdistributed.com/2012/01/amazon-dynamodb.html`

6. DB-Engines Ranking is a list of database management systems ranked by their current popularity, and is updated monthly:

    A. `https://db-engines.com/en/ranking/key-value+store`

    B. `https://db-engines.com/en/ranking_trend/key-value+store`

7. Redis – `https://redis.io/`

8. Gaurav Vaish, Getting Started with NoSQL, Packt Publishing, 2013 – `https://www.amazon.com/Getting-Started-NoSQL-Readiness-Review/dp/1849694982`

9. AWS blog mentioning DynamoDB with 1M+ customers – `https://aws.amazon.com/blogs/database/common-financial-services-use-cases-for-amazon-dynamodb/`

10. Amazon Herd: DynamoDB case study – `https://aws.amazon.com/solutions/case-studies/herd/`

11. Prime Day 2023 blog – `https://aws.amazon.com/blogs/aws/prime-day-2023-powered-by-aws-all-the-numbers/`

12. Disney+ press release – `https://press.aboutamazon.com/2021/4/the-walt-disney-company-uses-aws-to-support-the-global-expansion-of-disney`

13. AWS Global Infrastructure – `https://aws.amazon.com/about-aws/global-infrastructure/regions_az/#Regions`

14. Amazon DynamoDB Global Tables – `https://aws.amazon.com/dynamodb/global-tables/`

15. AWS re:Invent 2020: *How Disney+ scales globally on Amazon DynamoDB* – `https://www.youtube.com/watch?v=TCnmtSY2dFM`

16. Amazon OpenSearch Service – `https://docs.aws.amazon.com/opensearch-service/latest/developerguide/integrations.html#integrations-dynamodb`

17. Amazon OpenSearch DynamoDB integration developer documentation:

    A. Using DynamoDB Streams and AWS Lambda – `https://docs.aws.amazon.com/opensearch-service/latest/developerguide/integrations.html#integrations-dynamodb`

    B. Using zero-ETL integration – `https://docs.aws.amazon.com/amazondynamodb/latest/developerguide/OpenSearchIngestionForDynamoDB.html`

18. MongoDB – `https://www.mongodb.com/`

19. Global Secondary Indexes developer documentation – `https://docs.aws.amazon.com/amazondynamodb/latest/developerguide/GSI.html`

20. Amazon DynamoDB customer case studies – `https://aws.amazon.com/dynamodb/customers/`

2

The AWS Management Console and SDKs

Before we begin designing, building, and working with DynamoDB, we should understand what tools and options are available for us to interact and work with. Unlike some NoSQL databases that you connect directly to the primary server or host with (Redis or MongoDB, for example), DynamoDB is different—we interact with it through a series of stateless **application programming interfaces** (**APIs**).

The underlying service of those APIs corresponds to fleets of request routers and storage nodes, all of which take care of sending our requests to the correct storage nodes, managing any authentication and authorization alongside interfacing with any additional service features, such as **point-in-time recovery** (**PITR**) and read/write capacity modes.

Having most tasks abstracted away into an API means we can focus more on our application and data modeling. While, at first, querying a database through API calls may seem strange, it can really help with the simplification of performing queries and ensuring those queries are well-formed and validated before being sent to the database.

In this chapter, we will start off by taking a look through the AWS Management Console to familiarize ourselves with the layout and options that we have to work with in DynamoDB without needing to know any programming languages or install any software to get up and running. In fact, we can get fully started for free, as AWS offers a certain amount of use of DynamoDB under the AWS Free Tier.

We will then look at the AWS **software development kits** (**SDKs**) available that enable us to work directly with the DynamoDB APIs. Finally, we'll look at how to run our code natively in the cloud with AWS Lambda, rounding off with how we can make use of DynamoDB local to develop our application offline.

By the end of this chapter, you will understand how to access and work with DynamoDB to suit your preferred working environment and development needs.

In this chapter, we're going to cover the following main topics:

- Working with the AWS Management Console

- Navigating and working with items

- The AWS SDKs

- Using AWS Lambda and installing DynamoDB local

Working with the AWS Management Console

To get started with DynamoDB in the AWS Management Console, the first thing you will need is an AWS account. If you don't already have one, you can sign up for free and use a set number of services at no cost under the AWS Free Tier. There are three ways that the AWS Free Tier works—some services are offered as a free trial for a short term, others are free to use for 12 months, and lastly, some remain free forever and do not expire.

For DynamoDB, the AWS Free Tier offers the following:

- 25 GB of table storage

- 25 **write capacity units (WCU)** of provisioned capacity

- 25 **read capacity units (RCU)** of provisioned capacity

- 25 **replicated write capacity units (rWCU)**

- 2.5 million DynamoDB Streams read requests

- 1 GB data transfer out (15 GB for the first 12 months)

The benefits listed are given under the Standard table class.

Not only does this allow you to perform up to 200 million requests per month, but it also remains free forever and does not expire after 12 months.

To get started with a new account, and to take advantage of the AWS Free Tier, please visit the following URL: `https://aws.amazon.com/free`

> **Important note**
> As stated, some services in the AWS Free Tier are only valid for 12 months from sign-up, so ensure you fully understand what is and what is *not* covered during that period to avoid any excess charges during your usage period or at the end of it.

If creating an AWS account isn't something you're ready to do right now, but you still wish to carry on working with DynamoDB, then skip ahead to the *Using AWS Lambda and installing DynamoDB local* section, where you will find out how to download a version of DynamoDB that you can run locally instead.

Now that you have a new account (or an existing one to use), let's log in to the console. I will assume that you have followed the login process and are sitting at the **AWS Management Console** landing page:

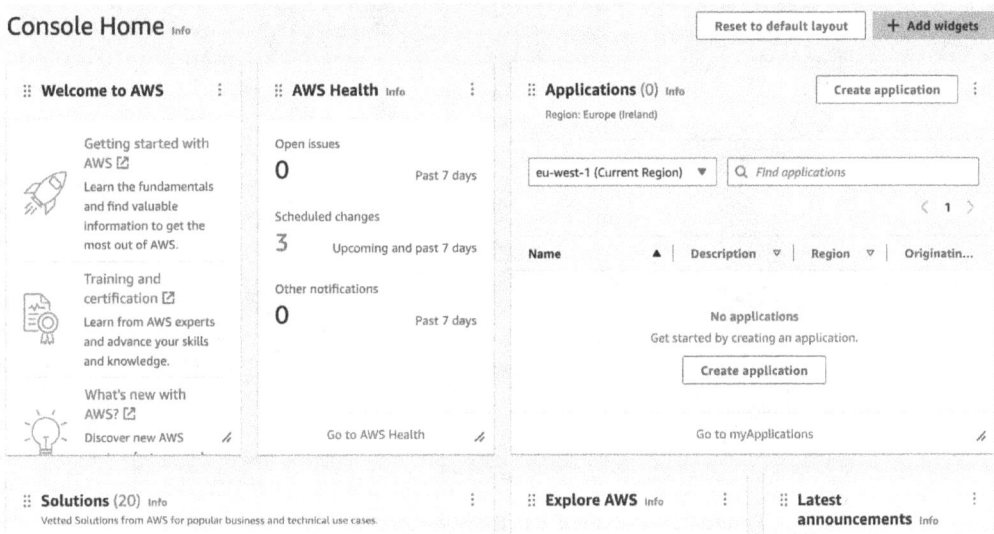

Figure 2.1 – Overview of the AWS Management Console landing page

When you first log in, you may be directed to the US East (N. Virginia) region. You are free to continue working in this region; however, I prefer to work in the region geographically closest to me (London). You can select your closest region from the drop-down list.

Feel free to explore the console in more depth and see what is available to you. When you're ready, simply enter DynamoDB into the main search box at the top of the page as shown in the following figure, and click on the title under the **Services** heading to be taken to the DynamoDB console:

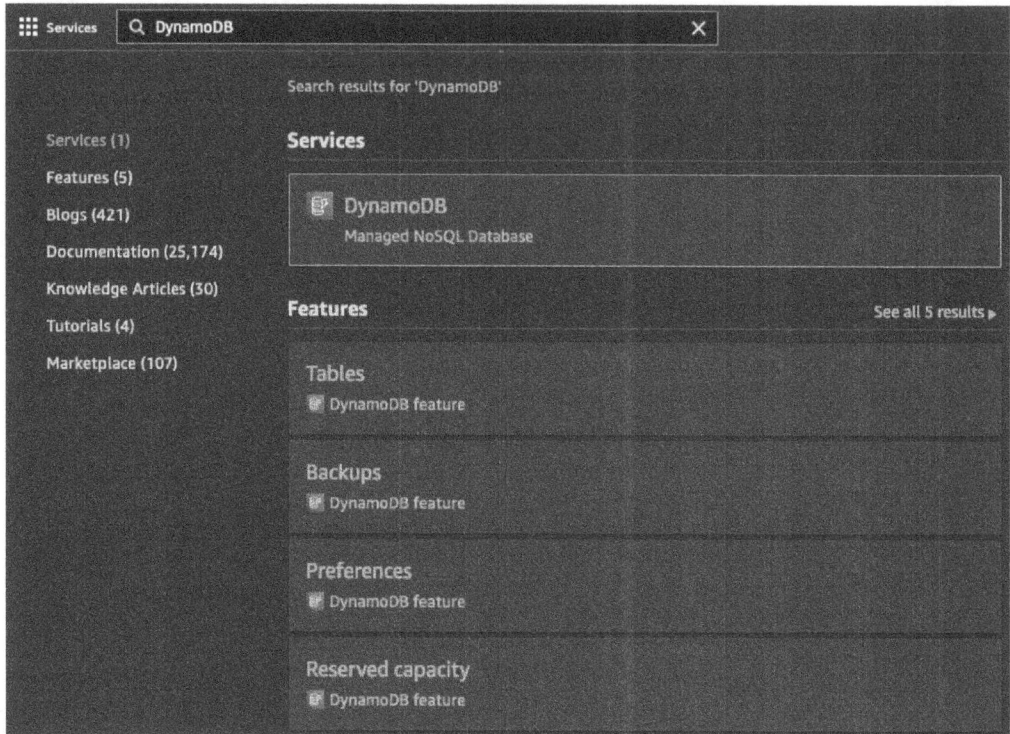

Figure 2.2 – DynamoDB in the Services search bar

Following this, you should be directed to the DynamoDB console. Next, let us use the console to create a DynamoDB table and play around.

Creating and using a DynamoDB table

As we're not yet building an application that connects to DynamoDB through the SDKs, but merely want to check out how DynamoDB works at a basic level and how we can enable and work with its supported features, this can all be done from within the DynamoDB console. From here, we can issue requests into DynamoDB and see the responses alongside being able to add, edit, and delete items in our table(s).

In this part of the chapter, we will start by creating a new table and then we'll add some items to it. We won't be issuing any advanced queries here, but we will be able to see how to quickly manage data in DynamoDB straight from the AWS Management Console.

Let's take our first jump into the world of DynamoDB and create our first table:

1. To start with, we're going to leave the majority of options at the default state when creating our table. Don't worry if you don't fully understand what each term or option means; we will cover these in depth later:

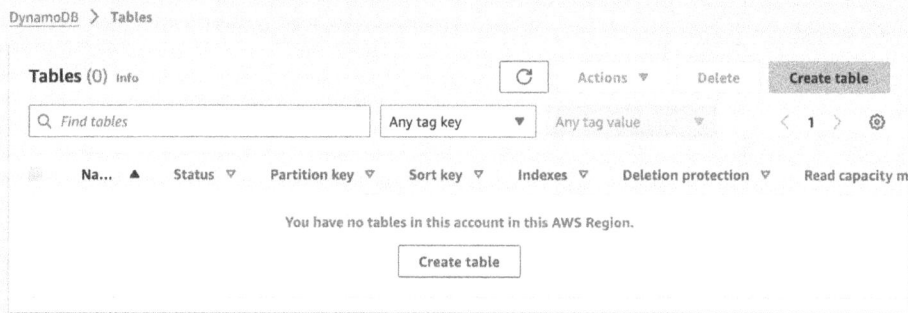

Figure 2.3 – DynamoDB console Tables landing page

2. Click on **Create table** and you will be taken to a relatively simple screen (*Figure 2.4*):

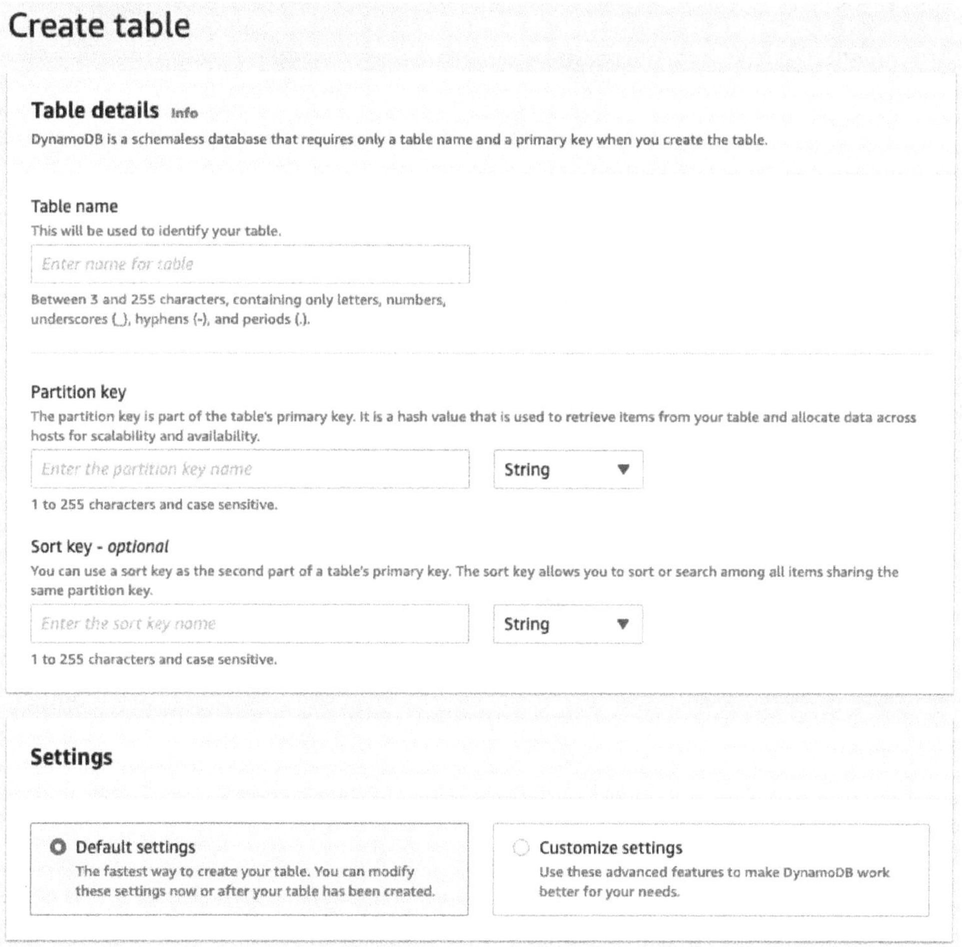

Figure 2.4 – Creating a new DynamoDB table

Here, we only need to enter information into two fields to create our DynamoDB table, **Table name** and **Partition key**. We are going to leave **Sort key** empty (it is optional anyway), and we will use **Default settings** for the rest of this table. For now, we do not need to add any tags to the table.

Once you have entered the table name (I'm using `DefinitiveGuide01`) and partition key value (I am using `id`), click on **Create table**. DynamoDB will now go ahead and make this table ready for you and will provision the necessary resources behind the scenes so that you are ready to go as soon as the table status changes from **Creating** to **Active** (this typically only takes a few seconds). The following figure shows the table in the **Creating** status:

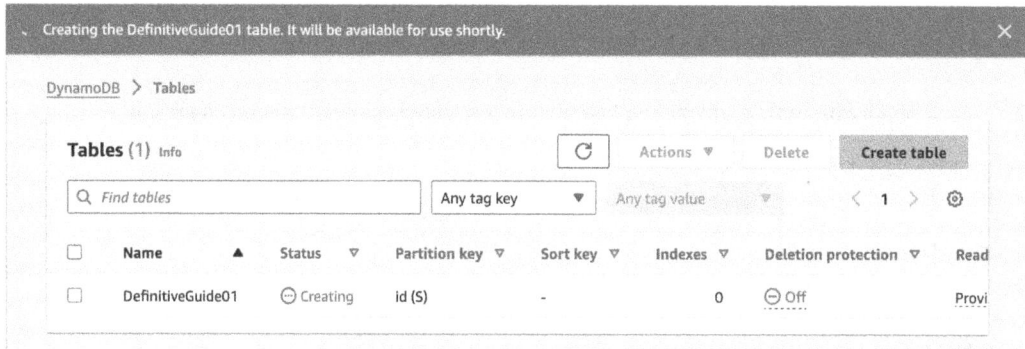

Figure 2.5 – Table in Creating status

As previously mentioned, the table should transition from the **Creating** to **Active** status within a few seconds. The following figure shows my table in the **Active** status and is ready to use:

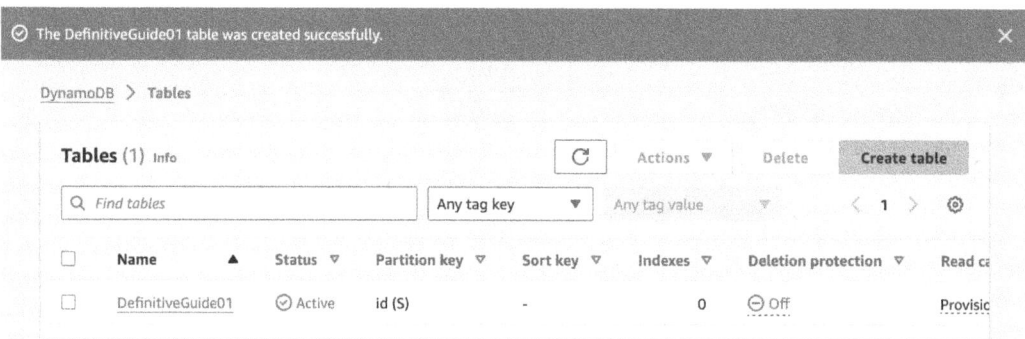

Figure 2.6 – Table in Active status

I've kept this initial table simple for now, including the partition key value. We will explore more advanced concepts in naming and how to use the sort key to our advantage in the next part of this book, *Core Data Modeling*.

Once the table is ready, the name of the table changes to a clickable link. Click on it and you will see a new screen that shows information about the table, including details such as **Capacity Mode** along with a section titled **Items summary**, which shows details about your table's data.

With the information screen about our newly created table open, clicking on the **View items** button at the top-right of the screen takes us to our item detail screen (*Figure 2.7*). At this point, our table is empty and you should see the **The query did not return any results** message onscreen, as shown in the following figure:

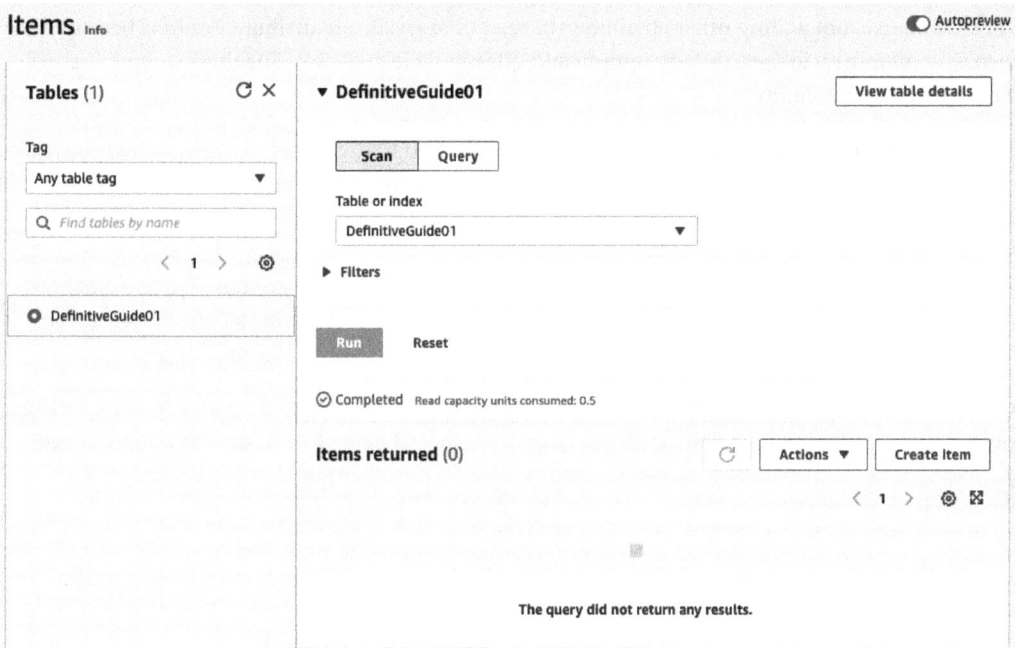

Figure 2.7 – Empty table results screen

This message is the result of the console running a **scan** on the table. A scan simply calls the DynamoDB API and effectively says *give me all the items that are stored in this table*. This can be useful at times, but as we will discover later, it is not always beneficial and can often end up costing a fair amount of RCUs when our table grows. For what we are doing presently in the console, it is fine and will not exceed our capacity throughput.

Our table is now ready, and we can start creating items in it. To summarize, we logged into the AWS Management Console and navigated our way through to the DynamoDB-specific console. From there, we created our first DynamoDB table and saw how the DynamoDB console can quickly show us the contents of our table. Next, we shall navigate through the console further and interact with our new DynamoDB table.

Navigating and working with items

You have seen how easy it is to create a table in DynamoDB. Now, we are going to populate that table with some items (an item in NoSQL is what we would call a row in a relational database), and I am pleased to tell you that it is just as easy.

On the right of the **Items returned** section, click on the **Create item** button and you will be taken to a screen that lets you create a new DynamoDB item (*Figure 2.8*). The first thing to notice here is that the only required attribute to create an item is the partition key. DynamoDB does not need anything further. Of course, not adding other attributes (in relational speak, an attribute would otherwise be known as a column) will mean a fairly empty and questionable database. For this example, we will add one attribute or column.

An important part to remember about DynamoDB is that no matter what, we must always have a partition key value—this is what forms the core concept of the **key-value** principle. It is this key that gets supplied to an internal hashing function and from there, DynamoDB determines which partition (or storage node) the item sits on. Whenever you perform a get_item or query operation, you have to supply the partition key regardless of whether you're using a sort key or not. The only time you do not need to know the item's partition key value is when you are performing a Scan operation on the table. So, it is vital that the partition key forms a solid and useful value and structure.

Sticking with keeping it simple, I have set my id, the partition key value to foo and added a **String** attribute (found under the **Add new attribute** drop-down menu) named title, with a value of bar:

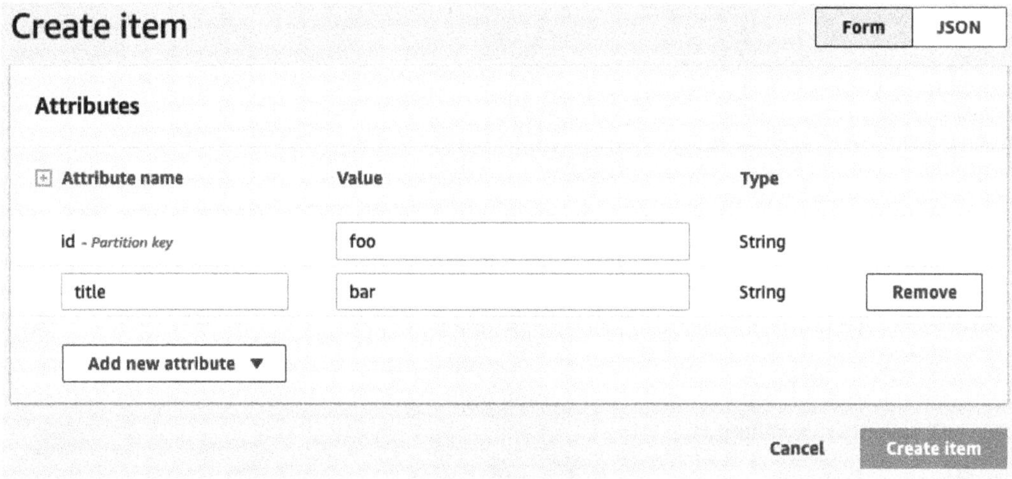

Create item

| Form | JSON |

Attributes

Attribute name	Value	Type	
id - *Partition key*	foo	String	
title	bar	String	Remove

Add new attribute ▼

Cancel   **Create item**

Figure 2.8 – Creating a new DynamoDB item

Once that is entered, click on **Create item** and that's it—your newly created item is now stored in DynamoDB: in the cloud, on three storage nodes, across multiple Availability Zones, and ready for retrieval with millisecond latency. How simple yet fantastic is that?!

Our **Items returned** list has now been updated (to **1**) and shows the newly created item along with its attribute names and values. The following figure shows this:

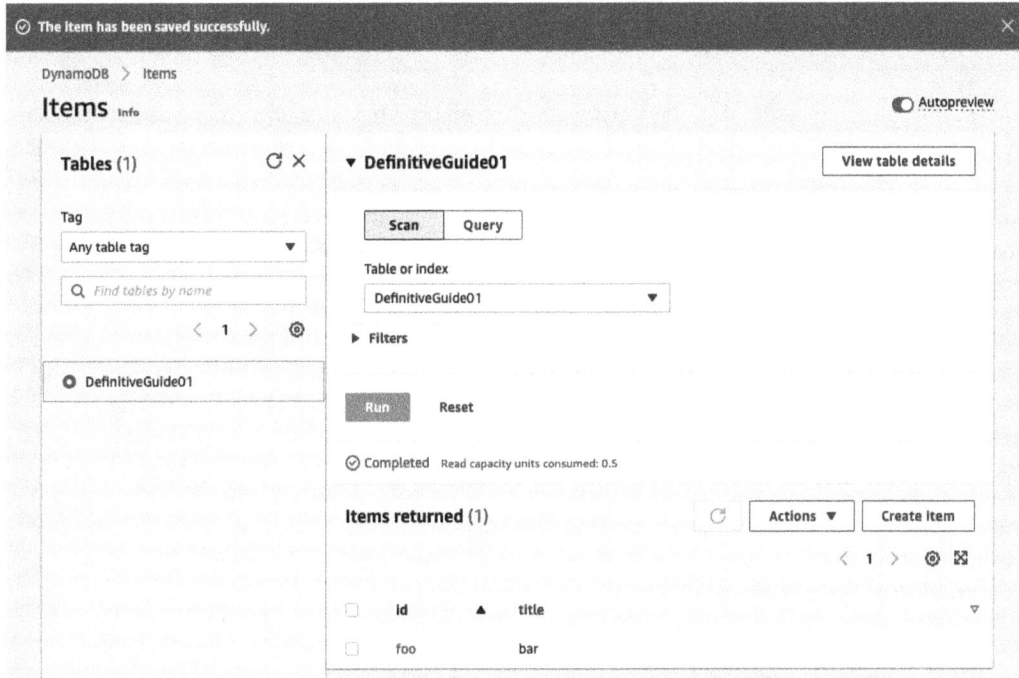

Figure 2.9 – A new DynamoDB has been created

To edit this item, we can either click on the partition key value (shown as a blue hyperlink) or we can select the checkbox next to it, then from the **Actions** drop-down menu, select **Edit**. Either approach will take you to a screen with the item value prefilled, allowing you to edit the attributes as necessary.

To delete an item, you must select it via the checkbox and use the **Actions** drop-down menu. Using this, you can also select and delete multiple items at once by selecting the relevant entries. Deletion only occurs once you have confirmed the action from the modal confirmation window, as shown in the following figure:

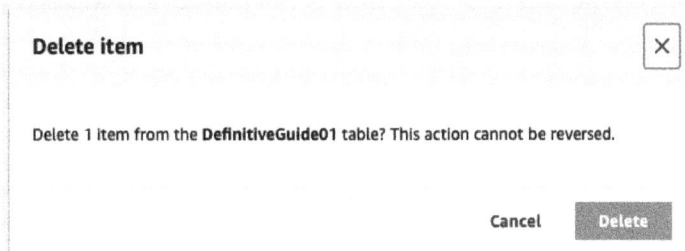

Figure 2.10 – Deletion confirmation modal

> **Top tip**
>
> The **Action** drop-down menu offers some very useful functions straight from the DynamoDB console. For example, it allows you to quickly and easily download the results of any query you see onscreen, or any selected items from the results, to a CSV file.

If you are not using the optional sort key attribute when you create your table, you must ensure that your partition key values are unique—you cannot have duplicate entries, otherwise DynamoDB will throw an error. If you are using a sort key, then the uniqueness is extended across to include the sort key attribute. This means you can have a duplicate partition key but the sort key (combined with the partition key) must be unique. This method forms the basis of *item collections*. We will explore this concept in more detail in *Chapter 4*.

So that we can explore how to perform a query with the console, let us add a few more items to our table. I am a huge electronic music fan, and one of my favorite artists is The Thrillseekers (1). For this example, I am going to add five entries from the *Adjusted Music* (2) record label, a label set up and managed by The Thrillseekers.

I will be using the catalog number (an identification number assigned to a music release by a record label) for each of the five songs for my partition key value as it represents a unique release. Feel free to replicate this yourself or create your own dataset going forward. The following figure shows my table with the five items I have created in it:

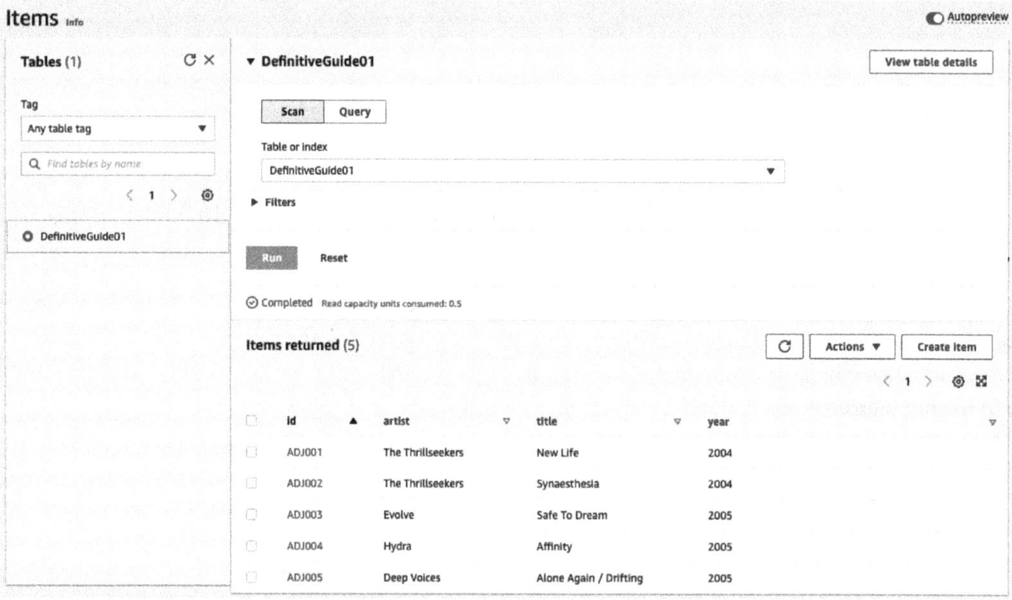

Figure 2.11 – A small number of items in our table

With all our newly added items in place, click on the **Query** button at the top of the **Items** screen, as marked in the following figure:

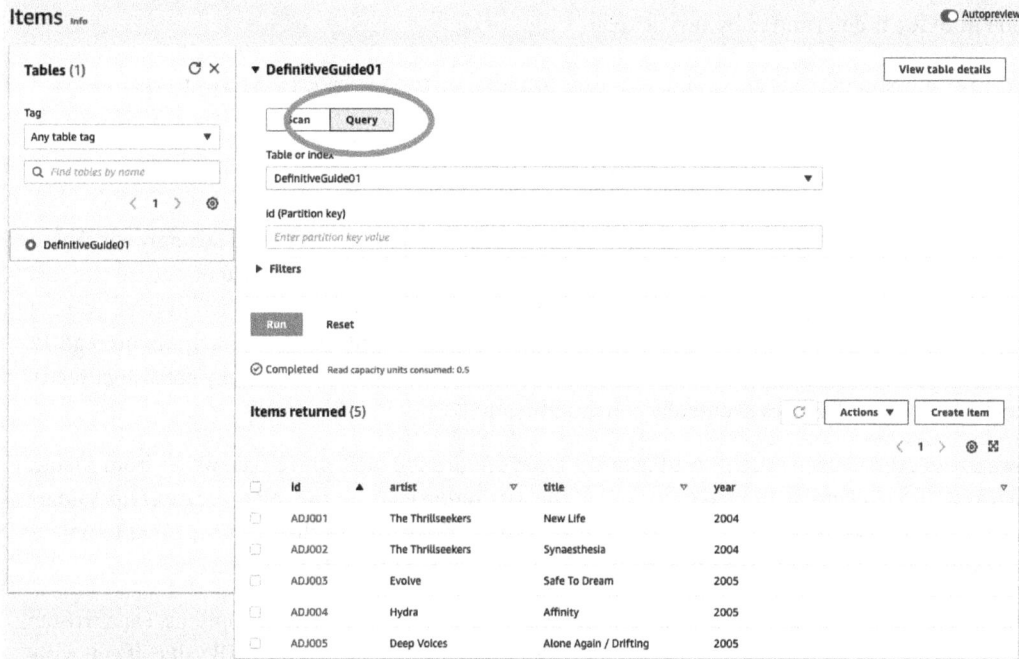

Figure 2.12 – The results of our DynamoDB query

It's really that simple. Essentially, this query returns any items matching the specified partition key value. The real power comes into play when you combine the sort key into the query as well, but even with our simple example, you can see how this could easily be built up into a powerful single-item retrieval system. Think about perhaps session IDs—typically, these are globally unique values and have a 1:1 mapping. The data stored against a session ID could be vital to how a user interacts with a website. All that's needed is a simple key-value system to power it, and you have just built one in the cloud using DynamoDB.

The last part of the console I want to highlight is that after performing a query or scan of your table, there is a small section added that shows the RCU consumption. The following figure shows this:

Figure 2.13 – Read capacity consumption displayed in the console

This value changes depending on how much data your read request consumes. With our small example table, reading one item has the same cost as reading all five items.

Let's take a short moment for an overview of how capacity charging works in DynamoDB.

A brief overview of how throughput is charged

When you perform any read action against DynamoDB, you are charged in RCUs. The cost of the action in RCUs is determined by two factors:

1. The amount of data read by DynamoDB, rounded up to the nearest 4 KB
2. The consistency of the read request

The basic formula is that a read charges 1 RCU per 4 KB of data read.

DynamoDB offers two types of read consistency—**eventually consistent reads** and **strongly consistent reads**. An eventually consistent read is charged at only half of a strongly consistent one, so the cost is 0.5 RCU per 4 KB. This flexible consistency model means that if you read 1 KB of data, you are charged for 4 KB (1 RCU, or 0.5 RCU with an eventually consistent read); similarly, if you read 10 KB of data then you are charged for 12 KB (3 RCU, or 1.5 RCU with an eventually consistent read). By default, DynamoDB uses eventually consistent reads.

However, an eventually consistent read may not always return the most up-to-date writes from a table, whereas a strongly consistent read ensures that the data returned from the call is the most up-to-date and reflects the latest write, thus no stale data is returned at the time of the call. One point to note is that if your read request returns zero results, you are still charged 0.5 RCU for the operation.

When writing into DynamoDB, it is straightforward as there are no consistency options. Data written into DynamoDB costs 1 WCU per 1 KB of data. This rounds up to the nearest 1 KB value. If you write 250 bytes, you are charged 1 KB (1 WCU); likewise, if you write 1.35 KB, you are charged 2 KB (2 WCU).

It's down to the owners and developers to determine the appropriate read consistency for their application. In reality, the odds of stale data in an eventual read are only one in three but must be catered for and factored in, although it is often worth the saving on cost. More details around consistency are covered in *Chapter 6, Read Consistency, Operations, and Transactions*. As of writing, the DynamoDB console does not allow for strongly consistent reads.

To summarize, we have covered a few core DynamoDB concepts in this section, from creating items to performing queries and understanding a little more about consistency and capacity. Leaving our example table in place (so we can reuse it), let us now explore the AWS SDK and learn how we can programmatically query our table using a choice of server-side languages.

Getting familiar with the AWS SDK

DynamoDB would be fairly limited if we could only interact with it via the AWS Management Console. To remedy this, AWS provides several helpful SDKS (3) that allow us to interact with AWS APIs to perform many service-based actions. These vary from control plane methods (to create new service resources, such as a new DynamoDB table) through to data plane actions that allow us to read or write items to and from DynamoDB.

AWS has developed the SDK for most of the popular programming languages in use today. At the time of writing, the list of available language-specific SDKs includes JavaScript, Python, PHP, .NET, Ruby, Java, Go, Node.js, C++, and Rust. That's quite an impressive list, not to mention that they are continually updated by the AWS SDK teams.

> **Important notice**
>
> To simplify my development and for the use of DynamoDB local, I'm using Docker to manage any environments in use on my local machine. This helps abstract multiple software installations into isolated containers. You can install and use Docker for free, available from `https://www.docker.com/`.

For the code examples used throughout this book, I will be using **Boto3**, the Python SDK. Feel free to use and adapt the code for your SDK language of choice.

To get started, and assuming that a recent (or the latest) version of Python and pip have already been installed, we first need to install the Python SDK—this is done with the following command:

```
pip install boto3
```

Once that has been installed, you are almost ready to go. When using an SDK or the AWS CLI, you will need to associate and authorize that software to call AWS services on your behalf. This is done via AWS **identity and access management (IAM)**. Here, we can create users in our AWS account that have access only to the services and functionality we grant them, a highly recommended practice for any AWS account. Let us do that now:

1. From the AWS Management Console search bar, enter IAM and click on the service icon, as shown in the following figure:

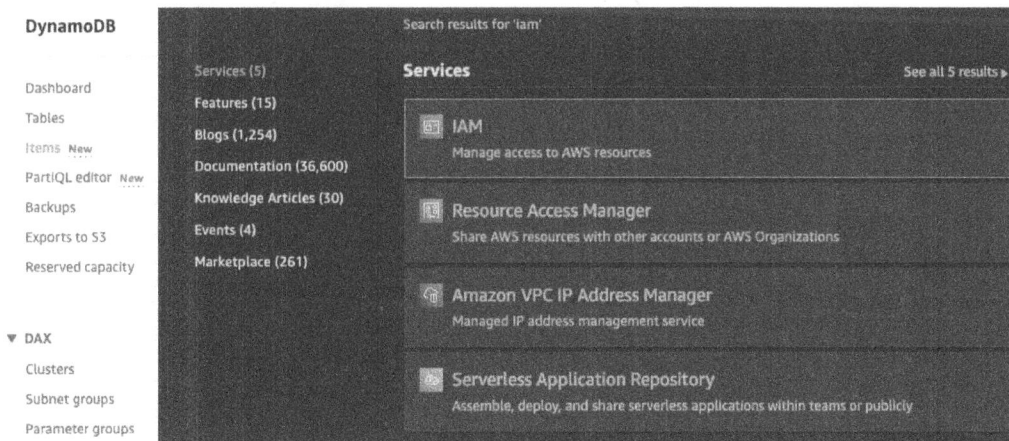

Figure 2.14 – IAM in the Services search bar

2.   You'll then be taken to the IAM dashboard, as shown in the following figure. From there, click on **Users** on the left-hand navigation, under **Access management**:

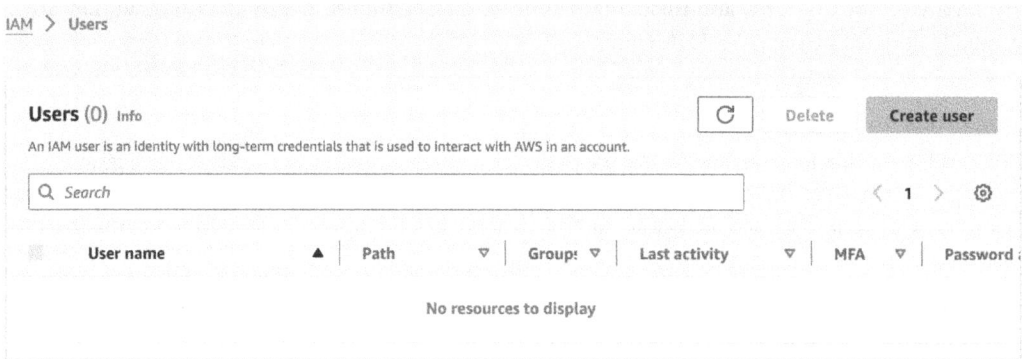

IAM > Users

Users (0) Info

An IAM user is an identity with long-term credentials that is used to interact with AWS in an account.

Create user

Delete

Q Search

	User name	▲	Path	▽	Group: ▽	Last activity	▽	MFA	▽	Password

No resources to display

Figure 2.15 – Users landing page

3.   We want to create a dedicated user purely for our Python SDK, so click on **Create user** to start the process. You will then be taken to the following page:

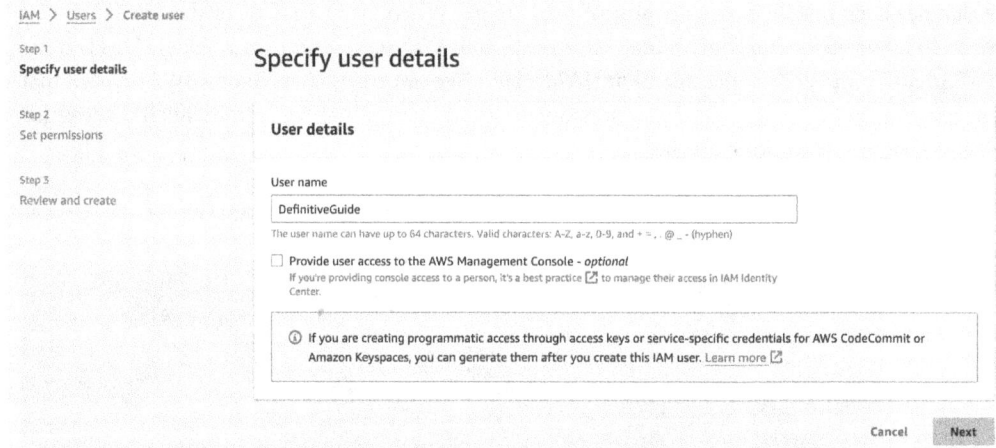

IAM > Users > Create user

Step 1
Specify user details

Step 2
Set permissions

Step 3
Review and create

Specify user details

User details

User name

DefinitiveGuide

The user name can have up to 64 characters. Valid characters: A-Z, a-z, 0-9, and + = , . @ _ - (hyphen)

☐ Provide user access to the AWS Management Console - *optional*
If you're providing console access to a person, it's a best practice 🗗 to manage their access in IAM Identity Center.

ⓘ If you are creating programmatic access through access keys or service-specific credentials for AWS CodeCommit or Amazon Keyspaces, you can generate them after you create this IAM user. Learn more 🗗

Cancel   Next

Figure 2.16 – Creating a new user within IAM with only programmatic access

Here, we give our user a recognizable name, such as `DefinitiveGuide`. This is useful as we can tell at a glance who this user is and what we can expect of their role. We do not want to grant them access to the AWS Management Console as it is not required for the job function, so we leave the **Provide user access to the AWS Management Console –optional** checkbox unchecked. Once the user is created, we will request creation of programmatic access keys to use it with the AWS CLI from our local machine. Programmatic keys give us a unique set of keys for the user that we store in our environment, allowing AWS to authenticate and help authorize our access. Click on **Next** to assign the relevant permissions to the user.

In the nature of least privilege, we do not want to assign admin permission to this user, or anything above what this user needs to achieve their job function. As we are only running a scan and query, we can assign them a basic level of DynamoDB—read-only access.

The IAM service also provides several standard sets of permissions called *policies*, which are managed by AWS on your behalf. You may also choose to create and maintain custom policies allowing desired actions with multiple AWS services. The following is a screenshot showing where I attach the **AmazonDynamoDBReadOnlyAccess** managed policy to my new IAM role:

Set permissions

Add user to an existing group or create a new one. Using groups is a best-practice way to manage user's permissions by job functions. Learn more 🗗

Permissions options

○ Add user to group	○ Copy permissions	● Attach policies directly
Add user to an existing group, or create a new group. We recommend using groups to manage user permissions by job function.	Copy all group memberships, attached managed policies, and inline policies from an existing user.	Attach a managed policy directly to a user. As a best practice, we recommend attaching policies to a group instead. Then, add the user to the appropriate group.

Permissions policies (1/1233)

Choose one or more policies to attach to your new user.

C⟳   Create policy 🗗

Filter by Type

Q DynamoDB   ✕ | All types   ▼ | 10 matches   ‹ 1 › ⚙

	Policy name 🗗 ▲	Type ▽	Attached entities ▽
☐ ⊞ 📋 AmazonDynamoDBFullAccess		AWS managed	0
☐ ⊞ 📋 AmazonDynamoDBFullAccesswithDataPipeline		AWS managed	0
☑ ⊞ 📋 AmazonDynamoDBReadOnlyAccess		AWS managed	1
☐ ⊞ 📋 AWSApplicationAutoscalingDynamoDBTablePolicy		AWS managed	1
☐ ⊞ 📋 AWSLambdaDynamoDBExecutionRole		AWS managed	0

Figure 2.17 – Assigning permissions only for the scope of DynamoDB

Feel free to skip adding any tags. Finally, review the user that you are about to create. Once you are happy, click on **Create user** and AWS will go ahead and create the user, as shown in the following figure:

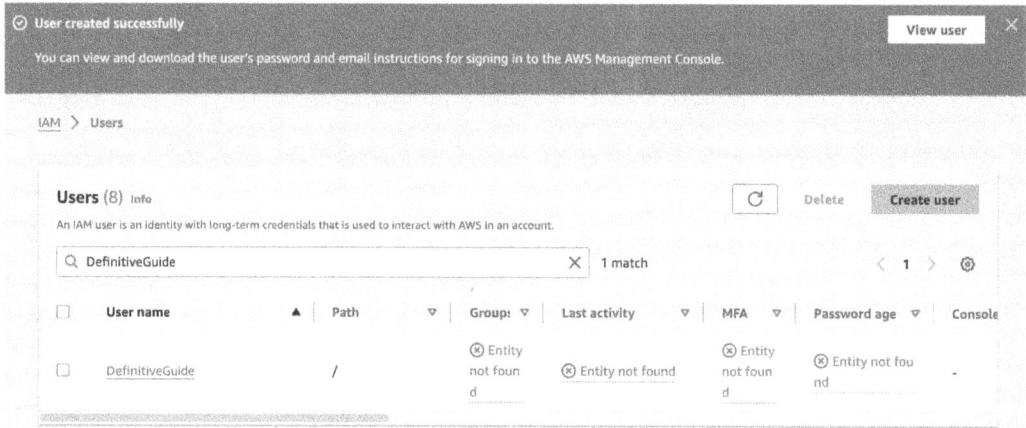

Figure 2.18 – Confirmation of user creation

Lastly, all that's required is to request a set of **access** keys for this new user, which can be done by following the **To create, modify, or delete the access keys of another IAM user (console)** section on the AWS docs (4).

Now that you have created your users and have downloaded the user credentials, you will need to store them in your environment so the SDK or CLI can access and use them. Full details on how to do this can be found in the *Configuration* section of the *Boto3 Quickstart Guide* (5).

Next, let us create and review AWS SDK-based scripts to interact with DynamoDB.

Using the AWS SDK

Let us replicate the scan and query calls we made in the DynamoDB console, but locally in a Docker container. If you do not want to set up Docker for isolating different runtime environments (in this case, Python), you can continue following this section, but run the scripts on your local machine without using Docker. Note that I am only using Docker for managing the runtime environments on my local machine, not deploying DynamoDB locally on a Docker container, which is also possible (6). For now, let us proceed with the AWS SDK-based scripts.

First, we need to write the Python code to run a scan. This is a relatively easy task, and the documentation is well written so we can easily discover how to write this using the SDK. We import the Boto3 client, specify the AWS resource we want to use (DynamoDB), give it the table name we wish to call against, and then call the `scan()` function as follows:

```
import boto3

dynamodb = boto3.resource('dynamodb', region_name='eu-west-2')
table = dynamodb.Table('DefinitiveGuide01')
response = table.scan()
```

```
items = response['Items']

print(items)
```

The `response` variable holds the full output of the scan function. You will notice that we are also creating a new variable called `items`; this holds the actual list of DynamoDB items from our table that we then print on the screen. If you print the `response` variable instead, you will see lots of additional information and metadata about the call such as `Count`, `ScannedCount`, `RequestId`, and so on. While this is interesting, we only want the list of items from the table—so for now, we just assign the value of the `Items` array to the python variable called `items` and print this variable..

The response of the Scan operation is as follows:

```
1 [{'year': Decimal('2005'), 'artist': 'Evolve', 'id': 'ADJ003',
  'title': 'Safe To Dream'}, {'year': Decimal('2004'), 'artist':
  'The Thrillseekers', 'id': 'ADJ002', 'title': 'Synaesthesia'},
  {'year': Decimal('2005'), 'artist': 'Deep Voices', 'id':
  'ADJ005', 'title': 'Alone Again / Drifting'}, {'year':
  Decimal('2005'), 'artist': 'Hydra', 'id': 'ADJ004', 'title':
  'Affinity'}, {'year': Decimal('2004'), 'artist': 'The
  Thrillseekers', 'id': 'ADJ001', 'title': 'NewLife'}]
```

Figure 2.19 – Output of Boto3-based script to perform a scan on DynamoDB

One thing to notice here is that the ordering of our items returned from DynamoDB does not necessarily match the ordering found in the console. This is because the DynamoDB console applies some ordering in the frontend automatically, along with additional formatting of the data. Ordering of data is applied when using sort keys, which we will see in later examples.

Let us now try our query to retrieve that single item in the same way that we did it in the console. Our Python code is similar, but instead of using the `scan()` function, we now call the `query()` function and specify the key condition using the property called `KeyConditionExpression` instead, as shown in the following snippet:

```
import boto3

dynamodb = boto3.resource('dynamodb', region_name='eu-west-2')
table = dynamodb.Table('DefinitiveGuide01')
response = table.query(
    KeyConditionExpression='id = :catnum',
    ExpressionAttributeValues={':catnum': 'ADJ004'}
)
```

```
items = response['Items']

print(items)
```

The result of the query is as follows:

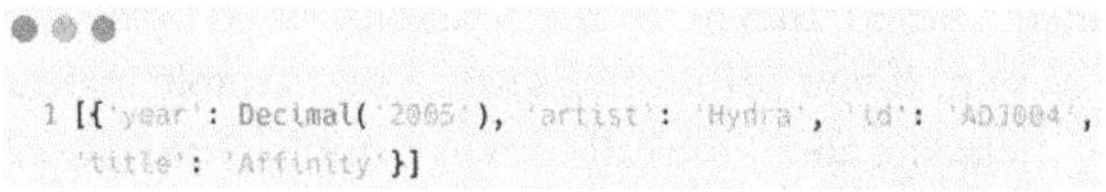

```
● ● ●
1 [{'year': Decimal('2005'), 'artist': 'Hydra', 'id': 'ADJ004',
    'title': 'Affinity'}]
```

Figure 2.20 – Output of Boto3-based script to perform a query on DynamoDB

This looks a lot more complex at first glance; however, all we are doing is building a query that says to return the item from DynamoDB where id (our partition key) equals ADJ004.

The first three lines are the same as the code used for the Scan operation, and again we are assigning the Items array response to be printed at the bottom. The difference in the way the query is constructed is to use KeyConditionExpression to specify the item selection criteria and then use placeholders (in this example, :catnum) to fill those values and generate the full query for the DynamoDB API.

DynamoDB has an API called GetItem that allows you to retrieve a single item based on the primary key value. Why didn't we use that here? We could have done it, but it is not available in the console, so to replicate the console behavior, we went with the query call. The code for getting a single item is simpler, so for comparison, the following is a version that uses the GetItem API, or the get_item method of the Python DynamoDB client to retrieve a single item:

```
import boto3

dynamodb = boto3.resource('dynamodb', region_name='eu-west-2')
table = dynamodb.Table('DefinitiveGuide01')
response = table.get_item(Key={'id': 'ADJ004'})
items = response['Item']

print(items)
```

The result of get_item() is as follows:

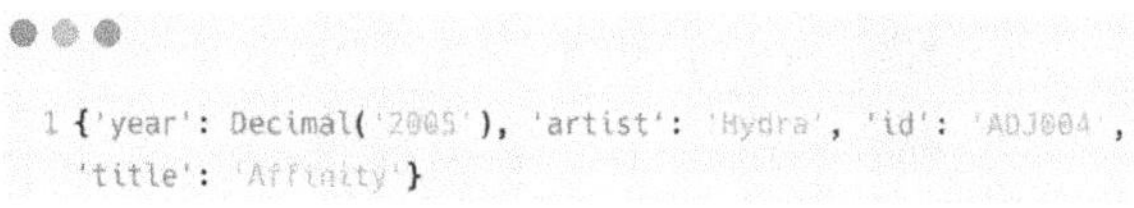

```
● ● ●
1 {'year': Decimal('2005'), 'artist': 'Hydra', 'id': 'ADJ004',
   'title': 'Affinity'}
```

Figure 2.21 – Output of Boto3-based script to perform get_item on DynamoDB

The output for both looks almost identical. The only difference is that `query()` can return multiple results, so it returns an array of items (`response['Items']`), whereas `get_item()` only expects at most one item to be returned, so simply returns that item object (`response['Item']`).

And with that, we have now successfully replicated both the scan and query calls of the console locally with the AWS SDK. We also learned how to set up an individual IAM user for the scope of running read-only DynamoDB calls, along with touching on how `KeyConditionExpression` works—we will use this a lot more in queries to come. For bonus points, we also looked at the code required to get a single item.

In the next section, we'll look at choosing whether to run code locally or in the cloud with AWS Lambda and how you can download and run a copy of DynamoDB locally for development purposes.

Using AWS Lambda and installing DynamoDB local

As we have explored, there are two straightforward methods to handle items in your DynamoDB table. The first, albeit basic, involves utilizing the AWS Management Console. While it has its limitations, it provides a means to oversee your table and its data. In the preceding section, we dove into programming, leveraging the AWS SDK to interact with our table. This method is prevalent, especially when developing a local application or one destined for hosting on an **Amazon Elastic Compute Cloud (Amazon EC2)** instance. However, there is another increasingly favored approach among developers—the realm of serverless compute using AWS Lambda.

What is AWS Lambda? In straightforward terms, AWS Lambda is a serverless compute service that enables you to upload and run your code in practically any programming language to support a wide range of applications or systems. The beauty of it lies in the fact that you do not have to manually provision or manage any servers; AWS handles that aspect for you. You just upload your code or write it directly in the console and you are good to go. AWS Lambda seamlessly integrates with numerous other AWS services, including DynamoDB, and you only incur charges for the executions of your function as they occur.

> **Important note**
> AWS Lambda is billed from the time your code begins executing until it returns or otherwise terminates, rounded up to the nearest 1 ms. Pricing is also dependent on the amount of memory that you allocate to the function. For our testing and to ensure we don't exceed the Free Tier allowance, we will stick with the lowest memory allocation of 128 MB.

Using our scan and query examples from the previous section, let us see how we can make use of AWS Lambda to query our DynamoDB data. If you have signed up for an AWS account using the AWS Free Tier, you have 1 million free Lambda executions per month, forever, so the examples we run here will not cost you anything to run.

1. In the AWS Management Console search bar, enter Lambda and click on the service icon, as shown in the following screenshot:

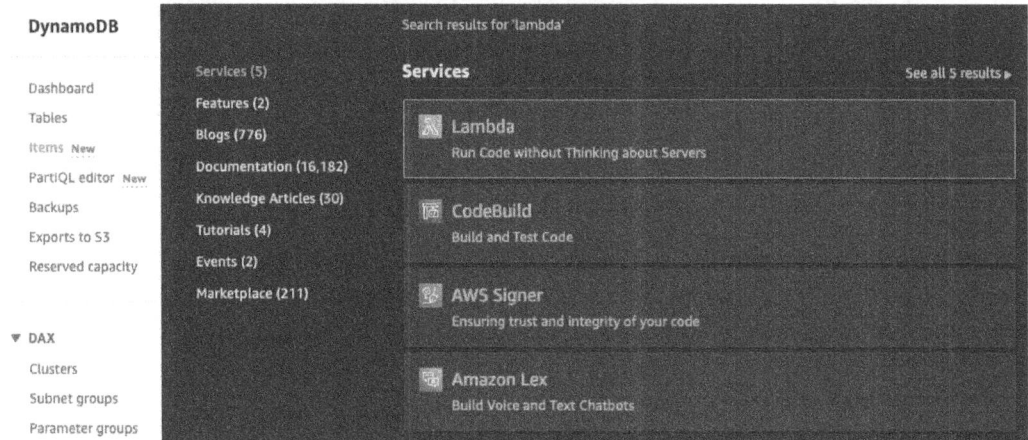

Figure 2.22 – Lambda in the Services search bar

2. By default, you should be taken to the **Functions** landing page. If not, from the left-hand navigation, click on **Functions**. We're going to stick with the selected default option of **Author from scratch** and we will give our function a name. Ensure you select **Python** (at the time of writing, the supplied Python version is 3.12), and leave the **Architecture** default as **x86_64**, as shown in the following figure:

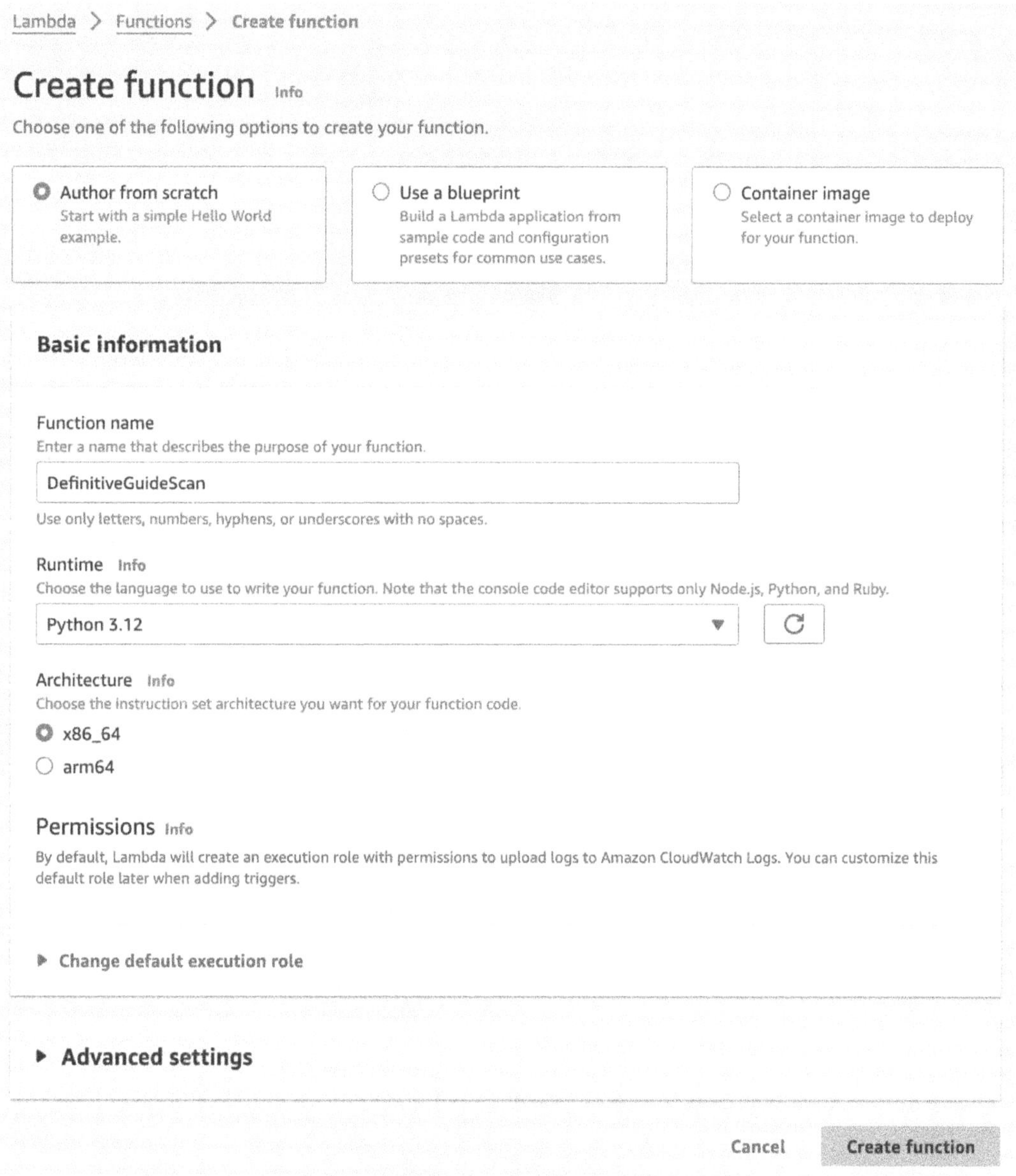

Lambda > Functions > **Create function**

Create function Info

Choose one of the following options to create your function.

● **Author from scratch**	○ **Use a blueprint**	○ **Container image**
Start with a simple Hello World example.	Build a Lambda application from sample code and configuration presets for common use cases.	Select a container image to deploy for your function.

Basic information

Function name

Enter a name that describes the purpose of your function.

```
DefinitiveGuideScan
```

Use only letters, numbers, hyphens, or underscores with no spaces.

Runtime Info

Choose the language to use to write your function. Note that the console code editor supports only Node.js, Python, and Ruby.

```
Python 3.12                                    ▼        ⟳
```

Architecture Info

Choose the instruction set architecture you want for your function code.

● x86_64

○ arm64

Permissions Info

By default, Lambda will create an execution role with permissions to upload logs to Amazon CloudWatch Logs. You can customize this default role later when adding triggers.

▶ **Change default execution role**

▶ **Advanced settings**

Cancel   **Create function**

Figure 2.23 – Creating a new Lambda function

3.   We don't need to change anything under **Advanced settings**, but we do need to work in the **Permissions** section. Here, we want to create a new **Execution role,** an IAM role that AWS Lambda will use to perform actions implemented in the function code, just for this function. So, under the **Change default execution role** menu, ensure that **Create a new role with basic Lambda permissions** is selected, as shown in the following figure:

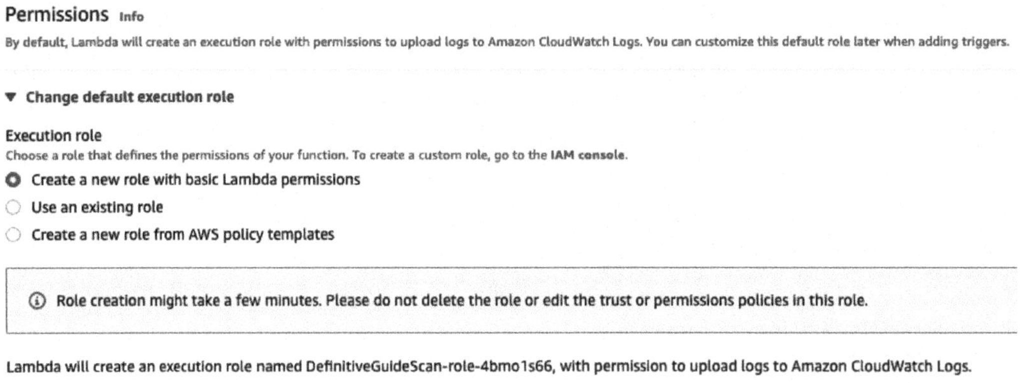

Permissions Info

By default, Lambda will create an execution role with permissions to upload logs to Amazon CloudWatch Logs. You can customize this default role later when adding triggers.

▼ Change default execution role

Execution role

Choose a role that defines the permissions of your function. To create a custom role, go to the **IAM console**.

◉ Create a new role with basic Lambda permissions

◯ Use an existing role

◯ Create a new role from AWS policy templates

ⓘ Role creation might take a few minutes. Please do not delete the role or edit the trust or permissions policies in this role.

Lambda will create an execution role named DefinitiveGuideScan-role-4bmo1s66, with permission to upload logs to Amazon CloudWatch Logs.

Figure 2.24 – Changing the default execution role

4.   Once that is done, click on **Create function**. Your function will now be created and once it's ready, you will be taken straight to the function screen with the source code available onscreen, as shown in the following figure:

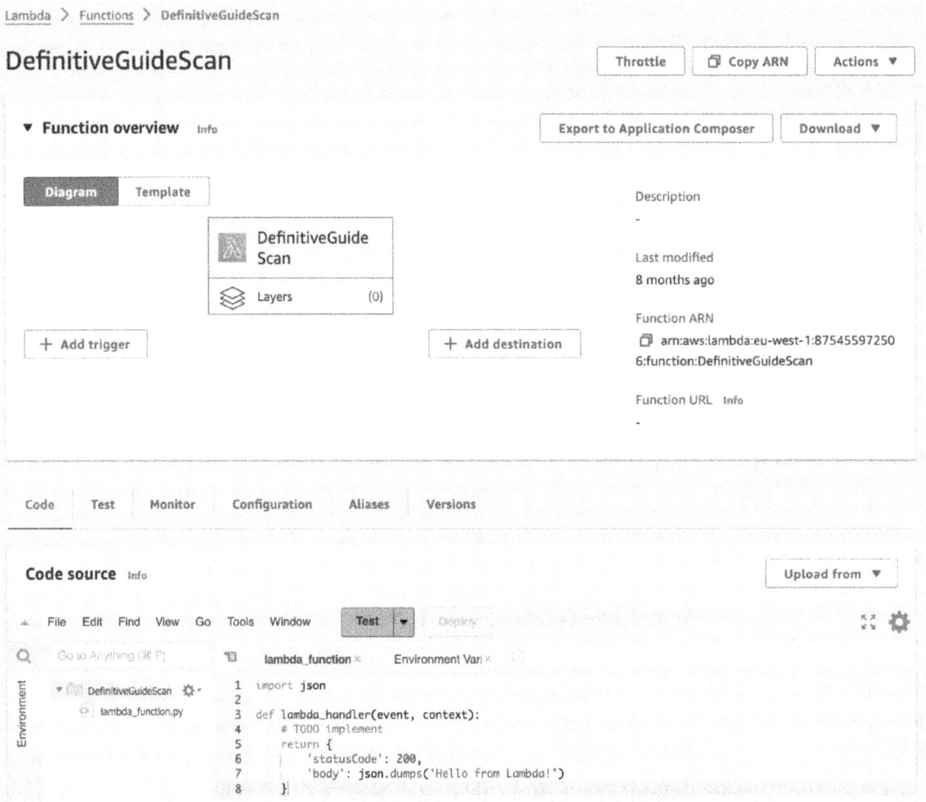

Figure 2.25 – Authoring a new Python function in Lambda

You will see the Lambda service has already pre-populated some basic Python code in the function for us, mainly a `lambda_handler` function. It is important to keep that in place as Lambda will call that code when the function is executed, and it is here that we will place our scan code. The following is the modified scan code to work with the Lambda function:

```python
import boto3

dynamodb = boto3.resource('dynamodb', region_name='eu-west-2')
table    = dynamodb.Table('DefinitiveGuide01')

def lambda_handler(event, context):
    response = table.scan(ReturnConsumedCapacity='TOTAL')
    items = response['Items']

return {
    'statusCode': 200,
    'body': items
}
```

As you can see, our code is fairly similar to before, except instead of printing the item to the screen, we are returning it from the function directly. Another added benefit of using AWS Lambda is that it automatically has the AWS SDK bundled with it, so there is no need to install anything for us to work with AWS services—it's all taken care of for us.

Before running any code that you have changed in the console, you need to deploy it to the Lambda function. This is done simply by clicking first on the **Deploy** button. If we now test our function, we immediately get an error in the **Executions** results tab:

```
[ERROR] ClientError: An error occurred (AccessDeniedException) when
calling the Scan operation
```

At first, this may be concerning, but what is happening here is that AWS is working on the same principle of least privilege that we talked about earlier. The execution role we gave our function allows it to be run by Lambda, but we have not allowed it to do anything else; specifically, we have not given it permission to communicate with DynamoDB. Let us fix that:

1. From the Lambda screen, click on the **Configuration** tab.

2. At the very top, we have the **Execution role** that was created for this function. If you click on the role name highlighted in blue, you will be taken to the IAM console in a new window. We're now going to follow the path we took earlier for our SDK user and give the Lambda function read-only access to DynamoDB.

3. With the **Permissions** tab open for this role, click on **Attach policies** under the **Add permissions** drop-down. Search for DynamoDB and tick the checkbox for **AmazonDynamoDBReadOnlyAccess**, as shown in the following figure:

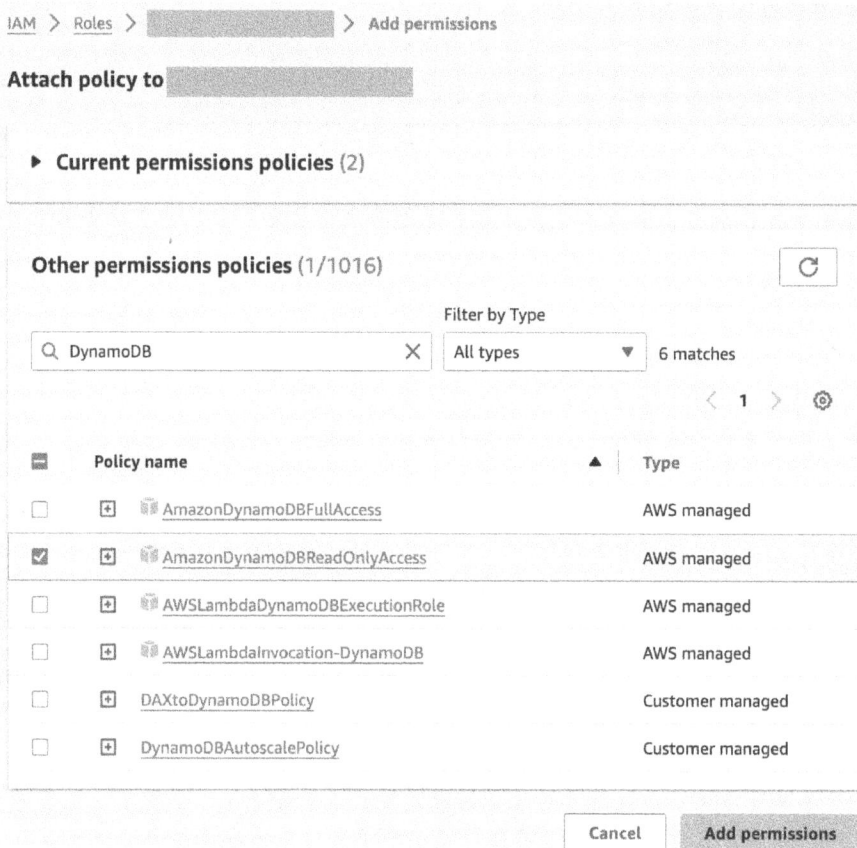

Figure 2.26 – Adding permissions to the Lambda service role

4. Once that is done, click on **Add permissions**. Once that's added, you can close this window and go back to the Lambda function.

5. Click on **Test** once more and in the **Execution results** tab, we should now see the following output (truncated in the figure):

```
 1 {
 2     "statusCode": 200,
 3     "body": [
 4         {
 5             "year": 2005,
 6             "artist": "Evolve",
 7             "id": "ADJ003",
 8             "title": "Safe To Dream"
 9         },
10         ...
11         {
12             "year": 2004,
13             "artist": "The Thrillseekers",
14             "id": "ADJ001",
15             "title": "NewLife"
16         }
17     ]
18 }
```

Figure 2.27 – Response of function test run

Our Lambda function has now successfully fired up, executed the Python code, connected to DynamoDB, performed a scan on our table, and returned the results.

We need to create another separate function for our query to get a single item, but the process is almost identical. The only change we need to make (apart from the different code that follows) is when we create our function—instead of creating a new execution role, we can reuse the scan role we created to automatically give us read-only access to DynamoDB. Following is the code for the query:

```python
import boto3

dynamodb = boto3.resource('dynamodb', region_name='eu-west-2')
table = dynamodb.Table('DefinitiveGuide01')

def lambda_handler(event, context):
    response = table.query(
        KeyConditionExpression='id = :catnum',
        ExpressionAttributeValues={
            ':catnum': 'ADJ004'
        }
    }
```

```
    )
    items = response['Items']
    return {
        'statusCode': 200,
        'body': items
    }
```

6. Click on **Deploy** first, then on **Test**, and in the **Execution results** tab, we should now see the following output:

```
 1 {
 2      "statusCode": 200,
 3      "body": [
 4          {
 5              "year": 2005,
 6              "artist": "Hydra",
 7              "id": "ADJ004",
 8              "title": "Affinity"
 9          }
10      ]
11 }
```

Figure 2.28 – Response of function test run

And that's all there is to serverless computing in the cloud! As you can see, using AWS Lambda is a lot simpler and quicker than setting up environments on your local computer.

Going forward with the further examples in this book, feel free to choose the preferred method for you. The examples we run will work both locally with the SDK and in the cloud using AWS Lambda. The code examples shown will be for locally executed Python code, so if you are running in AWS Lambda, you will need to modify them slightly to work inside the `lambda_handler` function, as we have done with the examples in this section.

Now that we have walked through some AWS Lambda function examples, let us look at installing DynamoDB Local—a Java-based application that emulates DynamoDB operations intended to be used for testing or development of DynamoDB-based systems.

Installing DynamoDB local

The final piece in getting up and running is the option to run DynamoDB locally. If you're not able to create an AWS account or simply don't want to, you can still follow along with the majority of examples and data modeling exercises in this book. DynamoDB local is a great option if you would like to develop offline, too, as it provides a local server version of the DynamoDB services. It is not recommended for production use and does not support all the service features the AWS-hosted version does, but it does provide a fantastic way to get started locally with the SDK and other tools we will be using.

DynamoDB local is available for download for free via a few options. The first option is an executable Java (.jar) file, which runs on Windows, macOS, Linux, and other platforms that support Java. Secondly, it is available as an Apache Maven dependency. You can also download and run it as a Docker image, and lastly, you can set up DynamoDB local with a flip of a button via the NoSQL Workbench—a free client-side tool that allows you to develop and visualize your data model for DynamoDB while also providing an interface to interact with the DynamoDB web service. We will learn more about the NoSQL Workbench in the next chapter, *NoSQL Workbench for DynamoDB*. All download and installation files can be found on the *Deploying DynamoDB locally on your computer* (7) URL. Given that I am using Docker for developing and running my Python SDK code, I find the Docker container easiest to get going with, but the NoSQL Workbench option is a good one too.

If you are using the SDK and want to use it in conjunction with DynamoDB local, all you need to do once you have it installed is specify the endpoint of the DynamoDB local server. This is done as follows:

```
import boto3

dynamodb = boto3.resource('dynamodb',
                          endpoint_url='http://localhost:8000')
```

This assumes that you have DynamoDB local running on your localhost loopback address (otherwise you can change it for the necessary hostname or IP address) and running on port 8000.

And with that, we now have many development and operational options available to use in order to start getting our hands on DynamoDB.

Summary

In this chapter, we covered the programming and access methods we'll use when working with DynamoDB. We started out by using the AWS Management Console to not only create our table but to add our first items into the database too.

We then expanded on our example table, introduced the AWS SDKs that can be used to start building real-world applications that utilize DynamoDB as the backend database, and showed how quick it is to get up and running with very few lines of code.

With this, we also introduced the concept of least privilege access and permissions inside AWS using the IAM service that fundamentally underpins everything in AWS.

Building on the SDKs, we explored how we can run code as functions in the cloud using a serverless compute model, AWS Lambda, that removes the need for any software installations or resources on our local machines, along with being able to download and run DynamoDB locally as a service—all for free.

In the next chapter, we will look at a final piece of software that can really help us with our data modeling exercises. It's a great, free tool from AWS that gives us a way to simplify and understand how our data models will be interpreted by DynamoDB.

References

1. *Wikipedia, The Thrillseekers*: https://en.wikipedia.org/wiki/The_Thrillseekers

2. *Discogs, The Thrillseekers discography*: https://www.discogs.com/label/24762-Adjusted-Music

3. *AWS Tools*: https://aws.amazon.com/tools

4. *AWS Docs, Managing access keys for IAM users*: https://docs.aws.amazon.com/IAM/latest/UserGuide/id_credentials_access-keys.html#Using_CreateAccessKey

5. *Python Boto3 Docs*: https://boto3.amazonaws.com/v1/documentation/api/latest/guide/quickstart.html

6. *DockerHub, DynamoDB Local*: https://hub.docker.com/r/amazon/dynamodb-local/

7. *AWS Docs, DynamoDB Local*: https://docs.aws.amazon.com/amazondynamodb/latest/developerguide/DynamoDBLocal.DownloadingAndRunning.html

3
NoSQL Workbench for DynamoDB

Having learned about the AWS Management Console, the SDKs, and how to interact with DynamoDB in the previous chapter, the next and probably the most important topic is **NoSQL data modeling**. A NoSQL database can easily appear to perform poorly or create more blockers than opportunities if the data model is not efficiently aligned with the application's access patterns.

Data modeling isn't a one-step process for an application that continues to grow and add more functionalities as it grows. In most cases, it is an iterative process whereby an initial data model is agreed upon to serve the initial set of business requirements. As these requirements are updated, corresponding changes to the data model may be necessary. This should be done with minimal disruption to the production environment. To design and iterate over the first, second, and Nth models, you need both a visual representation mechanism and a tool for the data model development.

For RDBMS in particular, the **Entity-Relationship** (**ER**) modeling process introduced in the 1970s (1) works well even in the 21st century. You know exactly which entities are established (which usually become individual RDBMS tables linked using foreign keys following Edgar F. Codd's relational model and normalization approach), as well as the nature of the relationship between those entities. Since this mechanism is widely used and understood, we will continue to use the representation part of it.

In this chapter, we will go through NoSQL Workbench for DynamoDB, a cross-platform client-side application for designing, creating, querying, and managing NoSQL databases. We will park the data modeling concepts for now, as they deserve its own set of chapters.

By the end of this chapter, you will be able to find your way around NoSQL Workbench for DynamoDB in particular, understand the functions of the tool, and use it for modern database development and operations.

In this chapter, we are going to cover the following main topics:

- Technical requirements
- Setting up NoSQL Workbench
- Viewing an existing data model
- Making changes to the `Employee` data model

Technical requirements

To follow along and use NoSQL Workbench in this chapter, we need a laptop or PC with a macOS, Windows, or Linux (Ubuntu 12.04, Fedora 21, Debian 8, or any newer versions of these distributions) operating system.

Setting up NoSQL Workbench

Let us start by downloading NoSQL Workbench and installing it on our local machines. The tool itself is free of cost and supports multiple platforms such as macOS, Windows, and major Linux distributions. First, you will download it from the AWS docs (2) and make sure to note the directory you download it to.

Downloading NoSQL Workbench

Depending on your operating system, you can download the appropriate package from the AWS docs (2). This process should be straightforward.

> **Fun Fact**
> The NoSQL Workbench installation packages are served off Amazon **Simple Storage Service (S3)**, an AWS object storage service with 99.999999999% (11 nines) of durability. An example from the public docs about S3: if you store 10,000,000 objects in it, you can on average expect to incur a loss of a single object once every 10,000 years.

Installing NoSQL Workbench

Yet again, the installation procedure may vary depending on your operating system. The corresponding operating system's installation wizard should guide you through setting it all up well. One thing to remember while installing NoSQL Workbench is that you must opt in for installing **DynamoDB local** bundle as well. This would be quite apparent and straightforward when you go through your operating system's installation wizard.

Once you fire up the application, the outcome should be the same irrespective of the operating system, that is, you should be able to see a landing page called the **AWS database catalog**, as shown in the following figure.

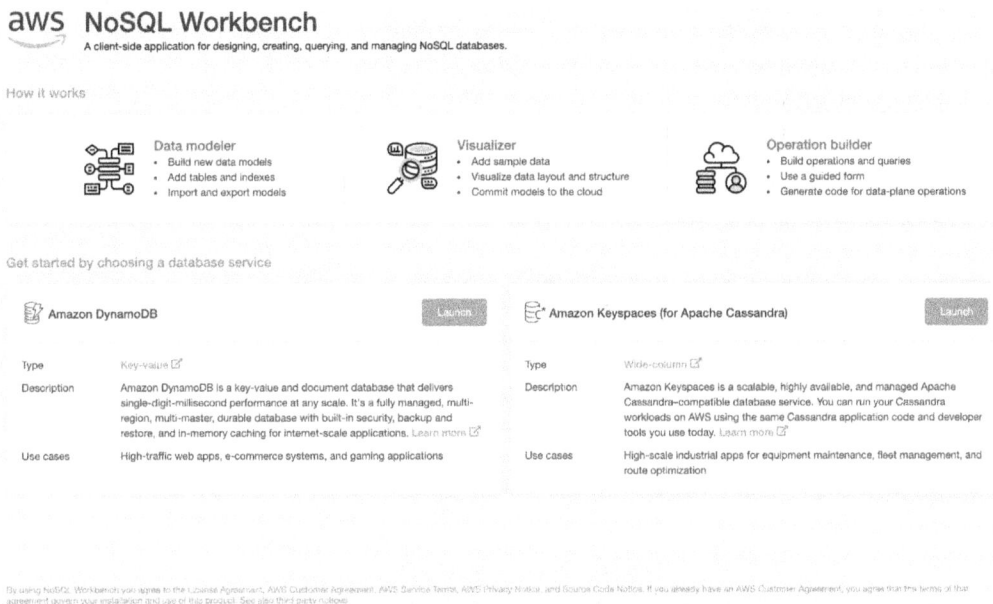

aws NoSQL Workbench
A client-side application for designing, creating, querying, and managing NoSQL databases.

How it works

Data modeler
• Build new data models
• Add tables and indexes
• Import and export models

Visualizer
• Add sample data
• Visualize data layout and structure
• Commit models to the cloud

Operation builder
• Build operations and queries
• Use a guided form
• Generate code for data-plane operations

Get started by choosing a database service

Amazon DynamoDB		Amazon Keyspaces (for Apache Cassandra)	
	Launch		Launch
Type	Key-value	Type	Wide-column
Description	Amazon DynamoDB is a key-value and document database that delivers single-digit-millisecond performance at any scale. It's a fully managed, multi-region, multi-master, durable database with built-in security, backup and restore, and in-memory caching for internet-scale applications. Learn more	Description	Amazon Keyspaces is a scalable, highly available, and managed Apache Cassandra-compatible database service. You can run your Cassandra workloads on AWS using the same Cassandra application code and developer tools you use today. Learn more
Use cases	High-traffic web apps, e-commerce systems, and gaming applications	Use cases	High-scale industrial apps for equipment maintenance, fleet management, and route optimization

By using NoSQL Workbench you agree to the License Agreement, AWS Customer Agreement, AWS Service Terms, AWS Privacy Notice, and Source Code Notice. If you already have an AWS Customer Agreement, you agree that the terms of that agreement govern your installation and use of this product. See also third party notices.

Figure 3.1 – The NoSQL Workbench database catalog page

> **Important note**
> The NoSQL Workbench application features, layouts, screenshots, and figures provided in this chapter are of version 3.12.0 (3). Like most things in the technology space, NoSQL Workbench can evolve. New functionalities and layouts may appear by the time you follow along with this content.

The database catalog page has some introductory information about the features of NoSQL Workbench. Apart from database development for DynamoDB, NoSQL Workbench also supports development and operations for **Amazon Keyspaces (for Apache Cassandra)** – a highly scalable, available, fully managed, serverless Apache Cassandra-compatible AWS database offering.

That's it for this section. Here, we learned how to go about setting up NoSQL Workbench on our local machines. This step is completely optional, but is needed if you prefer to follow along, as we will go through the different features using samples. Now, let us check out the DynamoDB part of the tool.

Viewing an existing data model

Thankfully, NoSQL Workbench also comes with sample datasets and data models that can help in the learning process. These sample data models are categorized into **Introductory** and **Advanced** skill levels. For the purpose of understanding, we shall review the different functionalities of the workbench using one of the **Introductory** sample data models: **Employee Data Model**. After selecting **Launch** that is next to **Amazon DynamoDB** in the initial window of NoSQL Workbench, use the **Import** option next to this model to start exploring, as shown in the figure that follows:

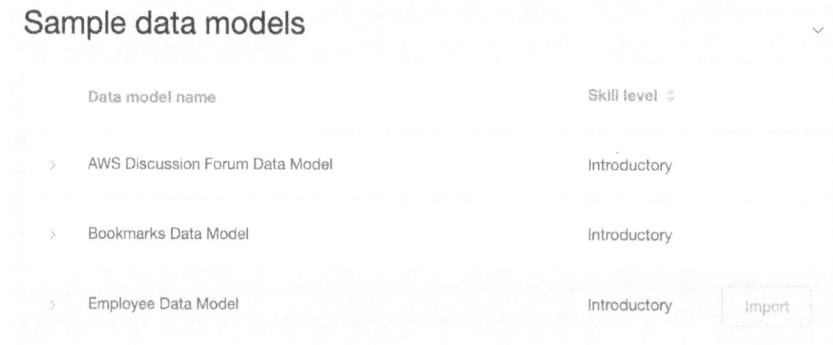

Sample data models

Data model name	Skill level
› AWS Discussion Forum Data Model	Introductory
› Bookmarks Data Model	Introductory
› Employee Data Model	Introductory Import

Figure 3.2 – Sample data models

In case the `Employee` data model is not available to you in the tool already, I have provided this in the public GitHub repository (4) containing assets for this book. All you would need to do to use the model is download the `Employee Data Model.json` file from the repo and use the **Import data model** option in the top right of the app in the **Getting started** section, followed by the **Import NoSQL Workbench model JSON** option, as shown in the following figure:

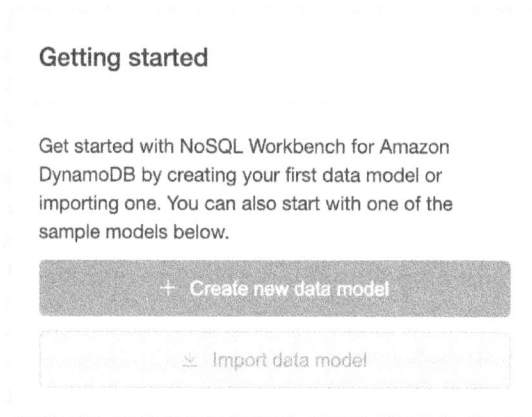

Getting started

Get started with NoSQL Workbench for Amazon DynamoDB by creating your first data model or importing one. You can also start with one of the sample models below.

+ Create new data model

⤓ Import data model

Figure 3.3 – The Getting started section

Next, let us look at the various tools that come with NoSQL Workbench and their applications in our database development.

Tool overview

NoSQL Workbench provides different subtools such as Data Modeler, Visualizer, and Operation Builder, as well as DynamoDB Local. Let us learn a bit about these subtools:

- The Data Modeler is where you define and iterate over the data model design. This includes defining almost all options you saw on the DynamoDB Management Console in the previous chapter and then some. In addition to defining tables, this subtool also allows you to define facets – basically, these are the access patterns that the particular data model must support.

- The Visualizer subtool allows you to see your model in practice. NoSQL modeling can appear quite messy at first and the Visualizer does an excellent job at making things pretty. For example, consider an e-commerce application that stores information about warehouses, shipments, orders, and products within each order, all together in a single table. This is potentially tricky to visualize without any help, right? With the Visualizer, you can also populate dummy data for your models to get the complete picture of what your model with data would look like and how it answers to the different facets or access patterns. You can visualize how the secondary indexes represent the same data as per their individual primary key schemas. Finally, you also have the option to commit the data model and sample data directly to the DynamoDB web service, say, to your development environment, to use it against your application and iterate over the model if needed. Both the Visualizer and the Data Modeler support the ability to add sample data, either managed by yourself or by NoSQL Workbench.

- The last subtool is the Operation Builder, which allows you to interact with existing DynamoDB resources in your AWS cloud environment or simply DynamoDB Local installation. Not only that; it also provides you with boilerplate code snippets for all DynamoDB data plane operations available specific to a particular table or secondary index using the respective AWS SDKs for different programming languages.

- In addition to the built-in subtools within NoSQL Workbench, the client application provides a convenient feature to effortlessly launch a DynamoDB Local instance. This feature enables you to develop and interact seamlessly with your established data model using DynamoDB operations. This hands-on experience allows you to gain insights into how the operations are executed on your implemented data model without having to use the DynamoDB web service and, therefore, pay for it.

Let us dive deeper into each subtool by reviewing the sample Employee data model.

Diving deeper into the tool

The navigation pane on the left-hand side of the workbench is your friend for going to different subtools that are part of NoSQL Workbench. Importing the `Employee` data model will take you to the first subtool: the Data Modeler.

Data Modeler

The Data Modeler is the model definition area. This is where you do the designing of the model. This includes creating tables and choosing the primary key schemas – be that a partition key alone or a combination of the partition and the sort key. This is also where you specify the capacity modes for each table. If you are following along with the tool at your end, select the `Employee` table from the **Tables** section to view the following screen:

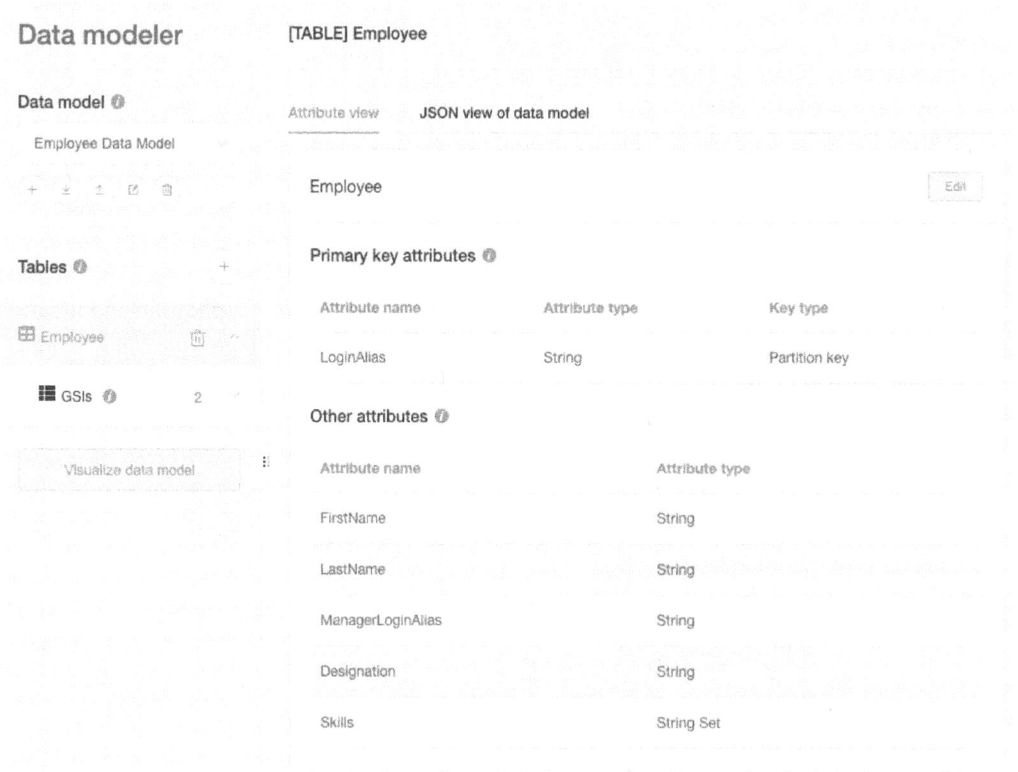

Data modeler   [TABLE] Employee

Data model ⓘ   Attribute view   JSON view of data model

Employee Data Model

+ ↓ ↑ ☑ 🗑   Employee   Edit

Tables ⓘ   +   **Primary key attributes** ⓘ

Attribute name	Attribute type	Key type
LoginAlias	String	Partition key

⊞ Employee   🗑

▦ GSIs ⓘ   2   **Other attributes** ⓘ

Attribute name	Attribute type
FirstName	String
LastName	String
ManagerLoginAlias	String
Designation	String
Skills	String Set

Visualize data model

Figure 3.4 – The Data Modeler view

From the preceding figure, you will notice that the `Employee` table has **LoginAlias** as its partition key, which is of the `String` type. This table does not have a sort key in its primary key schema. Along with the **Primary key attributes** section containing the primary key schema details, you can also see an **Other attributes** section, which contains what are called **non-key attributes** for the table. Non-key attributes are simply the attributes that are not part of the primary key schema for the particular table.

Just like the table, using the Data Modeler subtool you can also define the secondary indexes you need for the table and the primary key schema for each secondary index. When you clicked on the `Employee` table, you may have also observed that there were two **Global Secondary Indexes (GSIs)** listed for the table in the same **Tables** section. On hitting the tiny downward arrow next to the GSIs, you will see that these are named **Name** and **DirectReports**, respectively. Selecting a GSI displays its model definition, including the primary key schema, the non-key attributes, and the data types for each attribute.

For an advanced, curious user, you may also review the JSON view of the model, which is also what can be exported or imported for documenting data model revisions. As an architect or developer, I would export this JSON model and add it to a version control tool such as Git to document different revisions of a model that I can go back to at any given time. At the time of publishing, NoSQL Workbench itself does not provide native version control support. The Data Modeler also allows you to specify facets or the access patterns the model needs to support. The access patterns that the `Employee` data model supports are as follows:

- Get employee details for a given employee login alias
- Get employee details for a given first name
- Get all employees reporting to a given manager by manager login alias

Keep these access patterns in mind as we will come back to these after learning about the other subtools NoSQL Workbench provides. For now, hit the **Visualize data model** option or select the **Visualizer** option on the navigation bar.

Visualizer

The Visualizer or the Data Visualizer subtool is where your defined tables and secondary indexes start making more sense. This is where you can add dummy data to your defined tables and see how data would appear to your application interacting with the table. The data you add to the table will be represented in different ways for each secondary index according to key schemas defined for each index. Since the `Employee` data model comes with sample data, looking at the table via the Visualizer gives you a clear picture of the table and any secondary indexes.

The following is the view you will see on the Visualizer for the `Employee` data model:

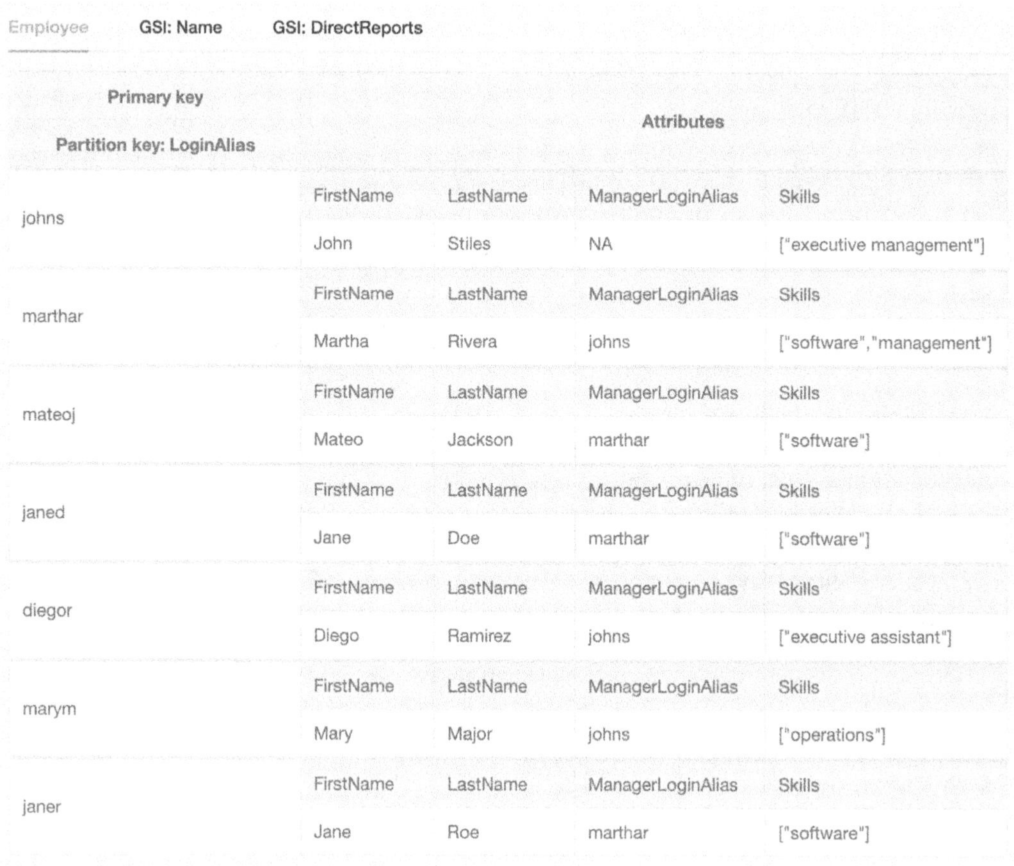

Employee	GSI: Name	GSI: DirectReports		

Primary key		Attributes		
Partition key: LoginAlias				
johns	FirstName	LastName	ManagerLoginAlias	Skills
	John	Stiles	NA	["executive management"]
marthar	FirstName	LastName	ManagerLoginAlias	Skills
	Martha	Rivera	johns	["software","management"]
mateoj	FirstName	LastName	ManagerLoginAlias	Skills
	Mateo	Jackson	marthar	["software"]
janed	FirstName	LastName	ManagerLoginAlias	Skills
	Jane	Doe	marthar	["software"]
diegor	FirstName	LastName	ManagerLoginAlias	Skills
	Diego	Ramirez	johns	["executive assistant"]
marym	FirstName	LastName	ManagerLoginAlias	Skills
	Mary	Major	johns	["operations"]
janer	FirstName	LastName	ManagerLoginAlias	Skills
	Jane	Roe	marthar	["software"]

Figure 3.5 – Visualizing the Employee data model

As you can see from the preceding figure, the `Employee` table shows well-structured sample data along with tabs for two GSIs. The secondary index view is probably my favorite part of the tool because of its capability to completely change the representation of the data as per the index key schema. Feel free to look at how the same data looks with each GSI based on their individual primary key schemas. We will dive deeper into secondary indexes in its dedicated *Chapter 8, Secondary Indexes*.

The Visualizer also allows an option to import data from a CSV file or add random sample data as per your specified data type automatically, which can be handy to improve productivity.

Once you are confident that your data model efficiently and appropriately addresses all the application access patterns, you can proceed with implementing the specific revision in several ways:

- You can export the model into its own JSON file, which will contain the model definitions as well as any dummy data you added to the table(s)

- You can also export the visual representation of the model with data as an image

- You can also choose to commit the data model to the DynamoDB web service or DynamoDB Local environments

Think about having a development environment in the AWS cloud or locally, where you would like to give the data model a test drive. The workbench makes it easy for you. In order to commit the model to the DynamoDB web service, you will have to establish a connection in the workbench with a valid **Identity and Access Management (IAM)** entity that is authorized to perform DynamoDB actions in the respective AWS account and region. The workbench is also able to recognize any IAM entities stored by default in the .aws folder of your local machine's home directory. For users of AWS CloudFormation (5) – an AWS service that allows you to model, provision, and manage AWS and third-party resources by treating infrastructure as code – I have good news! NoSQL Workbench also allows exporting JSON CloudFormation templates for the designed data model, as shown in the following figure:

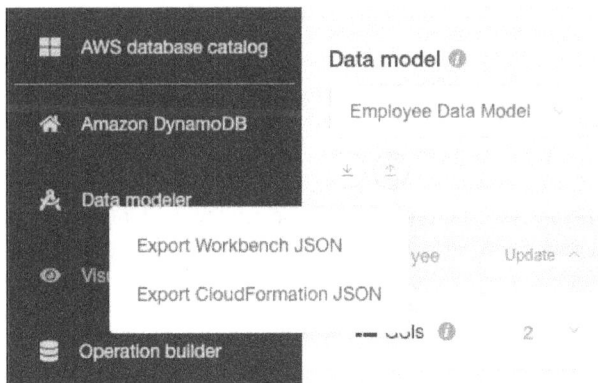

Figure 3.6 – Visualizer export options

Next, go to the **Operation builder** section via the navigation bar on the left of the app.

Operation Builder

The Operation Builder subtool is a graphical user interface that can be used to view, explore, and query DynamoDB resources in the cloud or in DynamoDB Local. It also enables you to perform and build DynamoDB data plane operations via your chosen connection, that is, the AWS cloud environment or DynamoDB Local.

I will create a connection using the `DefinitiveGuide` IAM user that Mike created in the previous chapter. Since I already configured the user credentials on my local machine in that chapter, the workbench will already show me an option for opening a connection with that user to connect to DynamoDB. I configured those user credentials as a named profile, `definitive-guide`, in my local machine, and so I see that reflected on the workbench as I click on **Operation builder**, as shown in the following figure:

Figure 3.7 – definitive-guide in the operation builder connections

On selecting the **DefinitiveGuide** table I created before, the workbench automatically attempts to perform a **Scan** operation to return a glimpse of the data that exists in the table. Speaking of building data plane operations, the UI allows you to use all DynamoDB data plane APIs available, including **PartiQL** – a SQL-compatible query language that can be used with DynamoDB. The following figure shows the output of the default **Scan** performed by the workbench and the available options around it:

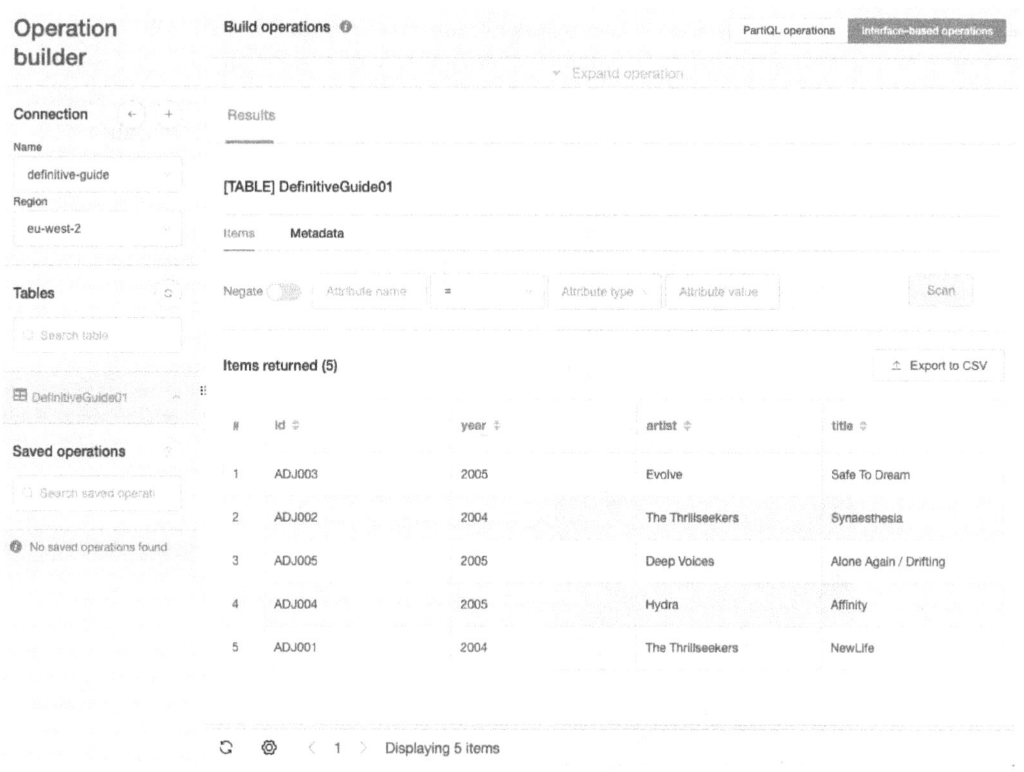

Figure 3.8 – The Operation Builder view

As mentioned previously, the Operation Builder allows the generation of boilerplate code for each of the supported data plane operations. For example, if I need to perform the `Query` operation on the `DefinitiveGuide01` table, where the partition key is `id = "ADJ001"`, I can generate the boilerplate code in Python, Node.js, or Java language and use it while developing my application. I tend to be a bit old school in that I often look up precise source code or extensive documentation, but this approach usually doesn't benefit my productivity. Using the Operation Builder's code generation capability, developer productivity will be high. That's one of the many benefits of NoSQL Workbench!

DynamoDB Local via NoSQL Workbench

In addition to the various built-in subtools of NoSQL Workbench, the free client-side application also offers a convenient one-click setup for DynamoDB Local. If you haven't explored the previous chapter, DynamoDB Local serves as the client-side, free version of the DynamoDB web service, designed to facilitate local development and testing of DynamoDB-based applications. DynamoDB Local, a Java-based application, can easily be set up on your local machine, enabling you to perform most DynamoDB operations as if interacting with the DynamoDB web service. It is important to note that this tool is specifically intended for development purposes and is not recommended for use in a production environment.

Traditionally, setting up DynamoDB Local involved downloading the DynamoDB Local JAR, configuring paths, and running it through a terminal window. Alternatively, you might choose to use Docker or Maven distributions. However, with NoSQL Workbench, configuring DynamoDB Local is simplified to a single switch button. This switch is conveniently located on the bottom of the left navigation pane, as illustrated in the figure that follows:

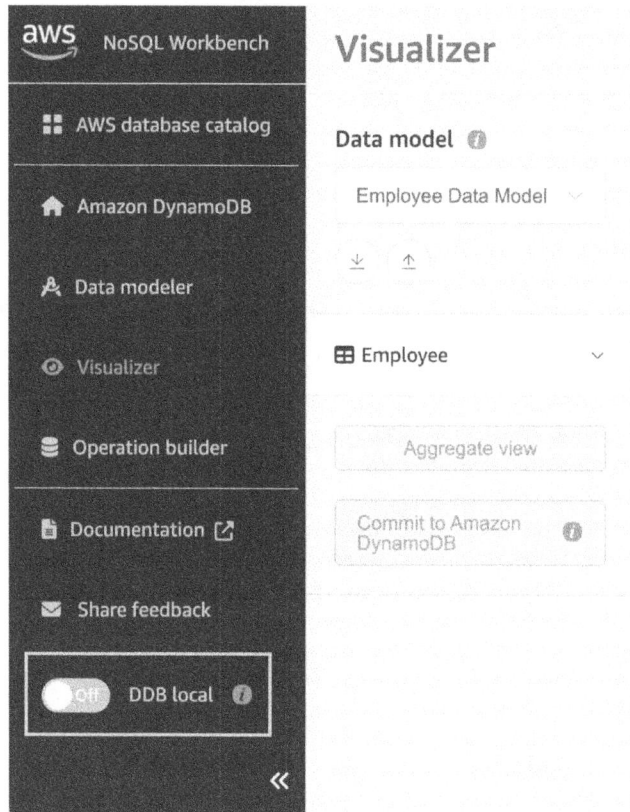

Figure 3.9 – DynamoDB Local from within NoSQL Workbench

Much like connecting to the DynamoDB web service via NoSQL Workbench, you can establish a connection to DynamoDB Local by activating the switch on the workbench. Within the Operation Builder subtool, there is a dedicated tab for connecting to DynamoDB Local. By default, it connects to localhost at port 8000, which is the default port for DynamoDB Local. The figure that follows illustrates the new connection dialog box:

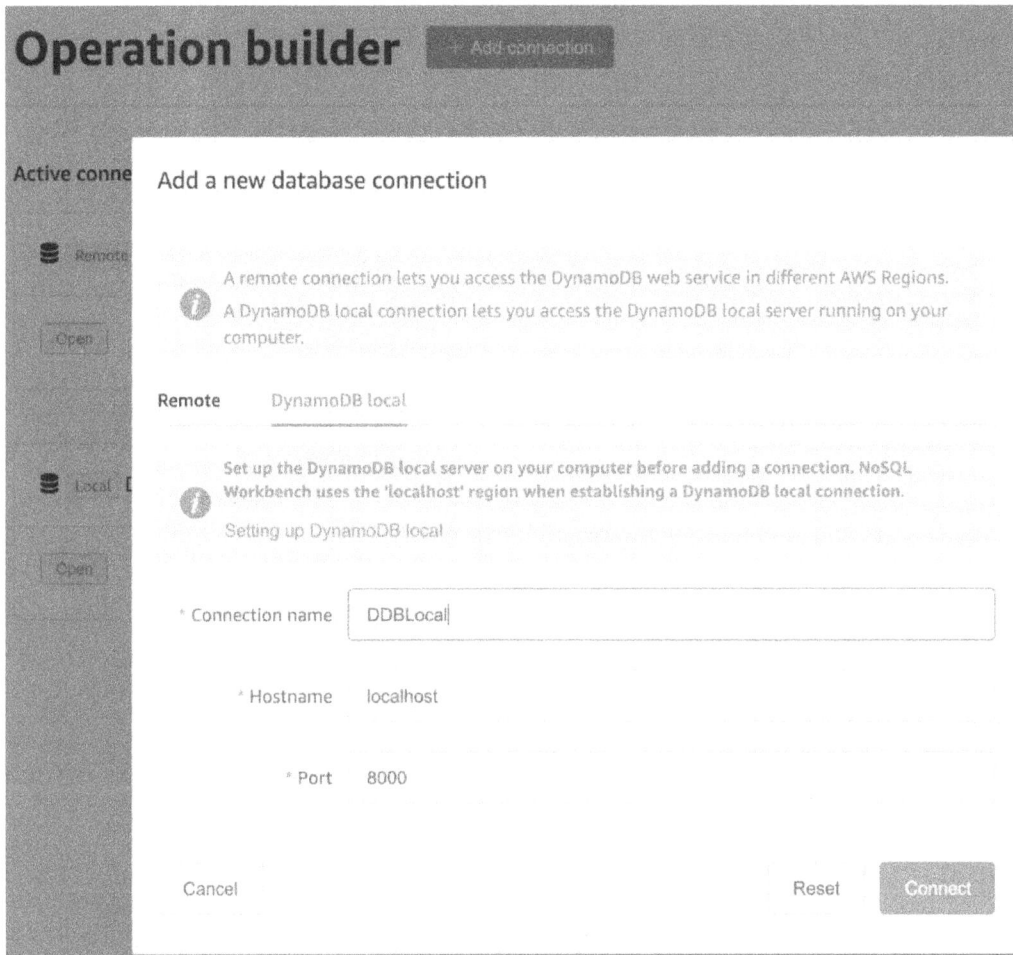

Figure 3.10 – The new connection dialog box for DynamoDB Local

After establishing the connection, you can utilize the **Operation builder** instance to seamlessly interact with the DynamoDB Local instance operating in the background. You have the flexibility to employ other subtools of NoSQL Workbench to work with the DynamoDB Local instance instead of the web service, enabling you to operate offline without the need to create an AWS account. Fantastic, right?!

With that, we have reviewed and learned about the different subtools available for DynamoDB database design and development using NoSQL Workbench. In the next section, we will apply this knowledge by making changes to the Employee data model and reviewing how those changes are reflected in the Visualizer.

Making changes to the Employee data model

After learning about the different capabilities of the workbench, let us put our data architect hat on and modify the `Employee` data model using the Data Modeler subtool. If you recall, the access patterns that the `Employee` data model supports are as follows:

- Get employee details for a given employee login alias
- Get employee details for a given first name
- Get all employees reporting to a given manager by manager login alias

The exact query conditions are not relevant for now as we focus on making changes to this model and visualizing those changes. Let us say that you, the architect, are sent an additional business requirement to support an additional access pattern: *Get all employees for a given designation*

If you observe the model definition for the `Employee` table in the Data Modeler, there is an attribute called `Designation` already, but when visualizing the model with dummy data, there is no item with a value for this attribute. Given that the existing data does not have values for `Designation`, in order to add support for this access pattern, we'll want to make two changes:

1. Review the `Designation` attribute and add designation data for each employee
2. Create a GSI to use `Designation` as its partition key attribute

To add the designation data for each employee, the designation attribute must exist in the data model. Let us go ahead and add this attribute and the data for it next.

Adding additional data

If a new attribute needs to be added to the data model, we need to use the Data Modeler subtool to first define this new attribute and its data type. With DynamoDB being NoSQL, and thus schemaless, not every item in the table would be expected to have this attribute, since it is a non-key attribute for the table. We see this in practice as the `Designation` attribute does not show at all in the Visualizer because none of the items in the table have a value for this attribute.

Before we populate this attribute with data, let us revisit the data modeler. Our goal will be to edit the model definition, specifically focusing on the `Designation` attribute:

1. Go to **Data modeler**, select the `Employee` table, and click **Edit**, which is on the right side of the app. You'll see that the `Designation` attribute has its data type as `String`. There is also an option to add additional attributes to the table. For each attribute added, you need to provide a name and its data type, as shown in the following figure:

Figure 3.11 – Adding new attributes

2. Now that we have reviewed the attribute information, let us populate data for this attribute for the employees. Head to the **Visualizer** subtool and click on the little **Update** option next to the Employee table.

3. Next, select the **Edit data** option on the top right of the app, under **Actions**. Give each employee item a value for the **Designation** attribute. Finally, save these changes. The following figure shows the edit view and I have added **Designation** values to every employee item:

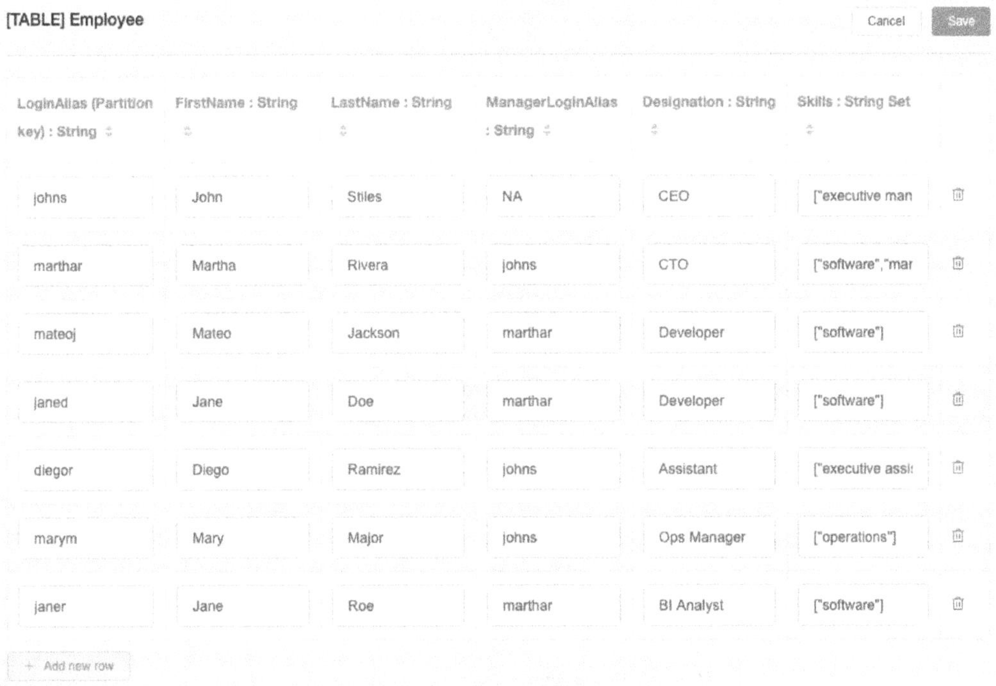

[TABLE] Employee   Cancel   Save

LoginAlias (Partition key) : String	FirstName : String	LastName : String	ManagerLoginAlias : String	Designation : String	Skills : String Set	
johns	John	Stiles	NA	CEO	["executive man	
marthar	Martha	Rivera	johns	CTO	["software","mar	
mateoj	Mateo	Jackson	marthar	Developer	["software"]	
janed	Jane	Doe	marthar	Developer	["software"]	
diegor	Diego	Ramirez	johns	Assistant	["executive assi:	
marym	Mary	Major	johns	Ops Manager	["operations"]	
janer	Jane	Roe	marthar	BI Analyst	["software"]	

+ Add new row

Figure 3.12 – Populating data

For change number two, which is creating a secondary index with the `Designation` attribute as the primary key, we will head back to the **Data modeler** to edit the `Employee` model:

4. By clicking **Edit model** and scrolling a bit, you'll find an **Add global secondary index** option. It requires basic information such as the name of the index (I've set `DesignationIndex`), the primary key schema where the partition key would be, and the `Designation` attribute. We'll add `LoginAlias` as the sort key for the index and leave the remaining configurations as is, like the projection type. The projection type essentially decides what information from the table needs to be propagated to the index for each item. We'll learn more about attribute projections in *Chapter 8, Secondary Indexes*. The following figure shows that I have configured a new GSI using the preceding parameter values:

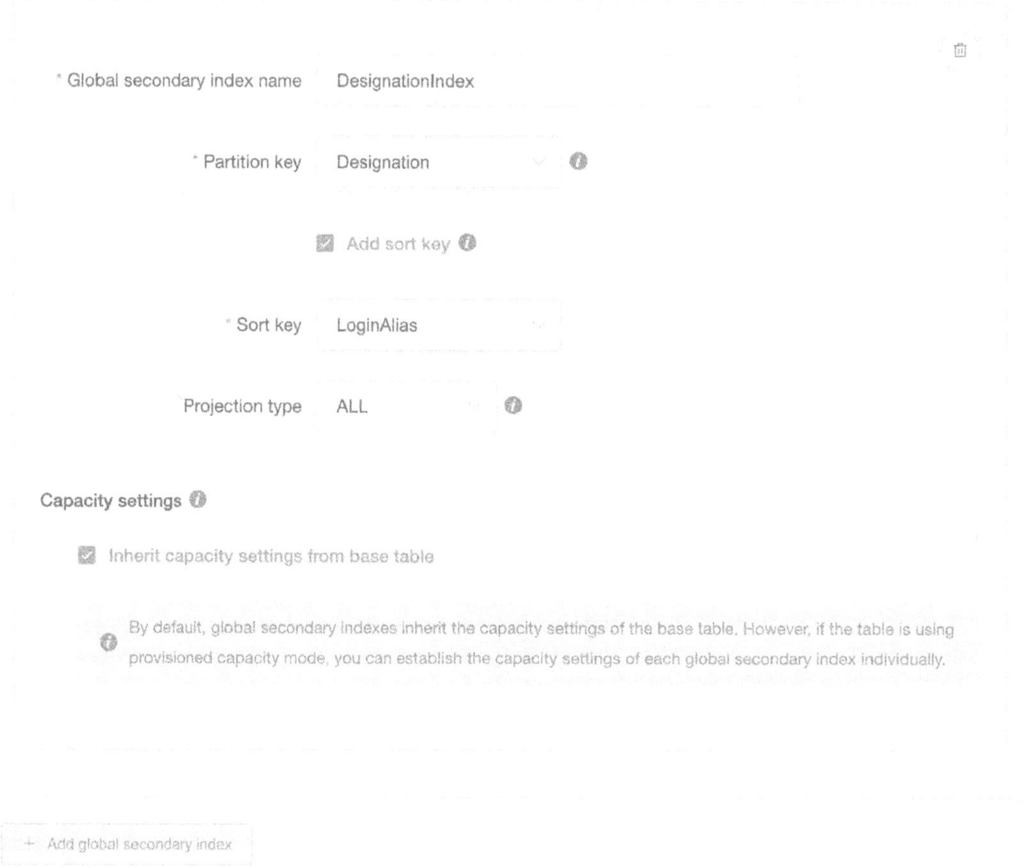

Figure 3.13 – Adding a secondary index

5. After hitting **Save edits** at the bottom of our page, we have now successfully added another global secondary index to the table. This completes change number two.

Let us visualize our changes using the Visualizer. You will find that our new secondary index now appears at the top ribbon of the Visualizer page when we select the Employee table. The following figure illustrates the same:

Employee	GSI: Name	GSI: DirectReports	GSI: DesignationIndex

Primary key		Attributes				
Partition key: Designation	Sort key: LoginAlias					
CEO	johns	FirstName	LastName	ManagerLoginAlias		Skills
		John	Stiles	NA		["executive management"]
CTO	marthar	FirstName	LastName	ManagerLoginAlias		Skills
		Martha	Rivera	johns		["software","management"]
Developer	janed	FirstName	LastName	ManagerLoginAlias		Skills
		Jane	Doe	marthar		["software"]
	mateoj	FirstName	LastName	ManagerLoginAlias		Skills
		Mateo	Jackson	marthar		["software"]
Assistant	diegor	FirstName	LastName	ManagerLoginAlias		Skills
		Diego	Ramirez	johns		["executive assistant"]
Ops Manager	marym	FirstName	LastName	ManagerLoginAlias		Skills
		Mary	Major	johns		["operations"]
BI Analyst	janer	FirstName	LastName	ManagerLoginAlias		Skills
		Jane	Roe	marthar		["software"]

Figure 3.14 – Visualizing the new secondary index

Finally, to serve the new access pattern of *Get all employees for a given designation*, I can simply perform the Query operation on the DesignationIndex secondary index and only pass the partition key value in the Query key condition.

For example, if I want to get all employees having a designation of a developer, I will perform the Query operation on the index, where Designation is set to "Developer". This would return two items from the index since I had set the designations of Jane Doe and Mateo Jackson as Developer. Finally, let us commit this model, along with the sample data, to DynamoDB.

Committing the model to DynamoDB

As the architect or developer, once I am happy with this model, I'd want to commit it to DynamoDB, so let us do exactly that. In the Visualizer itself, you will find the little **Commit to Amazon DynamoDB** option. You will be asked for a connection, that is, the IAM credentials. I'll be using the **definitive-guide** connection again. The following figure shows the commit option:

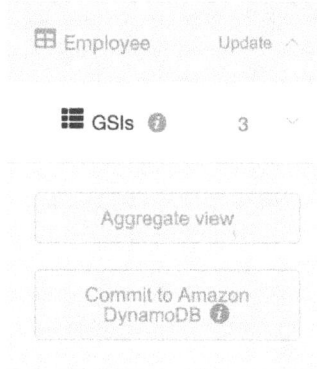

Figure 3.15 – Committing to DynamoDB

> **Important note**
>
> Be aware that committing a model to the DynamoDB web service can and will incur costs. Although some of the DynamoDB charges may be covered under the AWS free tier, the model created with default capacity settings also sets up AWS Application Auto Scaling alarms, which will not be covered by the AWS free tier. If you do not wish to use the webservice, you can use a DynamoDB Local instance through NoSQL Workbench and connect to it.

One step I am not covering here is that my **DefinitiveGuide01** IAM user only had the **AmazonDynamoDBReadOnlyAccess** policy attached to it, so in order to commit the Employee model to the DynamoDB webservice, I had to also attach the **AmazonDynamoDBFullAccess** policy to my user. Although I would never suggest using any of the **FullAccess** policies to my customers in their production applications (recall the least privilege concept from *Chapter 2*, *The AWS Management Console and SDKs*), since I am playing around in a non-production AWS account, I'll allow it.

Attaching the **AmazonDynamoDBFullAccess** policy would give my IAM user permissions to manipulate DynamoDB tables, as well as set up Application Auto Scaling – a feature that performs automatic scaling activities on the table and indexes based on throughput metrics for each table and index independently. We will dive deeper into Application Auto Scaling in *Chapter 9*, *Capacity Modes and Table Classes*. The policy also allows certain mutating actions on other AWS services that integrate with DynamoDB. You can find more information on attaching policies to IAM users in the AWS docs (5).

Once the Workbench creates service-side DynamoDB resources for me, I can review the table via the Operation Builder or via the DynamoDB console itself. The following figure shows the **Operation builder** view of my table and the results of a Scan operation on the Employee table:

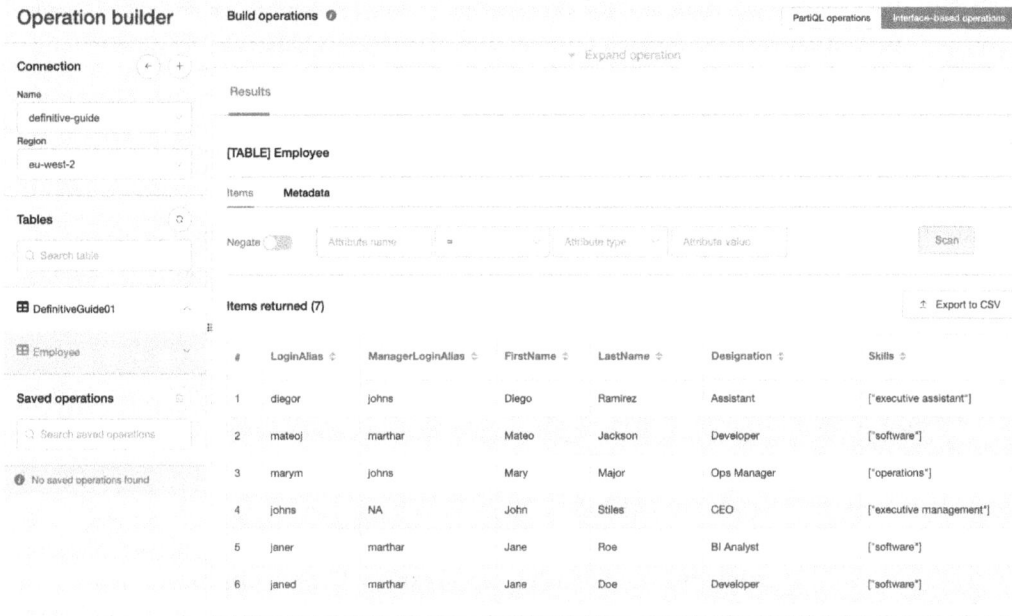

#	LoginAlias	ManagerLoginAlias	FirstName	LastName	Designation	Skills
1	diegor	johns	Diego	Ramirez	Assistant	["executive assistant"]
2	mateoj	marthar	Mateo	Jackson	Developer	["software"]
3	marym	johns	Mary	Major	Ops Manager	["operations"]
4	johns	NA	John	Stiles	CEO	["executive management"]
5	janer	marthar	Jane	Roe	BI Analyst	["software"]
6	janed	marthar	Jane	Doe	Developer	["software"]

Figure 3.16 – Interacting with the committed Employee table

With that, we have explored the most useful functions of NoSQL Workbench that will help you design and manage DynamoDB data models, as well as interact with DynamoDB resources in the AWS cloud or locally via DynamoDB Local.

> **Important note**
> Consider this a reminder to delete any resources you may have created while going through this chapter in your AWS account. We will continue using similar examples throughout the book wherever possible; you can always re-create the resources when you get to those parts and not have to pay for resources idling until then.

Summary

In this chapter, we introduced the NoSQL Workbench client application, a valuable tool for database development and operations. We explored the data modeler, focusing on its role in defining effective data models. The Visualizer was discussed for its powerful capability to represent data from different perspectives based on key schemas for tables and secondary indexes. Additionally, we dove into the functions of the Operation Builder, highlighting its role in interacting with DynamoDB services or locally, enhancing developer productivity by generating boilerplate code for various DynamoDB data plane operations. Apart from the subtools, we also learned about using DynamoDB Local via NoSQL Workbench.

We walked through a sample `Employee` dataset, demonstrating how to update data in the model, define a new secondary index, and visualize the resulting index data. The final step involved committing the model to DynamoDB, followed by verification through interactions with the table via the Operation Builder.

This concludes *Part 1: Introduction and Setup*, which aimed to introduce the potent cloud-native NoSQL database, Amazon DynamoDB. It familiarized you with SDKs, the AWS Management Console, and NoSQL Workbench.

Next, we proceed to *Part 2: Core Data Modeling*, where we will learn about various data modeling concepts, design patterns, and fundamental ideas. We will shift our perspective to *thinking* NoSQL, departing from the relational mindset. Additionally, we will explore DynamoDB capacity modes.

In the upcoming chapter, we will cover a fundamental data retrieval mechanism, **simple key-value**, through theory and examples. We will also dive into different data types supported by DynamoDB, understanding their benefits and limitations.

References

1. ACM Library: ERD Whitepaper – `https://dl.acm.org/doi/10.1145/320434.320440`

2. AWS Docs: Setting up NoSQL Workbench – `https://docs.aws.amazon.com/amazondynamodb/latest/developerguide/workbench.settingup.html`

3. AWS Docs: Release history for NoSQL Workbench – `https://docs.aws.amazon.com/amazondynamodb/latest/developerguide/WorkbenchDocumentHistory.html`

4. GitHub: Online assets for Chapter03 – `https://github.com/PacktPublishing/Amazon-DynamoDB---The-Definitive-Guide/tree/main/Chapter03`

5. AWS Docs: AWS CloudFormation – `https://docs.aws.amazon.com/AWSCloudFormation/latest/UserGuide/Welcome.html`

6. AWS Docs: Adding IAM Policies – `https://docs.aws.amazon.com/IAM/latest/UserGuide/id_users_change-permissions.html#users_change_permissions-add-directly-console`

Part 2:
Core Data Modeling

This part dives into the core principles of data modeling in DynamoDB. It starts with the basic key-value structure and discusses the transition from a relational database mindset to DynamoDB's schema-less model. It also explores essential topics such as read consistency, operations, and transaction management. Advanced data modeling techniques, including vertical partitioning and the use of secondary indexes are also covered. These concepts are critical for managing large datasets and ensuring efficient query performance. By the end of this part, you'll be well-versed in the fundamental and advanced techniques needed to model your data effectively in DynamoDB.

Part 2 has the following chapters:

- *Chapter 4, Simple Key-Value*
- *Chapter 5, Moving from a Relational Mindset*
- *Chapter 6, Read Consistency, Operations, and Transactions*
- *Chapter 7, Vertical Partitioning*
- *Chapter 8, Secondary Indexes*

4

Simple Key-Value

Now that we have gotten to know **DynamoDB**, as well as some of the tooling around its development and operations, let's dive into one of the most important parts of making the most of Amazon DynamoDB and NoSQL databases in general: **data modeling**.

Contrary to Mike, I (Aman) was introduced to DynamoDB while working within Amazon Web Services (AWS). I was an engineer in AWS support, focusing on DynamoDB and the big data/analytics suite of services that AWS provides. In this role, I primarily assisted customers who were already using DynamoDB and encountering issues with their implementations. Occasionally, I helped customers with data modeling, but most of my experience revolved around the operational aspects of the service.

After getting a good handle on the operational part, I decided to qualify as a DynamoDB SME internally, which meant that I would need to obtain deep-level knowledge of both the data modeling aspects, as well as the operational. It took me a while, but after working on DynamoDB for a few years, during one global but internal boot camp event in Sydney, I passed the SME board. Since then, I have been on the lookout for customers with data model-related conversations. I eventually found myself transitioning into a DynamoDB **Specialist Solutions Architect (SSA)**.

As an SSA, I have loads of exciting engagements with customers across industries and at varying stages of their journey in the cloud. Having only worked with relational databases before AWS, my experience with NoSQL and DynamoDB was no different than Mike's – I still have that *aha* moment each time I model a challenging multi-entity model on DynamoDB in a highly scalable and well-architected way.

Over the course of the next six chapters, we will learn how to *think* in terms of NoSQL – we will learn to think about horizontal scaling. This includes going over a few design patterns and practices specific to NoSQL databases and comparing them to traditional relational systems. We will also dive into DynamoDB-specific features such as read consistency, secondary indexes, and the two throughput capacity modes supported by DynamoDB tables and secondary indexes.

To start off, we will learn about the basic concept DynamoDB is tied to – **key-value**. This is the core mechanism of how data can be stored or read from the database. Store data associated with a key you know. Supply the key and DynamoDB will return the data corresponding to it, irrespective of the total number of these keys stored together. We'll also look into item collections and how they allow for super-efficient reads of multiple item result sets for related data.

By the end of this chapter, you will understand the key-value mechanism, know how to use it to interact with DynamoDB, know about the different data types that are supported, and understand the current limitations of the database.

In this chapter, we are going to cover the following main topics:

- What is key-value?
- An example database lookup
- DynamoDB data types
- DynamoDB limitations

What is key-value?

Academically, a **key-value datastore** would be a data storage design where information is stored in a dictionary or, in traditional computer science data structures, in a **hash table**. This information has an optimized, indexed part called the **key**, which is unique across a particular datastore. The **value** is the complementary part of the key and together, they are called a **key-value pair**. Essentially, a key-value datastore is used to store a bunch of key-value pairs, where each pair or object can be uniquely identified by its key. The value corresponding to each key could either be a single string of text, such as foobar, or could be a huge chunk of data with complex data types nested within each other.

The following table is an example of a key-value datastore with varying value types:

Key	Value	
K1	foobar	

K2	{
	 "first name": "aman",
	 "year": 2022,
	 "set_a": ["a", "b", "c"]
	 }	

K3	Example.com	12345
	Explore Inc	22222
	Exit Company	44444

Table 4.1 – A key-value example

To throw in an analogy here, a key-value datastore could be compared to a telephone directory (for those who were around that time; others – look up *yellow pages*), where keys are the letters of the alphabet themselves. Each letter can be uniquely identified. The value that corresponds to each key would be the huge chunk of several business names and their addresses, contact numbers, and so on. If I were to find information about Example.com, a company, I would know exactly which part of the telephone directory would have information about Example.com. You've got it – it would be under the section for the English letter E!

Since DynamoDB is a key-value datastore, looking up key-value pairs is a huge part of what it is good at. Every DynamoDB table item has its unique identifier as the primary key. As we saw in previous chapters, the **primary key** could be a **partition key** alone or a combination of a partition key and a **sort key**. Think of the primary key as the key, whereas the rest of the item collectively is the value in a key-value pair. To read a single item in the most efficient way, I would need to provide DynamoDB with the complete key. It would then return the whole item (the complete key-value pair) if the item with that key exists, or else it would return an empty set.

Single-item operations are also called **singleton** operations. Each singleton operation is expected to act on 0 or 1 items. For DynamoDB, in no particular order, some of the singleton APIs are as follows:

- `GetItem`
- `PutItem`
- `UpdateItem`
- `DeleteItem`
- `ExecuteStatement` (for writes only)

Each of these singleton operations expects the complete values of the primary key of the table. For example, if any of the preceding operations are executed on a table with a primary key that comprises a partition key and a sort key, the API request expects complete values of both. If the table's primary key was a partition key alone, then each of these operations would at least require the complete value of the partition key attribute in addition to their individual functionality parameters. For example, if my table's primary key consisted of a partition key called `PK` and a sort key attribute called `SK`, then for singleton operations, I would need to provide full values of both `PK` and `SK` to uniquely identify and operate on a single item. On the other hand, if the primary key of the table was simply a partition key in `PK`, then I would only need to provide the complete value of the `PK` attribute in order to perform the preceding singleton operations on any single item in the table.

In order to give a high-level overview of the preceding operations themselves, a `GetItem` API is a single-item read operation that has the ability to return the whole item on passing the primary key attribute values in the request. `PutItem`, `UpdateItem`, and `DeleteItem` are single-item write or mutating operations, which, like all other singleton operations, can perform an `insert`, `update`, or `delete` operation for exactly 0 or 1 item. I include 0 because if an item with the passed primary key-value does not exist, such as in the case of a `DeleteItem` API, no data is affected by the API call and yet a successful response, the 200 HTTP status code is returned.

Finally, the ExecuteStatement API is not only capable of performing a singleton write, update, or delete but also of performing a single or multi-item read! This API is part of the SQL-like query language supported by DynamoDB, which is called **PartiQL**. We will dive deeper into PartiQL and its operations in *Chapter 6, Read Consistency, Operations, and Transactions.*

The following figure visualizes the basics of how the preceding API calls would look, with the example of a GetItem operation to a DynamoDB endpoint in the AWS Cloud. It is absolutely okay to not know all about the figure at this time:

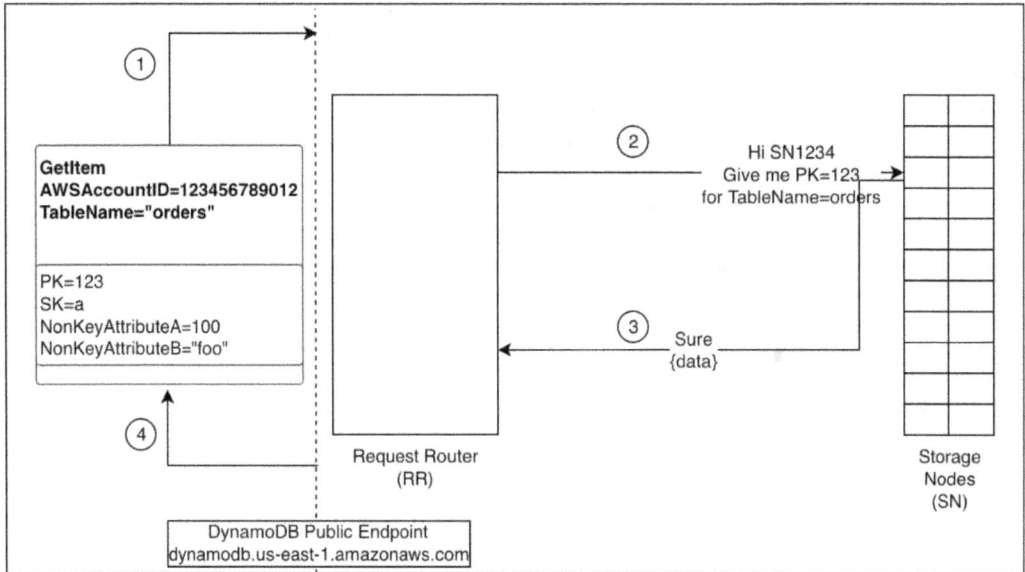

Figure 4.1 – GetItem and PutItem – a visual example

If you recall, DynamoDB partitions table data (items) across a horizontally scaled fleet of **backend nodes**, or **storage nodes**. To decide which node is responsible for what group of table items, DynamoDB uses **consistent hashing** (1) on the partition key-value. Imagine a key space of hexadecimal numbers from 0000 to FFFF (for example, DynamoDB might have a much larger range in reality) divided up into buckets of consecutive numbers. These buckets could be created for evenly distributed hex numbers or be uneven. The output of the hashing function specific to DynamoDB always returns a value between the hex ranges, and so whichever bucket or node this output belongs to, that node is responsible for the item.

For example, for an item with the id=ADJ003 partition key, the hashing function will always return the same hex value, such as 30F and for an item with the id=ADJ004 partition key, which seems like a sequential value after the previous item, the hashing function would return a completely different hex value such as 9A1. Thus, although both these items appear to be in sequence, they could land on completely different nodes. This is immensely beneficial when an application's access pattern is sequential in nature. This also helps the service do its best to distribute the request traffic evenly across the backend nodes, thus maximizing efficient utilization of the resources. We will learn more about these resources in *Chapter 10, Request Routers, Storage Nodes, and Other Core Components*. However, this brief overview should help us understand enough about the singleton operation behavior for now.

Coming back to our telephone directory analogy, if I tore the directory vertically into several smaller booklets, each booklet would still be useful in retrieving information about a particular company. If I needed to find information on Example.com among these booklets, I could look for the booklet that contains information on all companies starting with the letter E and then look up Example.com there due to the sorted nature of the company names within the section for the letter E. I could have torn the directory unevenly such that two or more booklets could contain information about company names starting with E. Even then, due to the lexicographical sort order of the company names, I could make an educated decision about which booklet might contain Example.com.

In the same manner, data could be partitioned across nodes for a DynamoDB table unevenly and DynamoDB would still be able to reach the right node responsible for a given item. As a reminder, the partitioning logic would involve the partition key of the item. This is how scaling horizontally across many booklets (or partitions) helps DynamoDB support millions of requests across table data. As the data volume or throughput requirements of the table increase, DynamoDB tears (or splits) the existing booklets (or partitions) into multiple smaller booklets to allow even higher throughput.

Performance with DynamoDB is at its best and most efficient if the access pattern is of the key-value kind and deals with the partition key in some regard. This becomes a best practice in terms of designing the table. Whether it is the table or the secondary indexes, if most of the access patterns for the application can be supported by the key schema, the most efficient reading and writing can be expected.

The key schema is the partition key alone if the table or the secondary index's primary key was the partition key only, and is the combination of partition key and the sort key if the table or index's primary key included both. It is very much possible that quite a few access patterns for your application may not be completely supported by the key schema of the table or any additional secondary index, and it makes sense to apply some filters to the non-key attributes on the items. DynamoDB APIs support using such additional filters, and this is fine. For this chapter, however, we will focus on the behavior with purely key-value access patterns where the primary key of the data can be fully utilized. This is also what formed the basis of why DynamoDB came into existence.

Now that we have covered key-value access on a conceptual basis, let's see how relevant this kind of access is in the real world.

Reviewing key-value use cases across industries

If you recall how DynamoDB service came into existence, it was about the Amazon.com retail platform at Amazon struggling to keep up with the growing scale that used traditional off-the-shelf commercial and open source relational database management system (RDBMS). To avoid outages of the platform due to database scaling issues, a group of distributed systems experts came together to design a highly scalable, highly available, horizontally scalable database system that catered to all of their requirements. This eventually came out as the Dynamo paper. The database system banked on the fact that 70-90% of the access patterns for the platform were of the key-value nature — about 70% of the simple key-value kind, 20% returning multiple rows but from the same table. This was the e-commerce industry.

No matter the industry, there are always use cases and workloads with mostly simple key-value access patterns. Common examples include storing user information, as well as authorizing data, session states, platform metadata, and configurations. These workloads benefit greatly from DynamoDB's efficient key-value store capabilities.

For industry-specific examples, consider game state and leaderboards in the gaming industry, product catalogs in e-commerce, user transactions and histories in finance, media metadata in media and entertainment, device statuses in manufacturing, and real-time impression tracking in ad tech. DynamoDB not only helps these workloads get up and running quickly but also supports their varying growth rates seamlessly. Using DynamoDB can lead to fewer late-night pages for engineers and a greater overall impact for businesses.

If you pick any use case you are currently working on or have worked on in the past, 70-80% of its Online Transaction Processing (OLTP) access may very well have been key-value-based. If not based on the key for the table itself, it may involve an index-like structure, which would allow reading the same data with a key schema. These would be decent candidates to evaluate DynamoDB as their primary datastore to realize the benefits the database has to offer.

Now that we have learned about the importance and behavior of the simple key-value pattern, let us put this into practice by briefly going over the employee data model and performing singleton read and write operations using different tools and SDKs for DynamoDB.

Example database operations of the simple key-value kind

Let us go ahead and perform read and write operations of the simple key-value kind. We will use the sample `Employee` table we used in the previous chapter. If you no longer have the table or would like to bootstrap the table to follow along, instructions to set the table up are in the GitHub repository (2) in the `Chapter03` folder.

If you recall or can observe, the `Employee` table key schema only has a partition key: `LoginAlias`. This means that each item in the table can be uniquely identified using the value of its partition key alone. For example, `LoginAlias=janed` would uniquely identify the employee details of `Jane Doe`, who is a developer as per the `Employee` table. Let us go ahead and make a key-value lookup to read this item programmatically.

If I were to read the item from the table, I would perform a `GetItem` operation against the table and pass the complete value of the `LoginAlias=janed` partition key.

Let us first see how this could be performed using Amazon Web Services Command Line Interface (AWS CLI). I will run the following in my terminal window of a macOS laptop that has AWS CLI installed. Refer to the AWS docs (2) for instructions on setting up the AWS CLI if you do not have it already:

```
aws dynamodb get-item \
--table-name "Employee" \
--key '{"LoginAlias": {"S": "janed"}}' \
--endpoint "https://dynamodb.eu-west-2.amazonaws.com" \
--region "eu-west-2"
```

In prettified JSON, the preceding code would return the following:

```
 1 {
 2     "Item": {
 3         "ManagerLoginAlias": {
 4             "S": "marthar"
 5         },
 6         "LoginAlias": {
 7             "S": "janed"
 8         },
 9         "FirstName": {
10             "S": "Jane"
11         },
12         "LastName": {
13             "S": "Doe"
14         },
15         "Designation": {
16             "S": "Developer"
17         },
18         "Skills": {
19             "SS": [
20                 "software"
21             ]
22         }
23     }
24 }
```

Figure 4.2 – AWS CLI GetItem example output

Notice that in my AWS CLI get-item command, I have also passed the --endpoint and --region parameters. Generally, you may not require these, unless you are operating in a different region than your default configured in AWS CLI locally. However, if you intend to point the command toward a **DynamoDBLocal** installation, you will need to pass the appropriate endpoint for DynamoDBLocal (for example, http://localhost:8000). You could also pass additional parameters to get more information apart from the data item itself. This includes information about the consumed read capacity units (RCU) for the `get-item` operation with the `--return-consumed-capacity` parameter, the `--output` parameter to instruct AWS CLI if you would like the result in text, JSON, or table format, among other parameters. Run `aws dynamodb get-item help` for more information on different supported parameters.

Next, we will perform the same operation using one of the AWS SDKs, specifically the AWS SDK for Python (Boto3). For this, you can also use the boilerplate script generated by the **operation builder** tool of the NoSQL Workbench tool. The following is how performing the same operation using a simple Boto3-based Python script would look:

```
import boto3
client = boto3.client("dynamodb", region_name="eu-west-2")
params = {
    "TableName": "Employee",
    "Key": {"LoginAlias": {"S": "janed"}}
}

response = client.get_item(**params)

print(response["Item"])
```

The following is the prettified output of the previous code snippet:

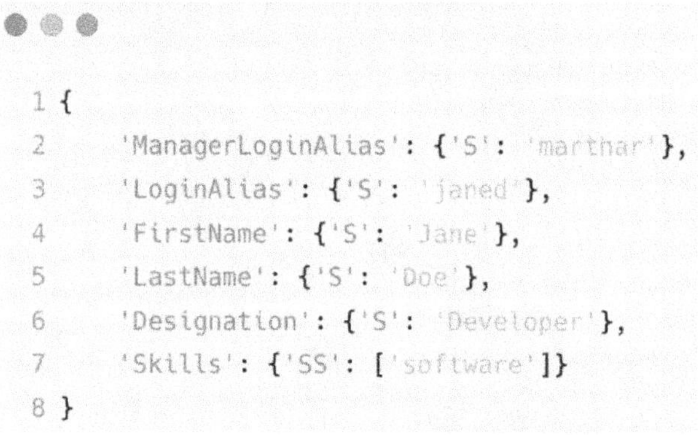

```
1 {
2       'ManagerLoginAlias': {'S': 'marthar'},
3       'LoginAlias': {'S': 'janed'},
4       'FirstName': {'S': 'Jane'},
5       'LastName': {'S': 'Doe'},
6       'Designation': {'S': 'Developer'},
7       'Skills': {'SS': ['software']}
8 }
```

Figure 4.3 – Python SDK (boto3) GetItem example output

The following example uses the PartiQL operation to read the same data:

```
import boto3
client = boto3.client("dynamodb", region_name="eu-west-2")
params = {
    "Statement": "SELECT * FROM Employee WHERE LoginAlias='janed'"
}
```

```
response = client.execute_statement(**params)

print(response["Items"])
```

The prettified output is as follows:

```
1 [{
2     'ManagerLoginAlias': {'S': 'marthar'},
3     'LoginAlias': {'S': 'janed'},
4     'FirstName': {'S': 'Jane'},
5     'LastName': {'S': 'Doe'},
6     'Designation': {'S': 'Developer'},
7     'Skills': {'SS': ['software']}
8 }]
```

Figure 4.4 – Python SDK (boto3) ExecuteStatement example output

Now that we have seen multiple ways of performing the singleton GetItem operation, let us see how we would perform simple key-value writes to the same Employee table. First, I would use the AWS CLI to perform PutItem using the following command:

```
aws dynamodb put-item \
--table-name Employee \
--item '{
    "LoginAlias": {"S": "amdhing"},
    "Designation": {"S": "Architect"},
    "FirstName": {"S": "Aman"},
    "LastName": {"S": "Dhingra"},
    "ManagerLoginAlias": {"S": "marthar"},
    "Skills": {"SS": ["software"]}
}' \
--region eu-west-2 \
--profile definitive-guide
```

In terms of writing to DynamoDB, if the response of a write request is an empty payload of HTTP 200 response, that means that the write has been persisted successfully. If the response code is anything other than HTTP 200, that would indicate an unsuccessful request. The actual HTTP status code would shed more light on the reason behind the request being unsuccessful. This may include authentication and authorization errors, request validation errors, throttling errors, or internal server errors, among others.

Let us perform the same PutItem operation using a Boto3-based Python script:

```python
import boto3

client = boto3.client("dynamodb", region_name="eu-west-2")

params = {
    "TableName": "Employee",
    "Item": {
        "LoginAlias": {"S": "ripani"},
        "ManagerLoginAlias": {"S": "marthar"},
        "FirstName": {"S": "Lorenzo"},
        "LastName": {"S": "Ripani"},
        "Designation": {"S": "Developer"},
        "Skills": {"SS": ["software"]}
    }
}

response = client.put_item(**params)

print(response)
```

The prettified output of the preceding script is as follows:

```
 1 {
 2     'ResponseMetadata': {
 3         'RequestId': XXXX,
 4         'HTTPStatusCode': 200,
 5         'HTTPHeaders': {
 6             'server': 'Server',
 7             'date': 'Sat, 09 Jul 2022 18:01:03 GMT',
 8             'content-type': 'application/x-amz-json-1.0',
 9             'content-length': '2',
10             'connection': 'keep-alive',
11             'x-amzn-requestid': XXXX,
12             'x-amz-crc32': 1234
13         },
14         'RetryAttempts': 0
15     }
16 }
```

Figure 4.5 – Python SDK (boto3) PutItem example output

All the DynamoDB APIs are documented in the *Reference* section (3), along with links to the developer docs for the programming language of your choice. Feel free to try out other data-plane operations or the same operations in other programming languages. You could also look at the samples repository by AWS in (4).

Note that simple key-value is not all that DynamoDB has to offer. You can also perform read and write operations in different ways that act on multiple items at the same time. You can perform multi-table multi-item operations in an all-or-nothing fashion with DynamoDB transaction APIs. You can also go through all the items in the table as part of a paginated read operation with Scan as we have seen in the previous chapters. You can also perform multi-item operations such as Query, which can operate on items with the same partition key-value but different sort key values – these are referred to as **item collections**. We will dive into item collections in *Chapter 7, Vertical Partitioning*.

After looking at some samples of interacting with DynamoDB with simple key-value operations, let us learn about the supported DynamoDB data types and naming rules.

Learning about DynamoDB data types

Let us now go over the different data types supported by DynamoDB. A lot of this would apply to databases in general. However, relational database systems generally support more data types that are very specific in terms of the operations they support. We will learn about how to map those most-used data types in relational systems onto the NoSQL and DynamoDB world.

The data types supported by DynamoDB are well documented (5), but I will aim to add value to this section by sharing practical insights from my experience. While you can find detailed specifications, such as the exact length of partition and sort key values, in the official docs, I will focus on practical usage. This approach is intentional. Additionally, I will cover optimizations and best practices that can enhance various use cases with these data types in real-world applications.

There are several data types supported in DynamoDB items, which are categorized into three groups:

- **Scalar types**: Scalar means a single value. These include binary, string, number, Boolean, and null values.

- **Document types**: These could be complex and nested and may contain multiple other same or distinct data types within, such as what you may find in JSON objects. Examples of document types include maps and lists. For example, a map attribute in a DynamoDB item could be a collection of lists, strings, numbers, and other maps.

- **Set types**: These contain multiple scalar data types but all values within a set would be of the same scalar data type. The set type includes binary sets, string sets, and number sets.

Let us look at each of these types in detail in the following sections.

Scalar types

Scalar data types such as string, number, and binary are the only data types allowed for key attributes, that is, for being partition keys, sort keys for tables, or secondary indexes. This makes them widely used and super beneficial. In addition to binary, string, and number, the scalar types also include Boolean and null. Let us dive into each of the scalar types and learn some best practices around using each data type.

Binary

This is your basic binary data type, which could be the binary form of compressed text, compressed files, or encrypted images, among other things. You must store binary data encoded in base64 when sending it to DynamoDB.

The most common use for binary data types is when you want to store complete programming objects as-is and interact with these objects only after retrieving all the data client-side. Another common use case is when the use case requires client-side encryption of data. By default, DynamoDB encrypts your data server-side. However, for use cases such as storing payment-related data for being **Payment Card Industry Data Security Standard (PCI DSS)** compliant or storing authentication-related data, you may need to encrypt the relevant attributes client-side before storing data in a database.

AWS has also open source client-side encryption libraries for DynamoDB (6) in Java and Python languages which use **AWS Key Management Service (AWS KMS)** for the encryption keys required to encrypt data client-side, before writing to DynamoDB. Similarly, the applications using these libraries also support decryption data after retrieving items from the table.

An important consideration with any form of client-side encryption is that the partition key and sort key data for the DynamoDB table and any secondary indexes must not be encrypted as this hampers any filtering ability for those keys. In other words, any attribute that needs to be encrypted in a DynamoDB item must not be part of the key schema of either the table or its secondary indexes, and must also not be subject to any filtering.

String

String attribute names and values are **Unicode** with UTF-8 binary encoding. Any data for a string attribute that is not UTF-8 encoded will automatically be converted to a UTF-8 encoded value by DynamoDB. For example, if I stored \uD83D\uDE42 as part of a string attribute in DynamoDB using any SDK, that part of the data will automatically convert to a smiley emoji, which is the UTF-8 form of the string. This can essentially corrupt the data as you may not know what the data was encoded as before.

> **Important note**
>
> If you write non-UTF-8 encoded data into DynamoDB, it will automatically convert that data into UTF-8 encoded values. You must ensure that the data being written to DynamoDB is UTF-8 encoded to prevent unexpected corruption.

In relational database systems, a timestamp is a common and widely used data type that supports different functions and operations such as changing time zones, changing formats, conversion to epoch, and extracting different components such as year, month, day of the week, and so on. Well, for DynamoDB, the two ways you can represent timestamps is either as a string or a number (in terms of storing epoch time – number of seconds elapsed since January 1, 1970, UTC/GMT time). Thus, it becomes the responsibility of the application to realize an attribute as a timestamp and then manipulate the value as required. Some may even convert the string timestamp to your programming language's native timestamp format to make use of all those functions and operations mentioned earlier.

A common practice and recommendation for using timestamps as a string is to have them in ISO 8601 format (8). The reason for this is that the format's well-defined standard and ability to be lexicographically sorted. Examples of such a format would include the following:

- 2022-07-09
- 2022-07-09T20:14:30Z (UTC)
- 2022-W27 (with the week number)

Using formats such as `2022-07-09T20:14:30Z` for the timestamp attribute can allow for additional filtering capabilities such as `begins_with()` or sorting methods. The hierarchical nature of `YYYY-MM-DDThh:mm:ss` can allow querying for questions such as *Get me all data for a particular year, for over a month within a particular year*, and so on. For example, if I need all data for a partition key where the timestamp was in the month of July 2022, I will use a `Query` API and have a condition on the sort key as `begins_with('2022-07')`.

Number

The DynamoDB number type maps to any kind of number-based data types you may be used to in the relational database or programming language worlds. The number format could be positive, negative, or zero. It can support 38 digits of precision and has the following bounds:

- **Positive range**: 1E-130 to 9.9999999999999999999999999999999999999E+125
- **Negative range**: -9.9999999999999999999999999999999999999E+125 to -1E-130

Apart from being just numbers, another use for the number data type is representing timestamps, such as epochs. This can be particularly beneficial when utilizing the **Time To Live** (**TTL**) (8) feature with DynamoDB that provides automatic, free-of-cost deletes for your items if the epoch time in a pre-configured number type TTL expiry attribute has passed the current time. TTL comes in handy in discarding data that has lost its relevance, in complying with data retention rules, or in simply controlling the storage size of the table for cost optimization purposes. We will learn more about the TTL functionality and its considerations in *Chapter 12, Streams and TTL*.

Boolean

This is a basic attribute that can store `true` or `false`.

Null

This is yet another basic attribute that indicates an unknown or undefined state of an attribute. However, with DynamoDB being schema-less, you may be better off skipping adding an attribute that has an undefined or unknown value – as NULL, in certain cases, it may not be appropriate to *assume* an unknown state simply by missing an attribute that could be NULL. There may be different interpretations of NULL in different systems, including empty, null, some state of the system, and the end of a page. Thus, having a proper definition and scope for a NULL attribute in your application is important if you are considering having NULL attributes.

Document types

Document types should be used for data that is mostly static or infrequently updated. This is because not all broad data manipulation functionalities are supported when interacting with document or map attributes. Although these types allow specific operations to be stored across nested levels using expressions (9), unless attributes need to be accessed together, they should not be stored together in non-scalar types such as maps, lists, and sets. Examples include storing strings in a list or parts of a postal address in a map. Document types such as maps and lists support up to 32 levels of nesting. Next, we will explore each of these document data types in more detail.

List

A list in DynamoDB is like a JSON array. You could have multiple scalar or document attributes nested within a list and elements in a list could be of different data types. You could access a list element like you would do a JSON array using `list [N]` notation, where N is the position of the element within the list starting with zero.

Map

A map in DynamoDB is like a JSON object. A common use for the map data type is when an application deals with JSON objects, but those objects are serialized into a DynamoDB map before storing the object in DynamoDB. Storing the object without serialization may require the binary data type. You can access and perform actions on elements within a map using the dot (.) notation such as `map1. nestedmap2.nestedStringElement`.

Set types

Sets always contain elements of the same type such as string, binary, or number. Set attributes in DynamoDB are unordered and each element within a set must be unique. Therefore, applications may not rely on the order of the elements within a set that they write the item with. Just like document data types, set data types cannot be part of the key schema for a table or the secondary index, and hence, these generally resort to static or limited access. Empty sets are not allowed; however, empty string or binary values within a set are allowed.

Now that we have reviewed the different data types supported by DynamoDB and how some conventional relational database system data types map or must be used when working with DynamoDB, we will next explore the limitations within DynamoDB.

Learning about DynamoDB data limitations

Predictable performance at any scale is a huge data point in favor of using DynamoDB. However, there are certain limitations that allow for such predictable performance. Some limitations are soft limits, meaning that they can be increased by requesting AWS, whereas other limits are hard limits that cannot be modified and may need working around using different design patterns and practices. Most of the limitations are well documented in the AWS docs (10) but I will attempt to add more value to this section by including practical information and best practices based on experience.

Throughput

DynamoDB supports two capacity modes for tables – **provisioned mode** and **on-demand mode**. While both modes have differences in the ways of metering, throttling, and scaling tables, both deal with units of throughput in similar ways. We will learn more about these capacity modes in *Chapter 9, Capacity Modes and Table Classes*. However, for this section, knowing that these two categories exist is enough.

In terms of provisioned mode, there are limitations applied on an AWS account-region level for the total throughput that can be utilized both for reads and writes. For example, the default DynamoDB throughput quota for a new AWS account would be 80,000 reads per second and 80,000 writes per second for every AWS region. Account-region level limits do not apply to on-demand mode tables.

Like account-region-wide quotas, DynamoDB limits reading and writing on a per-table level. At the time of writing, the numbers stand at 40,000 reads per second and 40,000 writes per second by default. Unlike account-region level quotas, the per-table-level quotas apply to tables in both provisioned and on-demand capacity modes.

Since we have already seen case studies about large customers making tens of millions of reads and writes per second using DynamoDB, these default quotas are adjustable by reaching out to AWS support, regardless of whether you subscribe to a Developer Support Plan or not. Next, let us dig into item sizes and their associated limitations and best practices.

Item sizes

One of the important limitations that you must be aware of with DynamoDB is the individual item size limit, which is 400 KB. You could have as many attributes in a DynamoDB item as you like, and they could be of whichever supported data type you'd like, as long as the cumulative size of the item does not exceed 400 KB. You could theoretically have an unlimited number of items stored in a DynamoDB table; however, each item cannot exceed 400 KB in size. There are multiple design patterns to use if you wish to store data in items that are beyond the 400 KB mark. These include breaking down single payloads into multiple smaller chunks in a meaningful way, compressing payloads, and offloading payloads to an object store such as Amazon S3 but referencing the object path in DynamoDB. We will look at these in depth in *Chapter 7, Vertical Partitioning*.

Another important fact about item sizes to keep in mind is that item size calculations take into consideration both the name of the attribute and the value. So, for an attribute such as `number_of_policies_for_user=1`, I would pay more for storing the attribute name than the attribute value. This is quite different from most relational systems.

Clearly, a good practice for DynamoDB would be using shorter attribute names that still make sense to the developer – or going one step further and storing a dictionary within the application code to map shorter attribute names present in the items to more meaningful, human-readable attribute names to optimize storage costs. From experience, customers have saved up to 30% on DynamoDB costs by simply shortening a few attribute names across tens of millions of items in their tables. The absolute value of cost savings in the case of the same customer was about tens of thousands of dollars each year.

This is a tip for you if you are involved in the designing of applications using DynamoDB – use short but meaningful attribute names right from the design phase. With that, let us learn about limitations associated with reserved keyword usage in DynamoDB.

Reserved words

Although this is a limitation of lesser importance, DynamoDB has a case-insensitive list of reserved words that cannot be used directly to reference attributes in expressions used to perform operations on items. You can use these as attribute names or values in the items themselves, but you cannot use them in expressions to make references to the items. You will find the list of reserved words within DynamoDB in (11) in the *Reference* section.

For example, `YEAR` is one of the reserved words in that list. Thus, to read an item where the partition key attribute name is `year` (case-insensitive), I will need to use a placeholder such as `#year_placeholder` for it in my **KeyConditionExpression** and pass the actual name within the **ExpressionAttributeNames** parameter. Not doing this would result in a failed request with a **ValidationException** error. The following is the correct way of referencing reserved words in operations:

```
dynamodb.query(
    "TableName" = "sample-table",
    "KeyConditionExpression" = "#year_placeholder = :year_value",
```

```
    "ExpressionAttributeNames" = {
        "#year_placeholder": "year"
    },
    "ExpressionAttributeValues" = {
        ":year_value": {"N": 2022}
    }
)
```

Next, let us learn about page sizes within DynamoDB.

Page sizes

DynamoDB supports requests with bounded pieces of work. If an operation supports acting on multiple items within a single request, there will be limitations on how many items it can interact with for that single request without compromising performance at scale.

In terms of multi-item read operations such as `Query/Scan` or PartiQL `ExecuteStatement`, a single API call can return up to 1 MB worth of items. If there are more items that may qualify the passed read parameters, DynamoDB would return a pagination token such as `LastEvaluatedKey` to indicate the same to the client. Clients, on encountering this parameter, must make successive API calls using such token received in the previous response to request DynamoDB for the next 1 MB page of results. A subsequent response that does not include this `LastEvaluatedKey` token would indicate that there aren't any more unread items matching the passed parameters.

There are some static limitations on payload sizes for specific DynamoDB APIs, which differ from each other. You will find the complete list of those in (10) in the *Reference* section. Specifically, you will find it under the *API-specific limits* section of the document that I highly recommend going through as part of learning.

Limitations on data types and dependent functionalities

We may have briefly touched upon this topic in previous chapters. However, based on the supported data types and operations, it may hint that DynamoDB does not support some advanced data types and functionalities when it comes to those advanced data types. It is that very reason why some use cases cannot be implemented without resorting to other purpose-built solutions. These include native geospatial lookups, weighted ranking, and full-text searching.

Take geospatial data types, for example. If I were to implement geo-based lookups for certain use cases, that would require some complex engineering such as performing spherical math, storing geohashes, and making use of libraries such as S2Geometry by Google (12) while managing all of this in the application. On the other hand, I could simply use a purpose-built solution such as Redis-compatible datastores such as Amazon ElastiCache for Redis or Amazon MemoryDB for Redis (13), both of which support geospatial lookups out of the box. This is not to say there aren't customers building interfaces on top of DynamoDB for such use cases, but that not providing this natively remains a limitation of the service.

Another example would be full-text searching. Even building interfaces on top of DynamoDB to achieve this would be an anti-pattern in terms of the functionality itself and not forgetting the impact on costs such a solution would have. Resorting to a purpose-built indexing solution such as Amazon OpenSearch (14) or Apache Lucene (15) based systems would prove much more beneficial instead. You could always use DynamoDB here as a persistent store to manage the source of truth or store additional information about data in these indexing systems.

> **Important note**
> Consider this as a reminder to delete any resources you may have created while following along with this chapter in your AWS account.

Summary

In this chapter, we built some foundation on the NoSQL data modeling theory by looking at the very basic, simple key-value mechanism. We learned how this works in the realm of DynamoDB and how DynamoDB can horizontally scale but still support consistent performance. We also covered how the mapping of keys across partitions works with a telephone directory analogy. We read about singleton operations and how they support the key-value mechanism within DynamoDB.

We covered how database access analyses within the amazon.com retail platform showed findings of 70-90% of access being of the key-value kind. We also learned that there are use cases across industries where the database access is of the key-value kind.

We walked through practical examples of performing singleton `GetItem` and `PutItem` operations on the sample `Employee` table using multiple means such as the AWS CLI, as well as Boto3, the AWS Python SDK.

Finally, we learned about the different data types supported by DynamoDB and how they can be used to implement certain advanced uses such as using ISO 8601 format timestamps as strings to query using `begins_with` operators. We also looked at the limitations of data types and operations supported by DynamoDB. This included missing data types and functionalities such as geospatial lookups, among others. We did establish the fact that in these use cases, it may be best to supplement DynamoDB with another purpose-built datastore.

In the next chapter, we will build further on the NoSQL theory by looking at how to implement relationships within NoSQL systems and DynamoDB. We will develop our understanding of NoSQL by covering de-normalization, exploring design to optimize on compute not on storage, and breaking down data based on different design aspects.

References

1. The Wikipedia page on consistent hashing: `https://en.wikipedia.org/wiki/Consistent_hashing`

2. AWS docs – install or update to the latest version of the AWS CLI: `https://docs.aws.amazon.com/cli/latest/userguide/getting-started-install.html`

3. DynamoDB API reference: `https://docs.aws.amazon.com/amazondynamodb/latest/APIReference/API_Operations_Amazon_DynamoDB.html`

4. GitHub aws-samples: `https://github.com/aws-samples/aws-dynamodb-examples/tree/master/DynamoDB-SDK-Examples`

5. Supported data types and naming rules in Amazon DynamoDB: `https://docs.aws.amazon.com/amazondynamodb/latest/developerguide/HowItWorks.NamingRulesDataTypes.html`

6. DynamoDB encryption client: `https://docs.aws.amazon.com/dynamodb-encryption-client/latest/devguide/what-is-ddb-encrypt.html`

7. The ISO 8601 format: `https://en.wikipedia.org/wiki/ISO_8601`

8. DynamoDB TTL: `https://docs.aws.amazon.com/amazondynamodb/latest/developerguide/TTL.html`

9. Expressions on non-scalar data types: `https://docs.aws.amazon.com/amazondynamodb/latest/developerguide/Expressions.html`

10. Service quotas for DynamoDB: `https://docs.aws.amazon.com/amazondynamodb/latest/developerguide/ServiceQuotas.html`

11. Reserved words in DynamoDB: `https://docs.aws.amazon.com/amazondynamodb/latest/developerguide/ReservedWords.html`

12. Google S2 Geometry: `https://s2geometry.io/`

13. AWS in-memory databases: `https://aws.amazon.com/nosql/in-memory/`

14. Amazon OpenSearch: `https://aws.amazon.com/opensearch-service/`

15. Apache Lucene: `https://lucene.apache.org/`

5
Moving from a
Relational Mindset

We hope that the previous chapter, *Simple Key-Value*, helped you understand a bit about the NoSQL world and particularly how DynamoDB is good at what it was built for – consistent performance at scale. We will build on the same in this chapter, where I'll attempt to connect DynamoDB data modeling with related aspects from the **relational database management systems (RDBMS)** world. Although these aspects would be related, they may not necessarily be similar as some of the concepts are more likely the opposite of those in the RDBMS paradigm. For example, we have touched upon the fact that normalization of data with primary and foreign keys is important across relational databases to optimize storage and that is the industry norm, but, with NoSQL, denormalization is strongly encouraged and can very well be the only way to optimize compute as well as support horizontal scaling and the goodness that comes with it.

In the technology domain, many of us are accustomed to thinking in terms of relationships and normalization concerning our databases. While you might have limited knowledge about the application accessing the database, new information tends to have minimal impact on the conventional process of listing entities, creating an **entity-relationship diagram (ERD)**, and normalizing the data model extensively. Typically, changes occur in the **data manipulation language (DML)** when major functionalities are added to the application supported by the database.

In contrast, with NoSQL, the process takes a different trajectory. The efficiency, performance, and scalability of a data model using a NoSQL database hinge on tailoring the model to specific access patterns of the intended application. This involves understanding the entities, their relationships, and approximately 70-80% of the access patterns the database needs to support. It's noteworthy that even in the context of a NoSQL database such as DynamoDB, relationships between entities play a crucial role. In this chapter, you will dive into the nuances of modeling in DynamoDB while preserving these relationships.

Just like the preceding examples, we will unpick those relational database concepts and see why they couldn't apply to NoSQL and build on them by learning the DynamoDB way.

In this chapter, you will learn about the following topics:

- Realizing and understanding relationships in data

- Denormalization – storing multiple copies is OK

- Multi-table versus single-table design

- Breaking down your data

Realizing and understanding relationships in data

No, you did not make a mistake in reading the heading – in this section, we will learn about relationships that exist in data, regardless of the database technology or anything beyond. Let's dig right in.

All data is relational

While this statement may initially seem unconventional, it holds true. Regardless of the type or nature of data, there are always relationships among different data components, and these relationships can manifest in various ways. It is crucial to grasp that NoSQL does not imply non-relational data. Instead, it refers to data stores that diverge from the traditional SQL language-based RDBMS in terms of data storage and management. NoSQL databases adopt diverse approaches to store and access data while preserving relationships between entities. Although DynamoDB may employ a unique method for storing and accessing data compared to other NoSQL databases, the core concept remains. The representation of relationships in NoSQL systems differs from SQL-based RDBMS in that the arrangement of data in NoSQL systems must be based on the relationship between entities in the data, something that is not essential in RDBMS.

Therefore, adopting NoSQL databases requires a shift in mindset – one that moves away from the familiar territory of SQL, normalization, and RDBMS concepts that have been integral to our practices for decades.

In the preceding chapters, we explored potential use cases across industries that heavily rely on the simple key-value access pattern. Now, let us dive into various use cases and industries to examine examples illustrating how the data they use or store may inherently contain relationships.

Use case – monitoring

In various industries, robust logging within applications plays a pivotal role in detecting anomalies, monitoring system health and performance, and ensuring overall observability across systems. In the context of logging, each entry has a one-to-many (*1:m*) relationship with the timestamp, signifying that multiple log entries may share the exact timestamp. Furthermore, if stack traces are present in the logs, each stack trace establishes a *1:m* relationship with its corresponding calling method/class. Retrieving and analyzing these logs often involves filtering based on time ranges, status codes, method names, or specific content within the log body. These operations heavily rely on the relationships that exist between timestamps, status codes, method names, and log body entities. Storing logs in a semi-structured, queryable manner aligns well with the capabilities of a NoSQL datastore.

Use case – social media

Social media applications exemplify the utilization of diverse relationships within their data structures, often showcasing many-to-many (*m:n*) relationships among different entities. Consider a popular professional networking application, where a single user can foster numerous connections. Each user connection, in turn, may have its own set of connections – distinct or shared connections. Users might also belong to multiple groups based on their interests, associations, and various technical or non-technical certifications they've acquired.

When you search for a specific user or group on this application, it might reveal the degree of connection you have with other users in the search results. This information is derived from the relationships that exist between users or the communities you both are associated with. While graph traversal may be involved, it is likely facilitated by a NoSQL database, representing relationships between user entities in a manner distinct from traditional RDBMS. NoSQL databases designed for efficient traversal of entities and relationships are commonly categorized as **graph databases**.

Fun fact, in the realm of social network traversal, the concept of *six degrees of separation* suggests that any two people in the world are connected by six or fewer relationships. This intriguing idea underscores the interconnected nature of social networks (1).

Use case – authentication

Yet another use case that spans all industries is **authentication**. Applications often incorporate multiple authentication methods to interface with diverse third-party systems. In this scenario, a single user may possess various modes of authentication, and the user experience should seamlessly grant access to the same account regardless of the chosen authentication method. This establishes a one-to-many (*1:m*) relationship between the user entity and the various authentication modes. DynamoDB finds application in this context across a broad spectrum of industries, be it software and internet, media and entertainment, manufacturing, gaming, or others.

Use case – internet of things (IoT)

IoT stands out as a prominent use case for NoSQL databases, particularly when there is a demand for time-series-based access. In IoT scenarios, thousands or even millions of devices continually contribute telemetry data to a database. Access patterns in this context commonly involve retrieving a historical view of a specific device over time (*1:m* relationship) or analyzing data within a particular period (*1:n* relationship, often utilizing a global secondary index), among other scenarios.

Although the data being written may have a shorter shelf life, the challenge arises from the unpredictable number of devices generating data. To effectively manage the varying throughput, the data store must be capable of scaling out or in. Traditional databases may prove unsuitable due to the sheer volume and velocity of data generated in IoT environments. NoSQL databases not only support these intricate relationships but also provide the scalability needed to adjust to the dynamic demands, influencing the cost profile based on actual usage.

Use case – finance

Finance is yet another industry where NoSQL databases are growing in popularity because of the benefits they come with. Within fintech organizations, NoSQL databases play a pivotal role in various solutions such as e-payments, delivering enriched market index data, core banking, fraud detection, and credit scoring. Despite the finance sector historically having a slower adoption rate for modern technologies, the adaptability of NoSQL databases, allowing for access to data with maintained relationships and accommodating unprecedented growth, positions them prominently in mission-critical scenarios.

Notably, institutions such as Capital One publicly acknowledge their reliance on DynamoDB, sharing insights and guidance with the community through blogs (2). DynamoDB, in their context, supports customers using mobile applications, ensuring low-latency access to their systems. This stands as one among many systems backed by DynamoDB and other NoSQL databases in the financial domain.

Use case – user profile

User profiles form a cornerstone across diverse industries, especially in applications that cater to end users. In the gaming sector, user profiles encompass storing players' game states, earned awards, statistics for leaderboards, and custom game configurations. This results in the modeling of data with *1:1*, *1:m*, and *m:n* relationships. Transitioning to your favorite food delivery app, user profile data extends to include preferred restaurants, recent order history, location preferences, and saved payment options, often modeled with *1:1* and *1:m* relationships. Similarly, user profile data for ride-hailing apps, banking apps, and messaging apps may leverage NoSQL databases beneath the surface to deliver a familiar user experience.

While traditional RDBMS have been historically prevalent in these industries and use cases, the surge in data volume and the demand for low-latency access have steered applications toward NoSQL databases. This shift allows for increased reliability and availability, all while preserving essential relationships between entities. Although these relationships are maintained, the representation differs from traditional SQL databases.

In these diverse applications, sticking to a traditional relational database could lead to higher latencies due to scaling characteristics, concurrency control mechanisms, and the compute-intensive nature of RDBMS. NoSQL databases, designed to address these challenges and performance bottlenecks, offer the same data representation and access in a manner that boasts orders of magnitude in performance improvements and nearly infinite scalability in most cases.

Use case – e-commerce (online shop)

While there are various use cases discussed earlier, let's highlight another critical one – **e-commerce**. Previous chapters shed light on the significance of the e-commerce use case for Amazon itself, showcasing how a NoSQL system such as DynamoDB drives the retail platform comprehensively, from managing product catalogs to handling financial transactions. To illustrate the modeling of multiple entities related to each other using NoSQL principles, let's consider an example from a GitHub repository (3) shared by a colleague. This resource is openly available, aiding us in guiding customers on effective modeling practices for DynamoDB.

An online shop may have multiple entities to manage – products, orders, customers ordering the products, warehouses stocking these products, invoices for products ordered by customers, and so on. A typical ERD for such a use case would look like the following:

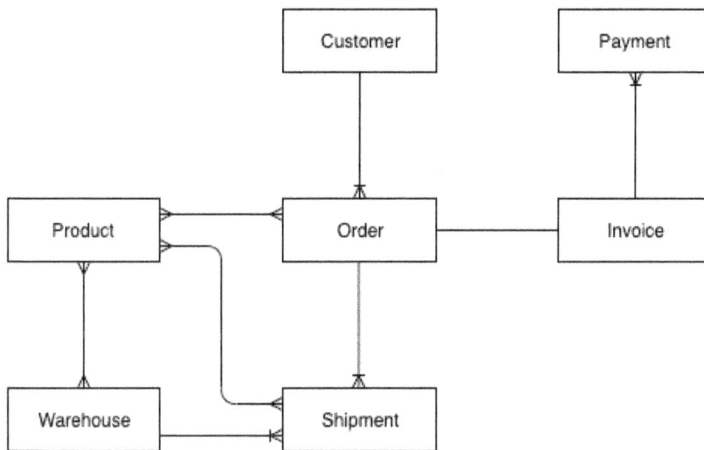

Figure 5.1 – ERD for online shop

If we were to model the described structure in a relational database, it would require a minimum of seven tables, with all access patterns reliant on JOIN operations across these distinct tables. Let's narrow our focus to a specific entity: orders. In this context, a single order can be linked to multiple individual products, be part of a single invoice, and be associated with a specific customer. This establishes a *1:m* relationship with products, a *1:1* relationship with an invoice, and another *1:1* relationship with a customer. Leveraging DynamoDB, the sort key proves to be an efficient mechanism for modeling these diverse relationships associated with an order. For this example, the access patterns we consider at this point are as follows:

1. Get all products for a given order.
2. Get an invoice for a given order.
3. Get all order details for a given order.

For the preceding access patterns, a key hint of the DynamoDB **partition key** (**PK**) is that they all deal with a unique identifier for an order, such as `orderID`. Modeling for these access patterns would mean the PK for our table must store the `orderID` value. Now, for representing the other entities and their relationship with an order, we can use the DynamoDB sort key. The following is how the preceding access patterns could be modeled using a NoSQL database such as DynamoDB, keeping the relationships between different entities intact.

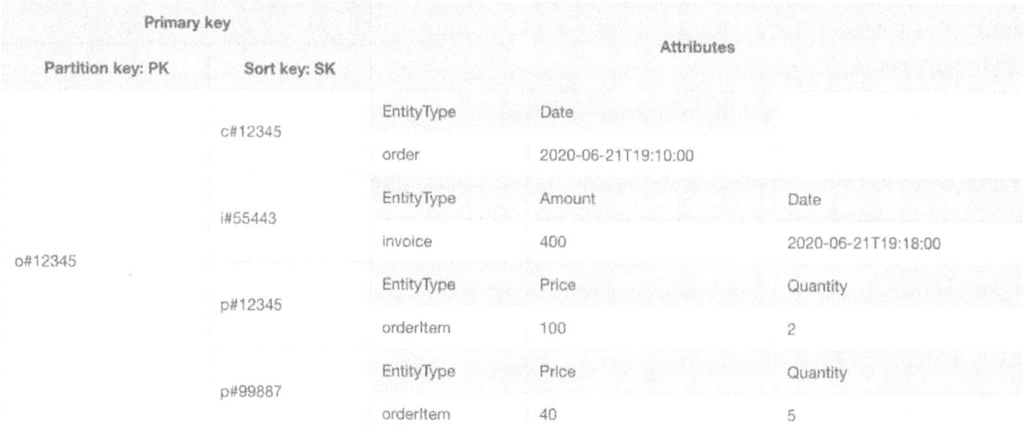

Primary key		Attributes		
Partition key: PK	Sort key: SK			
o#12345	c#12345	EntityType	Date	
		order	2020-06-21T19:10:00	
	i#55443	EntityType	Amount	Date
		invoice	400	2020-06-21T19:18:00
	p#12345	EntityType	Price	Quantity
		orderItem	100	2
	p#99887	EntityType	Price	Quantity
		orderItem	40	5

Figure 5.2 – Data model

As you can see, all the items have a PK value of `o#12345`, indicating a relationship with the order entity. For all four items, the sort key values have a prefix indicating their entity types as customer, invoice, and product. Having such prefixes can allow efficient targeted querying ability using the `begins_with()` functionality of the DynamoDB Query operation. Each item also has an `EntityType` non-key attribute that represents the entity the data belongs to. For the given list of access patterns and based on the preceding data model, the query patterns would look like the following:

Access pattern	Table/Index	Key condition	Example

Get all products for a given order	Table	`PK = orderID`
		
		`SK begins_with("p#")`
			`PK = "o#12345"`
			
			`SK begins_with("p#")`

Get invoice for a given order	Table	`PK = orderID`
		
		`SK begins_with("i#")`
			`PK = "o#12345"`
			
			`SK begins_with("i#")`

Get all order details for a given order	Table	`PK = orderID`
		
		`SK begins_with("c#")`
			`PK = "o#12345"`
			
			`SK begins_with("c#")`

Table 5.1 – Access/query patterns

In the example, you can observe that all relationships between the order, product, and invoice entities are preserved when accessing data from DynamoDB – a NoSQL database. It's the representation that is different from how a RDBMS model would look. The outcome is improved performance, scalability, and freedom from compute-intensive JOIN operations when accessing data for the specified patterns. The purpose of this example was to bust the myth about NoSQL meaning non-relational data.

Next, let us learn the rationale behind another design choice that is common for NoSQL databases: denormalization.

Denormalization – storing multiple copies is OK

Another modeling practice that would be considered a strong anti-pattern in the RDBMS world is **denormalization,** a practice where storing multiple copies of data is considered acceptable. By multiple copies here, I mean occasionally duplicating attributes across items if that helps answer specific access patterns well! In the RDBMS realm, normalization, often up to the **third normal form (3NF)**, is a standard practice aimed at minimizing data duplication, preventing anomalies, ensuring data integrity, and simplifying data management. A RDBMS data model is deemed to be in the 3NF when all rows are functionally dependent solely on the primary keys. This requires combining or joining data elements at runtime to fulfill data access requirements, making traditional RDBMS heavily reliant on JOIN operations.

In the e-commerce scenario discussed earlier, a critical access pattern involves retrieving all products associated with a specific order. *Figure 5.2* illustrates the orderItem entity, where each product is uniquely represented. However, the items primarily feature attributes such as productID, quantity, and price.

In a real-world application, if a user wishes to view detailed order information, relying solely on this design would require multiple database queries. Initially, the app would fetch productIDs, quantities, and prices. Subsequently, to provide a more user-friendly experience, the application must make additional calls to obtain essential product details such as names and image links. This multiple-query approach not only increases database read operations but also introduces latency, potentially affecting user satisfaction. Not forgetting yet another request will be made per product to an object/blob store to get an image of the product, for a decent user experience.

To optimize in such scenarios, consider embedding additional product details such as names and image links alongside productID, quantity, and price within the orderItem entity. This pre-built approach minimizes database lookups, enhancing efficiency by reducing read operations. While this increases storage requirements marginally due to the added metadata, it significantly cuts down database read traffic in half – a noteworthy optimization. The following figure illustrates the new model that includes these duplicated attributes in product name and product image metadata.

Primary key		Attributes					
Partition key: PK	**Sort key: SK**						
o#12345	c#12345	EntityType	Date				
		order	2020-06-21T19:10:00				
	i#55443	EntityType	Amount	Date			
		invoice	400	2020-06-21T19:18:00			
	p#12345	EntityType	Price	Quantity	ProductId	ProductName	ProductImgUrl
		orderItem	100	2	12345	Raisins Can Prepacked Tins	https://example-blob-store/path/to/raisins.png
	p#99887	EntityType	Price	Quantity	ProductId	ProductName	ProductImgUrl
		orderItem	40	5	99887	Crispy Bites	https://example-blob-store/path/to/bites.png

Figure 5.3 – Data model with duplicated product information

Furthermore, avoiding separate queries for product catalog data helps mitigate potential hotspots. Without efficient caching, high-demand items could strain the database, compromising user experience for new customers. Thus, this design balances storage costs against performance gains, ensuring a smoother user experience.

While performing joins or multiple queries in such scenarios may seem straightforward in a small-scale example, the computational overhead of similar joins at scale, especially for latency-sensitive modern applications, can be substantial. Introducing write activity into such models may cause issues, leading to blocking queries, deadlocks, and timeouts. These challenges can even contribute to resource exhaustion, potentially causing the database server to experience downtime. The root of these issues lies in the compute-intensive nature of runtime JOIN operations.

Going back a few decades when RDBMS was considered modern, efficient, and scalable, both compute and storage were considerably expensive. Storage devices, predominantly magnetic tapes, could only support **kilobytes** (**KBs**) to single-digit megabytes. Compute power, represented by processors, was also costly, though server workloads seldom demanded high-computing capacity. During this era, sensitivity to storage costs surpassed that of compute costs, leading to the widespread adoption of a high degree of normalization – a practice optimized for storage. This, however, came at the expense of optimizing workloads for compute. RDBMS gained prominence for effectively addressing prevalent challenges.

Fast forward to the 21st century, where storage costs have significantly decreased, albeit with a relatively slower rate of decrease in compute costs. You may have a 2-terabyte external storage device lying around, which may seem insignificant but the same cannot be said about a 2-gigahertz computer processor. The surge in the volume and velocity of data generated today is remarkable compared to a few decades ago. Performing runtime joins on exponentially larger datasets and complex schemas may no longer yield efficient results. A major reason for performing JOIN operations is the normalization of data. With NoSQL, the optimization of compute is achieved by storing pre-built results together, enabling access without the need for complex JOIN operations.

Next, let us learn about a modern-day NoSQL rule of thumb associated with duplication and denormalization of data to optimize for expensive runtime compute.

Data that needs to be accessed together must be stored together

The concept of denormalization involves pre-building and storing data in the manner it needs to be accessed. If an application requires access to all employee data simultaneously, different entities must be stored to allow comprehensive access, ideally in a single request without requiring a complex JOIN operation. In scenarios where your use case incorporates multiple entities with diverse relationships between them, the principle remains: If those entities need to be accessed together, you store them together in NoSQL databases.

In many instances, this would mean duplicating certain parts of the data, such as employee roles or projects, a practice encouraged in NoSQL data modeling if it aligns with the access patterns. Since, in most cases, an application's read activity surpasses its write activity, the advantages of duplicating information across multiple objects outweigh the drawbacks. These benefits contribute to a highly performant application, positively impacting customer experience and business outcomes.

Addressing concerns about managing multiple copies of data during updates or deletions, DynamoDB offers the capability to update multiple items across the table (or tables) using transactional **application programming interfaces** (**APIs**) in an all-or-nothing operation. Either all updates succeed, or none do. While there may be scenarios where a substantial amount of data requires an update following a system change, the approach taken should be based on the potential impact of serving stale data on the business. Such cases are generally limited to extremely low frequencies – perhaps occurring once a month or a couple of times a year.

> **Important note**
> Research (4) shows the critical impact of site speed on user engagement. Even a slight delay of a few seconds can lead to customer loss. As a page's load time extends from 1 second to 5 seconds, the potential for customers to leave the site increases by 90%. Moreover, 53% of mobile users are likely to abandon a page if it takes longer than 3 seconds (5).

In today's landscape, customers no longer tolerate extended wait times. These times demand optimal service responsiveness. To meet the challenges of this environment, sustain business growth, and expand the customer base, it is important to ensure that customers encounter minimal wait times when accessing and purchasing services, products, or offerings. A direct correlation exists between prolonged wait times and potential revenue loss.

The key takeaway is to design your data access with a focus on optimizing CPU performance, even if it means not prioritizing storage optimization to the same extent. Pre-build your data by having relationships between several entities in a way that the system can reliably fetch it with low latency, irrespective of scale. This approach ensures overall page load times remain exceptionally low, contributing to enhanced business performance.

As we dive into the prospect of storing data for multiple entities together, this often involves storing such pre-built data within a single table. As a general practice, initiating the data modeling process in DynamoDB with a single table is almost always advisable. Of course, exceptions may exist, and we will explore the nuances of making such decisions next.

Multi-table versus single-table design

Given DynamoDB's lack of support for runtime joins (which is by design, by the way), a common but potentially naive design choice for those new to NoSQL might involve normalizing data across multiple DynamoDB tables akin to relational databases. The intention might be to perform joins at runtime in the application layer. While DynamoDB could still maintain performance in this scenario, the application may spend more time engaging in **input/output** (**I/O**) round trips with the DynamoDB service than executing meaningful business logic.

Denormalization and occasional data duplication across items offer a strategy to circumvent the need for runtime joins, even at the application layer. Conversely, adopting a single table for all data without a clear requirement or data access pattern that requires all data together could introduce unnecessary complexity in development and troubleshooting, and may, in general, do more harm than good.

In the upcoming sections, we will explore when to opt for multiple tables to store application data and then dive into considerations for a single-table approach. Finally, we will provide a comprehensive summary for a thorough comparison.

Using a single DynamoDB table for your application

For users accustomed to relational database systems transitioning to DynamoDB without delving into NoSQL data modeling, the initial instinct might be to replicate structures from the old system – each table managing data for a specific entity. This could lead to the application making multiple requests to various tables, such as orders, customers, and invoices, to fetch and aggregate data required for display on the web store. The architecture might involve substantial data manipulation, putting strain on compute and memory resources on the application host.

This scenario represents a case where using multiple tables seems inappropriate, impacting overall application performance despite the seemingly straightforward database design and application development. The latency of the application becomes contingent on the number of network calls needed at the database level for each end-user-initiated request through the web store. It results in spending a considerable amount of time, especially in website page load terms, fetching data from individual tables for different entities and processing it. Such situations might lead to the perception that DynamoDB is slow or unsuitable for the application.

In my interactions with customers, encountering scenarios like the one described was not uncommon. This prompted engagements where professionals like me offered guidance on NoSQL design patterns, aiming to assist customers in understanding how to model for NoSQL systems, specifically DynamoDB. Following these initial discussions, collaborative efforts often involve multiple design reviews and workshops tailored to the specific use cases. The outcome of this iterative process typically resulted in the development of a robust NoSQL design that the team could seamlessly adopt. In the given example, this often translated into opting for a single-table design tailored to the specific entities.

Hopefully, the preceding example helps you understand when to use a single-table design. Going back to the earlier mantra, data that needs to be accessed together must be stored together (in a single table).

Apart from facilitating easier access to data for related entities, which by itself is a massive benefit, adopting a single-table design can offer additional benefits. It can result in cost savings, especially when using a provisioned capacity mode with autoscaling. In provisioned capacity mode, users often provision capacity with a buffer to prevent throttling during occasional traffic surges. With autoscaling, there is an inherent buffer of at least 20% over the configured scaling threshold, leading to cumulative inefficiencies when using several DynamoDB tables in a multi-table design.

While, if suitable, a single-table design offers advantages, there are scenarios where it might not be the most suitable choice. In the following sections, we will explore instances when opting for multiple tables is a more appropriate solution.

Using multiple DynamoDB tables for your application

Consider an application managing user profiles for a game that also facilitates peer-to-peer chat messaging. While both access patterns are within a single application, they represent distinct use cases at a lower level. Importantly, there is no scenario where simultaneous access to both user profiles and chat messages by a user to other users is required within the same scope. In this context, implementing a single table could introduce unwanted complexity in development, and using separate tables for managing data for different entities would be a more appropriate solution. With a multi-table design in this use case, developers would not need to contend with chat messaging-related data while working on user profile services, and vice versa.

Using the same DynamoDB table for different entities, as described in the example of the gaming use case, could lead to unwarranted costs in certain aspects of the architecture. Let's consider this scenario:

- **Analytical use cases**: If the application requires generating recommendations, lists of users, or unique marketing workflows based on user profile data, having a large dataset of chat messages in the same table might result in exporting unnecessary data. This can lead to increased costs, as exporting data for analytical purposes would involve exporting all data, including potentially irrelevant chat messages, which may be most of the table data.

- **Asynchronous workflows**: Setting up asynchronous workflows using DynamoDB Streams for both chat messages and user profiles can introduce unwanted processing overhead. If the application is only interested in reading profile updates, having stream events generated for chat message writes would lead to unnecessary computational efforts.

- **Backup and restore**: In scenarios where backup and restore operations are necessary, using the same table means that restoring data involves bringing back both user profile and chat message data. This may mean you will need to export the vastly irrelevant chat message data to get the user profile and pay for it in money and time.

- **DynamoDB table classes**: DynamoDB offers different table classes, such as Standard and **Standard Infrequent Access (Standard-IA)**. If user profile data is considered infrequently accessed at scale, separating it into its own table allows for optimization of storage costs by moving it to the Standard-IA class. This optimization becomes impractical when user profile data shares a table with more frequently accessed and voluminous chat message data. You might still opt to benefit from the Standard-IA table class and expire data in the Standard class table using **time to live (TTL)** and transfer it to a Standard-IA class table through a DynamoDB Streams and AWS Lambda-based pipeline (6). However, in this scenario, you are still handling two tables and managing a stream processing pipeline, and the considerations mentioned here still apply.

In situations like the one mentioned, where an application manages user profiles and supports peer-to-peer chat messaging, using multiple DynamoDB tables is highly beneficial. This approach offers cost efficiency, faster backup restoration, flexible data export, and efficient asynchronous processing. It allows for better control over costs, quicker recovery of specific data, and streamlined processes for exporting and processing individual datasets. Overall, using multiple tables is an optimal solution in such specific cases, which enhances both cost-effectiveness and system performance. Of course, one size does not fit all, and using multiple tables for every use case is not appropriate.

Next, let us explore some concepts that might be new for those new to NoSQL or DynamoDB. These concepts will assist in moving from a relational mindset about organizing data to a NoSQL database.

Breaking down your data

This section of the chapter, like the others, assumes you are relatively new to NoSQL data modeling and may have some proficiency with modeling databases in RDBMS. Going from the RDBMS-based model to a NoSQL-based one, you may need to move your mindset from normalization to denormalization with potentially some duplication as well; however, the denormalization here could itself be done cleverly. Simply dumping all the columns/attributes into a single large JSON blob and retrieving that blob for any data access request that may not warrant reading the 100 other attributes within the blob is not quite what would work. If it does, that would not be efficient anyway.

This section will showcase a few design patterns and thought-provoking concepts that would help you think about the denormalization of data coming from the relational world itself, but also doing it cleverly, by identifying opportunities to break up the data as per the data access patterns, to be highly efficient, both cost and performance wise.

First, let us review the denormalization part of the journey, and apply all that we have learned about it so far in this chapter.

Single table – swapping extra tables for items

For use cases where related data needs to be accessed together, you now know that they must be stored together. To use this concept for transitioning from a traditional relational database paradigm to DynamoDB's NoSQL approach, let us revisit the e-commerce example introduced earlier in this chapter. The subsequent figure depicts the entities from the ERD: customer, order, and product.

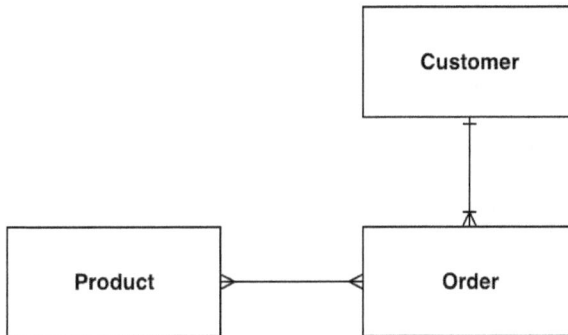

Figure 5.4 – ERD with customer, order, and product entities

In an RDBMS context, these entities would exist as distinct tables within a database, each equipped with its primary key, namely, `customerID`, `orderID`, and `productID`. These tables might have foreign keys to denote relationships with other entities. Moreover, the design could see further normalization, such as introducing an order status dimension table for the order entity and an address table associated with both the order and customer entities. Data retrieval here would require executing joins and unions across multiple tables to assemble the data for presentation on the e-commerce website's user interface.

In DynamoDB, a NoSQL database, the approach differs. To maintain relationships between these entities and facilitate efficient data retrieval based on those relationships, we consolidate the data for these entities within a single table. Instead of deploying separate tables, we opt for a single table where items are organized based on the table's primary key and based on sort keys representing foreign keys for related entities. Typically, you would want to go for the ID type attribute from each entity to link the data together, as shown here:

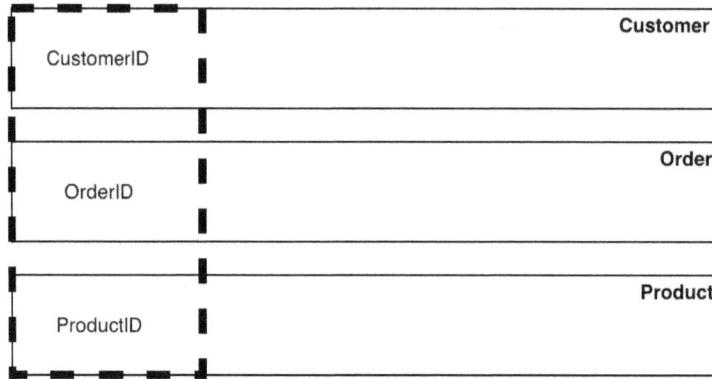

Figure 5.5 – NoSQL way of aligning entities into a single table

The ID type attributes are generally good candidates to be part of the primary key in some way. Whether these become the PK or the sort key depends on the ERD, the relationships between the entities, and the access patterns that need to be answered. Let us consider that the model we want here must be able to get data based on the customer entity, for example:

- Get all orders for a customer

- Get customer information by `customerID`

- Get product information by `productID`

With the preceding access patterns in mind, `customerID` could potentially be a suitable candidate for the PK, whereas the other entities that are related to the customer entity could become part of the primary key, serving as the table's sort key. The following is how the model could come about:

Primary key		Attributes				
Partition key: PK	Sort key: SK					
	!	EntityType	customerID	timestamp		
		customer	12345	2023-02-19T19:07:40.178Z		
c#12345	o#234	EntityType	orderID	timestamp		
		order	234	2024-02-02T12:51:09.694Z		
	o#555	EntityType	orderID	timestamp		
		order	555	2021-05-02T12:51:09.694Z		
p#99887	!	EntityType	productID	productName	productImgUrl	timestamp
		product	99887	Raisins Cans Prepacked Tins	https://example-blob-store/path/to/raisins.png	2023-07-13T04:19:10.990Z

Figure 5.6 – Data model to address access patterns linked via customer entity

In the previous figure, observe the presence of an item collection encompassing both customer-related static information and customer order details. By linking the order information using the PK of `customerID`, I can efficiently retrieve all orders associated with a specific customer, alongside the customer's information, in a single request. The result can be readily displayed on the user account page of the e-commerce web store. Additionally, if the access pattern is to exclusively list the orders, notice the order prefix associated with each order entity item. This enables the use of the `begins_with` operator within the Query API, streamlining the retrieval of order-specific data only.

Also, note that the table design in *Figure 5.7* differs from the one previously presented in this chapter, despite being based on the same ERD. This is because earlier, we were modeling access patterns for orders, invoices, and customers, all centered around an `orderID` value. In contrast, the current scenario is based on the same entities but addresses a different set of access patterns, with each pattern linked to a `customerID` value.

In practical scenarios, these two views of data, based on the same entities, might represent the schemas of a table and its **global secondary index (GSI)**. Determining which view serves as the table's schema and which constitutes the GSI's schema depends on additional factors such as write access patterns, read-after-write consistency requirements, and other considerations. If you are not concerned about strong read-after-write consistency, my recommendation is to select the view that aligns with most of your primary access patterns as the base table's schema. For example, if listing orders by `customerID` is expected to be one of the most frequent access patterns, then it becomes a part of the base table's schema. As a bottom line, most of your database schema modeling choices should cater to specific access patterns, whether for reads or writes.

This design also offers future-proofing for your application. If new requirements emerge to associate the customer entity with additional entities such as wish lists, shopping carts, or payment options, the current model can be effortlessly extended. This involves adding new items to the model, each sharing the same PK (`customerID`), while the sort key-value indicates different entities with specific prefixes. For instance, the sort key for wish lists could be `WISH#<wish_list_item>`, for payment options, it could be `PAY#<payment_option_id>`, and so forth. The use of generic names for the PK and sort key proves beneficial in this context. For other access patterns involving these new entities, you can either overload the GSIs like what was done with the table here or create new GSIs if extending the existing ones is impractical for any reason.

Another notable point is the separate modeling of the product entity with its own PK. This decision is influenced by the specific use case as it does not require accessing product catalog data along with customer data in, say, a single request. Product catalog data is owned by the organization, not the customer, and thus does not need to be intricately linked with customer data. While the order entity may contain information about products within an order, having a distinct product catalog microservice can efficiently provide detailed product information when a user accesses the product page for any item from their orders.

Taking this analysis a step further, it may not be logical to store the product catalog data in the same table as other entity data. Hence, a preferable approach would be to maintain the product catalog data in a dedicated, separate table, allowing independent management of this data distinct from customer-specific information.

In short, leveraging related items and querying them based on these connections can streamline your DynamoDB table design. Instead of adding multiple tables as you would in a traditional relational database, you can use items with the same PK but varying sort keys to manage related data efficiently. This approach is also called vertical partitioning, and you will see more examples of this in *Chapter 7, Vertical Partitioning*. However, some entities might still warrant their own DynamoDB table if they do not need to be accessed alongside other data. Moving forward, we will learn how breaking down items further can optimize costs while still retaining performance efficiency.

Breaking down items

In the NoSQL realm, another strategic approach involves breaking down data based on attribute aspects in addition to access patterns. Rather than consolidating all attributes into a monolithic JSON blob with numerous fields, prioritize structuring the data into logical groupings. This segmentation facilitates separate access, individual updates, and independent indexing, optimizing both performance and manageability.

Thoughtful breaking down of data can yield significant benefits in terms of both performance and costs. Consider a scenario where you have a substantial 20 KB JSON blob stored as a single item or row in a database. In many cases, for serving an application request, you likely only need a portion of that 20 KB, not the entire blob. For example, in a gaming use case, you might only require inventory data stored along with the user profile and game progress, and in an email service provider use case, fetching the sender and subject line may be sufficient without needing the entire email body. The following figure illustrates the gaming use case:

game-data

Primary key					
Partition key: PK			**Attributes**		
	EntityType	userID	timestamp	inventory-blob	profile-blob
u#Johnnie.Roberts	game-blob	Johnnie.Roberts	2024-03-03T17:24:25.744Z	{ "inventory": { "potions": { "life-potion": 5, "heal-potion": 0, "craft-potion": 22 } }, "other-inventory-attributes": "<large blob data>" }	{"user-name":"Johnnie.Roberts", "email":"Johnnie.Roberts@example.com", "home-server":"IAD", "other-attributes":"<large blob data>"}
	EntityType	userID	timestamp	inventory-blob	profile-blob
u#Edythe_Hermann7	game-blob	Edythe_Hermann7	2023-03-24T07:26:44.694Z	{ "inventory": { "potions": { "life-potion": 11, "heal-potion": 10, "craft-potion": 27 } }, "other-inventory-attributes": "<large blob data>" }	{ "user-name": "Edythe_Hermann7", "email": "Edythe_Hermann7@example.com", "home-server": "DUB", "other-attributes": "<large blob data>" }

Figure 5.7 – NoSQL model of game use case with large data blobs

The following figure illustrates the email use case:

mailbox

Primary key		Attributes				
Partition key: PK	Sort key: SK					
		email-sender	userID	mail-metadata	mail-blob	timestamp
u#Noel89	2023-03-30T22:44:30.845Z	Jean_OKeefe@yahoo.com	Noel89	{ "server": "mail-1234@example.com", "origin-id": 3216532068892672, "network-id": 2811475563380736, "other-mail-metadata": "\<large blob data>" }	"Hello, I hope this message finds you well. I wanted to share some exciting news and updates with you. Please stay tuned for further details. Best regards, Jean O"	2023-09-16T07:29:04.625Z
		email-sender	userID	mail-metadata	mail-blob	timestamp
	2024-10-25T05:27:38.531Z	Allan_Klein11@yahoo.com	Danielle.Pfannerstill	{ "server": "mail-4@example.com", "origin-id": 21312399867, "network-id": 888756565655, "other-mail-metadata": "\<large blob data>" }	"Dear Noel, I trust this email finds you in good health. I wanted to discuss the upcoming project and gather your thoughts on the proposed timeline. Your input is highly valued. Regards, Allan K"	2023-07-19T19:08:55.302Z

Figure 5.8 – NoSQL model of email use case with large data blobs

Breaking down these bulky blobs into smaller, distinct chunks that can be queried individually and retrieved together can result in substantial savings on read IOPS (input/output operations per second) if we think in terms of raw disk access in operating systems. Instead of storing the entire 20 KB together when it is not always needed together, you can create separate items or rows for specific chunks of data. For instance, in the gaming use case, you might have distinct items for user profile, recommendations, and inventory, as shown in the following figure.

Primary key		Attributes			
Partition key: PK	Sort key: SK				
u#Johnnie.Roberts	inventory	timestamp	inventory-blob		
		2024-03-03T17:24:25.744Z	{ "inventory": { "potions": { "life-potion": 5, "heal-potion": 0, "craft-potion": 22 } }, "other-inventory-attributes": "\<large blob data>" }		
	profile	userID	timestamp		profile-blob
		Johnnie.Roberts	2014-12-03T17:24:25.744Z		{"user-name":"Johnnie.Roberts", "email":"Johnnie.Roberts@example.com", "home-server":"IAD", "other-attributes":"\<large blob data>"}
u#Edythe_Hermann7	inventory	timestamp	inventory-blob		
		2023-03-24T07:26:44.694Z	{ "inventory": { "potions": { "life-potion": 11, "heal-potion": 10, "craft-potion": 27 } }, "other-inventory-attributes": "\<large blob data>" }		
	profile	userID	timestamp		profile-blob
		Edythe_Hermann7	2023-03-14T17:00:44.694Z		{ "user-name": "Edythe_Hermann7", "email": "Edythe_Hermann7@example.com", "home-server": "DUB", "other-attributes": "\<large blob data>" }

Item Collection

Figure 5.9 – NoSQL model for gaming use case post-breaking down of large items

Similarly, in the email use case, separate items for email metadata and the email body would make for efficient reading and perhaps updating. The following is what a version of the model could look like:

Primary key		Attributes				
Partition key: PK	Sort key: SK					

Partition key: PK	Sort key: SK					
u#Noel89	ksuid#2b0fWuvPaw jE1Wtd34DzNpdisO l#body	mail-blob "Hello, I hope this message finds you well. I wanted to share some exciting news and updates with you. Please stay tuned for further details. Best regards, Jean O"				
	ksuid#2b0fWuvPaw jE1Wtd34DzNpdisO l#meta	email-sender Jean_OKeefe@yahoo.com	userID Noel89	mail-metadata { "server": "mail-1234@example.com", "origin-id": 3216532068892672, "network-id": 2811475563380736, "other-mail-metadata": "<large blob data>" }		timestamp 2023-09-16T07:29:04.6 25Z
	ksuid#2b0fY4c9GZl 6WU5X8SKHfCpW 9G0#body	mail-blob "Dear Noel, I trust this email finds you in good health. I wanted to discuss the upcoming project and gather your thoughts on the proposed timeline. Your input is highly valued. Regards, Allan K"				
	ksuid#2b0fY4c9GZl 6WU5X8SKHfCpW 9G0#meta	email-sender Allan_Klein11@yahoo.com	userID Danielle.Pfan nerstill	mail-metadata { "server": "mail-4@example.com", "origin-id": 21312399867, "network-id": 888756565655, "other-mail-metadata": "<large blob data>" }		timestamp 2023-07-19T19:08:55.3 02Z

Figure 5.10 – NoSQL model for email use case post-breaking down of large items

Fetching these smaller, individual chunks at the lowest level in a database translates to fewer IOPS on the disk. This approach is advantageous both from a cost perspective and an overall latency perspective. The key is to link these individual chunks together using the same PK so that they can be retrieved individually or together, mimicking the behavior of a single blob. This supports access patterns that require the complete set of data while providing flexibility and efficiency for scenarios that only demand specific chunks.

When considering the breaking down of data, another valuable approach involves assessing the anticipated IOPS that individual data chunks might generate. Let us dive into another e-commerce scenario where the product catalog encompasses detailed product descriptions, names, brand information, and dynamic attributes such as inventory levels and user feedback.

Primary key		Attributes						
Partition key: PK	Sort key: SK	name	brand_name	category	others	description	inventory	user_feedback
0079893601113	!	Raisins Can Prepacked Tins	O Organics	packed foods	other attributes	\"**O Organics Raisins Can: Nature's Sweetness in Prepacked Tins**\r\n\r\nWelcome to the world of O Organics, where we celebrate the pure goodness of nature in every bite. Our O Organics Raisins Can is a tribute to the simplicity and purity of organic ingredients, thoughtfully prepacked in tins.....	224	<user feedback blob>

Figure 5.11 – NoSQL model for e-commerce use case with large items of mixed IO

Given that a product's popularity can influence its data access patterns – more sales might lead to frequent inventory updates and increased user feedback – it only makes sense to segment this data thoughtfully. Firstly, you might isolate user feedback and comments to separate items or even a separate NoSQL table, considering they may not always need to be accessed along with primary product attributes.

Regarding inventory details, their dynamic nature may demand special consideration. While product descriptions and other static attributes remain relatively unchanged, inventory data would typically experience frequent updates due to sales or restocking. This dynamic nature of inventory can result in high read and write operations, generating significant IOPS compared to the more static product details.

In the context of DynamoDB, updating a portion of a larger data item incurs throughput charges for the entire item size, which can be inefficient. Most database technologies out there perform additional IO operations to handle updates to small parts of the data, since the data on disk may be read and replaced by the updated image of the data. By breaking down the data based on anticipated IOPS, you can create distinct items or rows. For instance, one item could focus on high-velocity inventory attributes, while another captures static product details. Crucially, linking these items with a common PK would ensure cohesive retrieval when necessary.

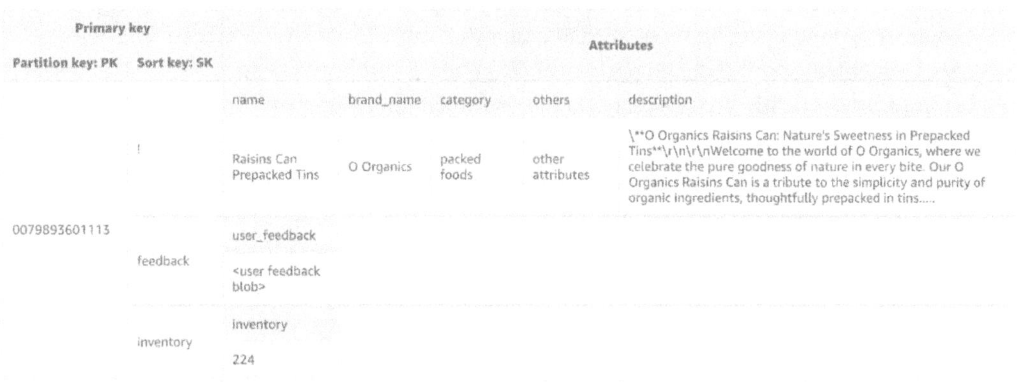

Primary key		Attributes				
Partition key: PK	Sort key: SK					
		name	brand_name	category	others	description
	!	Raisins Can Prepacked Tins	O Organics	packed foods	other attributes	**O Organics Raisins Can: Nature's Sweetness in Prepacked Tins**\r\n\r\nWelcome to the world of O Organics, where we celebrate the pure goodness of nature in every bite. Our O Organics Raisins Can is a tribute to the simplicity and purity of organic ingredients, thoughtfully prepacked in tins.....
0079893601113	feedback	user_feedback				
		<user feedback blob>				
	inventory	inventory				
		224				

Figure 5.12 – NoSQL model for e-commerce use case post-breaking of items based on expected IO

This strategy not only aligns with the data's static or dynamic characteristics but also complements earlier approaches focused on logical data segmentation. By blending both methodologies, you achieve a balanced, efficient data model tailored to your application's unique needs.

That was all about breaking down items for now. More examples and guidance about breaking down data further can be found in *Chapter 7, Vertical Partitioning*. Next, let us learn about moving from a relational mindset about storing extremely **large object (LOB)** data within the **online transaction processing (OLTP)** database.

Handling large items

Over the decades, RDBMS has evolved into versatile solutions, attempting to cater to various data types rather than being exclusive to transactional access. In contrast, NoSQL databases opt for a streamlined approach, sacrificing the Swiss Army knife-like flexibility of databases for consistent and scalable performance. Although organizations often require functionalities beyond transactional data access, clever design patterns can address these needs within DynamoDB by utilizing strategic data storage methods or integrating purpose-built data stores alongside the primary DynamoDB table.

Addressing the storage of LOBs in DynamoDB involves thoughtful consideration of factors such as filtering, update frequency, and performance requirements. It requires choosing a strategy tailored to the specific demands of your use case. It is crucial to be mindful of the limitations of each DynamoDB item. Each item can only hold up to 400 KB of data, encompassing both attribute names and their corresponding values. This constraint guides the decisions regarding how to efficiently manage LOBs within DynamoDB:

- **Compression for fast access**: To ensure rapid access to LOBs without adding complexities to your application stack, consider utilizing efficient data compression techniques before storing data in DynamoDB. Employing libraries, such as Snappy or zlib, can achieve significant compression ratios, optimizing both cost and performance. While decompression post-retrieval increases compute usage slightly, the benefits generally outweigh this minor overhead.

- **Object store integration**: If immediate DynamoDB access for LOBs is not a priority and you are open to integrating additional dependencies, leveraging an object store such as Amazon S3 can be cost-effective. Store a reference link to the S3 object within DynamoDB, turning it into a metadata store. For enhanced speed, introduce a caching layer in front of S3, especially beneficial for frequently accessed objects. This approach is particularly effective when LOBs undergo infrequent updates.

- **Uncompressed storage with vertical partitioning**: If feasible, and if you require rapid access to uncompressed data within DynamoDB, consider storing the LOB data as is. However, to sidestep the 400 KB item size limit, segment the LOB into multiple DynamoDB items. Ensure these items share a common PK for seamless retrieval. Although this method might elevate costs, it promises optimal DynamoDB performance, especially when compressed storage is not suitable.

The uncompressed storage within the DynamoDB strategy is a byproduct of implementing vertical partitioning, and you can learn more about this design pattern in *Chapter 7, Vertical Partitioning*. All the previously mentioned strategies for handling LOBs within the NoSQL and DynamoDB realm are also explained in an AWS blog (8).

> **Important note**
>
> Consider this as a reminder to delete any resources you may have created while going through this chapter in your AWS account.

This marks the end of the guidance on handling LOBs and how it is different from the traditional RDBMS.

Summary

In this chapter, we expanded our understanding of NoSQL, emphasizing the distinctive design patterns between RDBMS and NoSQL, with a specific focus on DynamoDB. We explored strategies such as duplication and denormalization, which optimize for compute over storage. This approach minimizes expensive runtime joins and prioritizes efficient data retrieval.

While discussing denormalization, we delved into strategies for efficient data modeling in DynamoDB. We highlighted scenarios favoring a single DynamoDB table for related entities versus multiple tables, emphasizing that there is no one-size-fits-all approach. Sometimes, segregating data into separate DynamoDB tables proves more efficient.

Furthermore, we examined the intricacies of breaking down data in NoSQL databases, emphasizing smart partitioning strategies. This section also covered handling LOBs in DynamoDB, detailing strategies based on factors such as latency, update frequency, and application dependencies.

In the upcoming chapter, we will pivot to reading data from DynamoDB. We will explore various APIs and functions for data retrieval, delve into DynamoDB's read consistency options, and provide best practices for efficient data access, especially in OLTP scenarios. Additionally, we will demystify DynamoDB's transactional APIs and its Atomicity, Consistency, Isolation, and Durability (ACID) compliance mechanisms.

References

1. Six degrees of separation: `https://en.wikipedia.org/wiki/Six_degrees_of_separation`

2. Understanding the basics of using DynamoDB: `https://www.capitalone.com/tech/software-engineering/introductory-guide-to-dynamodb/`

3. Online Shop: `https://github.com/aws-samples/amazon-dynamodb-design-patterns/tree/master/examples/an-online-shop`

4. Find out how you stack up to new industry benchmarks for mobile page speed: `https://www.thinkwithgoogle.com/marketing-strategies/app-and-mobile/mobile-page-speed-new-industry-benchmarks/`

5. How Page Load Speed Affects Customer Behavior: `https://www.business.com/articles/website-page-speed-affects-behavior/`

6. Archive to cold storage with Amazon DynamoDB: `https://aws.amazon.com/blogs/database/archive-to-cold-storage-with-amazon-dynamodb/`

7. Compression libraries:

    I.   Snappy: `https://en.wikipedia.org/wiki/Snappy_(compression)`

    II. zlib: `https://en.wikipedia.org/wiki/Zlib`

    III. Large object storage strategies for Amazon DynamoDB: `https://aws.amazon.com/blogs/database/large-object-storage-strategies-for-amazon-dynamodb/`

6

Read Consistency, Operations, and Transactions

Continuing the momentum from the previous chapters of *Part 2* of the book, which focuses on data modeling, let us maintain this thematic flow but pivot purely into the read aspect of DynamoDB for this chapter.

If you are used to **Relational Database Management Systems** (**RDBMS**) and have been following along with the earlier parts of this book, you know by now that NoSQL or DynamoDB works differently. It is about organizing data in a new way, and you need to be open-minded about learning this.

The main idea is to make your data model work well for the specific ways in which you want to get information. This means giving up some flexibility to make things work smoothly and consistently. Your workload, being read-heavy or write-heavy, is one of the aspects that decides how you should set up your data.

For example, in a read-heavy scenario, you want your model to work efficiently from the beginning. You don't want the database to fetch a ton of data – just what's needed – and do most of the filtering and sorting on its side. It might be okay to spend more time on writing and duplicating attributes across items, so that data is pre-built to be read quickly.

On the flip side, if you have a write-heavy workload, you want your model to focus on finishing the writing process reliably and quickly. Duplicating, pre-building, or doing a ton of processing before writing to the database might not be the best in this case. The key is to adapt your data model based on what your workload needs, whether that's more reading or more writing.

If you're dealing with a read-heavy workload, which is the case for most situations, it's crucial to understand how reading works best in the database you're using. In DynamoDB, this involves knowing about the various ways you can retrieve data (read APIs) and what each is good for. Also, it would help to know how the system ensures consistency when reading data.

In this chapter, we'll kick things off by examining DynamoDB's read consistency model. We'll dive into the design choices made within the database and offer guidance on incorporating read consistency into your application. Moving forward, we'll explore each of DynamoDB's read APIs, which include `GetItem`, `BatchGetItem`, `Query`, `Scan`, `ExecuteStatement`, and `BatchExecuteStatement`. Our aim is to evaluate the strengths of each API while staying alert to any potential downsides.

After exploring the consistency model and the read operations, we'll introduce you to **Atomicity, Consistency, Isolation, and Durability** (**ACID**) transactions in DynamoDB, focusing on the transaction read and write APIs. We'll emphasize the differences between these and other DynamoDB APIs, along with the types of consistency and isolation they deliver. Importantly, we'll provide guidance on when it's suitable to employ these transactional APIs.

By the end of this chapter, you'll have a strong understanding of DynamoDB's read consistency model. You'll be well-versed in the various supported read APIs, knowing the appropriate use cases for each. Furthermore, you'll be familiar with the ACID-compliant transactional APIs within DynamoDB, gaining insights into how they differ from transactional operations you might be used to, particularly if you're transitioning from the RDBMS world.

In this chapter, we are going to cover the following main topics:

- Understanding the read consistency model
- Reviewing read APIs
- Transactions and atomicity

Understanding the read consistency model

In traditional RDBMS, significant emphasis is placed on preventing various forms of data corruption through mechanisms such as data locks, waiting queries, and ensuring read consistency. These systems provide a fixed set of support for all use cases and workloads. However, in scenarios where only a small portion of data needs updating, database locks can still occur, impacting other incoming requests.

In contrast, NoSQL systems such as DynamoDB have evolved to adopt different measures tailored to specific use cases, focusing on flexibility and scalability.

For instance, in SQL Server databases, data updates could result in the locking of one or more pages of the table, even if the update only involved one specific data row on the page. Similarly, in Postgres, access-exclusive locks might be used by applications to lock the entire table during a particular read operation, causing any other concurrent operations on the table to wait until the lock is released.

Turning the spotlight to read consistency, most NoSQL systems, being distributed in nature, adhere to the principles of the **Consistency, Availability, and Partition Tolerance (CAP)** theorem. Revisiting computer science lessons, the CAP theorem states that distributed systems, such as NoSQL databases, can provide only two out of the following three guarantees:

- Consistency

- Availability

- Partition tolerance

During a network partition in a distributed database system, a crucial trade-off arises between prioritizing availability or consistency. When a system gives precedence to availability over consistency, it can continue serving data even during a network partition, but it cannot guarantee that it can provide the most up-to-date data. Conversely, prioritizing consistency over availability ensures that the system provides the most current state of the data, even though it might experience failures when it is unable to guarantee this service. In situations without a network partition, the system can support both availability and consistency concurrently.

While all database systems can support AC, NoSQL databases, especially those designed for horizontal scaling, typically do well in supporting AP. DynamoDB stands out by supporting all three aspects: AP, CP, and AC. While DynamoDB doesn't defy the CAP theorem by simultaneously supporting AP and CP, it allows you to choose between the two on a per-request basis using a simple parameter. This behavior is closely tied to DynamoDB's read consistency, offering two levels for reads:

- Strong consistency

- Eventual consistency

Let us dive a bit deeper into each of these modes and understand when it's suitable to use one over the other.

Strong consistency

As outlined in earlier chapters, DynamoDB ensures high availability by replicating table data across at least three AWS Availability Zones. Across these Availability Zones, the replicas of the same data slice (partition) operate in a leader-follower topology, with one replica serving as the leader and the other two as followers. This topology is established because all write requests are exclusively processed by the leader, and the leader then propagates the write intent to both followers simultaneously. A customer write is considered successful when, apart from the leader, persistence acknowledgment from at least one follower is received. While the second follower provides its acknowledgment shortly after, the response confirming the customer write success is already sent to the client after the first follower acknowledgement is received. This is all the background we need to learn about read consistency in this chapter. For more information about the leader-follower topology and how DynamoDB is set up under the hood, head to *Chapter 10, Request Routers, Storage Nodes, and Other Core Components*.

Given that either of the two followers could have the latest state of the data, achieving read-after-write consistency requires requesting a strongly consistent read. In this case, the read is consistently processed by the leader, ensuring access to the latest data state.

A strongly consistent read from the leader supports the CP aspect of the CAP theorem. Even during a network partition, DynamoDB can authoritatively serve the most up-to-date data through the replication leader. However, it is important to note that with CP, there may be a trade-off in terms of lower availability, relative to AP. If DynamoDB encounters difficulties establishing a connection with the replica leader or recognizing a replica as the leader, a strongly consistent read request may result in a server error (**HTTP 500**) until a leader is recognized or a successful connection to the leader is established.

In DynamoDB, even simultaneous write and strongly consistent read requests toward a table item are processed with serializable isolation. This ensures that each request is processed as if no other request is being handled concurrently. If the strongly consistent read is processed before the write, the most up-to-date state of the item at that moment is returned to the client. Conversely, if the write is processed before the strongly consistent read, the read returns the complete item state after the write request.

DynamoDB allows you to set the read consistency level on a per-request basis using a simple Boolean flag. By default, DynamoDB performs eventually consistent reads, a concept we will dive into next.

Eventual consistency

Understanding eventual consistency in DynamoDB follows logically from our exploration of strong consistency. In the context of the same replication group for a DynamoDB table partition, when a read request is made with eventual consistency, DynamoDB accesses any of the three replicas to fulfill the read. Given that at least two out of the three replicas, including the leader, would have persisted and acknowledged the most recent write request for the same DynamoDB item, there is a two-out-of-three chance that the eventual consistent read request will be served with the most up-to-date state of the item. However, since there's a possibility of one out of three requests being served with a stale copy of the item, eventual consistent reads should only be used if your application can tolerate a brief period of receiving potentially outdated data shortly after a write request.

In terms of the CAP theorem, eventual consistent reads support the AP aspect. During a network partition, DynamoDB can support eventual consistent reads with higher availability, compared to strongly consistent reads. However, it cannot guarantee that the response will always contain the most up-to-date state of the data. If high availability takes precedence over strong read-after-write consistency in your application, it's appropriate to use eventual consistent reads within DynamoDB.

That was all the important bits about read consistency fundamentals. More about read consistency can also be found in the AWS docs (1). Next, let us learn about when to choose one or the other, as well as how to think about the read consistency model in the context of your application.

Choosing the right read consistency mode

Understanding the choice between eventually consistent reads and strongly consistent reads prompts the question: why not always choose strongly consistent reads since they guarantee the most up-to-date data? Well, the decision hinges on the priorities set by the specific access pattern. Eventually, consistent reads offering AP tolerance guarantees in the CAP theorem become favorable when availability takes precedence. Additionally, eventual consistent reads cost half as much as strongly consistent reads – they are a more economical choice.

Lastly, a lesser-known insight is that with eventually consistent reads, since the read could be served by any of the three partition replicas, the overall read throughput that can be consumed on the overall data held by the partition is higher than the 3000 **Read Capacity Units** (**RCU**) per second per partition documented limit. It's crucial to evaluate each access pattern individually to determine whether strong read-after-write consistency is necessary.

For those transitioning from a traditional RDBMS mindset, where every read is strongly consistent, there's a need to shift thinking. In DynamoDB, considering access patterns and aiming for the highest availability possible becomes crucial.

Consider the example of a banking application in the financial services industry. For real-time transaction verification, where users make a transaction and immediately check the transaction history, strongly consistent reads are crucial, despite potential read failures during network partitions. However, for non-financial data updates such as personal profile details, strongly consistent reads may not be necessary.

On the flip side, in a social media application tracking likes, comments, and follower counts, eventual consistency may be entirely acceptable, with users viewing the latest information within milliseconds to a few seconds.

When deciding between eventual and strong consistency for an access pattern, consider the following:

- The business impact if the information returned could be stale
- Whether you prioritize strong read-after-write consistency over occasional availability during network partitions

> **Important note**
>
> Despite the read consistency mode chosen for your access patterns, DynamoDB SLA (2) would apply to the table's monthly availability at 99.99% for single region tables and 99.999% for global tables.

To sum up, DynamoDB supports both CP and AP modes as per the CAP theorem for reads. It's essential to think about read consistency on a per-access-pattern basis, ensuring a cost-optimized approach with the right balance of availability and consistency in your application. In the next section, let us review the different read APIs supported by DynamoDB and when to use each.

Reviewing read APIs

Now that you have a good idea of the read consistency model within DynamoDB, let us review the different read APIs offered by DynamoDB and their best practices. At the time of writing, DynamoDB supports the following APIs to retrieve data:

- `GetItem`
- `BatchGetItem`
- `Query`
- `Scan`
- `ExecuteStatement`
- `BatchExecuteStatement`
- `ExecuteTransaction`
- `TransactGetItems`

If you have followed the earlier chapters, you are likely familiar with the `GetItem` and `Query` operations in DynamoDB. These operations are fundamental for retrieving single and multiple items, respectively, and are among the most commonly used. Let us review all these read operations with insights and best practices based on my experience interacting with DynamoDB and its customers.

For detailed information, including request/response structures, limits, and SDK reference pages, you can refer to the AWS docs API reference pages (3). Instead of duplicating the documentation, I will focus on key aspects and crucial best practices.

Let us kick off the review with the `GetItem` operation.

GetItem

The `GetItem` operation in DynamoDB provides a straightforward key value lookup. When provided with a primary key value, it returns at most a single item from a DynamoDB table. Depending on the table's primary key schema, which can consist of a partition key alone or both a partition key and a sort key attribute, the `GetItem` request must include the complete values of these keys for a successful retrieval:

- If the primary key schema includes only a partition key, each partition key value corresponds to a unique item in the table
- If the primary key schema includes both a partition key and a sort key, each item is uniquely identified by a combination of these keys

If the values provided in the GetItem request are partial or incorrect, DynamoDB does not raise an error but responds with an empty list. The GetItem API supports various parameters, including a Boolean flag for the read consistency level, which defaults to eventual consistency.

GetItem is ideal for single-item lookups with the lowest expected latency. Due to the horizontal partitioning of DynamoDB table data, a GetItem operation targets exactly one partition responsible for the item, making it highly performant. The typical latency for such retrievals is in the single-digit milliseconds. If the end-to-end latencies significantly exceed this range, troubleshooting on the client side may be necessary.

For use cases that require batching singleton GetItem requests, DynamoDB offers the BatchGetItem API, which we will explore next.

BatchGetItem

As the name implies, BatchGetItem enables the execution of a batch of individual GetItem requests. This batch processing results in a single network request going to and from DynamoDB, reducing the number of network round trips and optimizing crucial milliseconds, especially when dealing with high-traffic applications. Each individual GetItem payload within a BatchGetItem request resembles a standalone GetItem request.

Here are some key details about BatchGetItem:

- A BatchGetItem request can contain up to 100 individual GetItem requests
- The response from BatchGetItem is limited to returning no more than 16 MB as a collective response size
- Considering that the DynamoDB maximum item size is 400 KB, the BatchGetItem response can return a maximum of just under 40 items, each 400 KB in size, or all 100 items where each item is less than 160 KB
- UnprocessedItems is an array returned for keys that were not processed successfully in the batch, indicating the need for the application to include them in another BatchGetItem request

The decision to use BatchGetItem depends on your application's request rate and architecture. It is beneficial when your application can aggregate data access requests for short periods before sending them to DynamoDB. Many customers with high application traffic leverage BatchGetItem for its efficiency in reducing network round trips.

It is important to note that, apart from the network time, there is essentially no difference between a single BatchGetItem request and 100 individual GetItem requests from the database's perspective. Read throughput costs are computed per item, not based on the cumulative size of the response.

With `BatchGetItem`, you can retrieve data with different pairs of partition keys and sort key values. Additionally, you can retrieve data from different DynamoDB tables within the same AWS account and region. For use cases where you need to fetch one or more items from a single table linked with the same partition key value, the `Query` API is what you would want to use, and we will explore it next.

Query

The `Query` API in DynamoDB allows you to retrieve one or more items with the same value for the partition key, but with potentially differing in sort key values. This operation is especially useful when dealing with items that share a **1:m** relationship with the partition key value. The items returned by a `Query` API response are sorted by their sort key values, supporting both ascending and descending order.

Here are key aspects of the `Query` API:

- **Needs complete partition key value**: You can pass the complete value of the partition key, and DynamoDB will return all items in the table with that partition key value in a paginated fashion.

- **Response size limit**: Each `Query` API response is capped at a maximum of 1 MiB worth of items. If more items fall outside this limit, DynamoDB provides a `LastEvaluatedKey` element in the response as a continuation token for additional requests.

- **Metering for RCU**: Each `Query` API is metered for RCU based on the size of data read by DynamoDB to return the result set to the client. This is different from `GetItem` and `BatchGetItem`, where the RCU is charged on a per-item per-4 KB basis for strongly consistent reads, and a per-item per-8 KB basis for eventually consistent reads.

- **Sorting and optimized functions**: The `Query` API supports sorting by sort key values, both in ascending (default) and descending order. Additionally, it offers optimized functions such as top or bottom N results, item count, and the crucial `begins_with` function.

- **The begins_with function**: `begins_with` is essential for retrieving data for multiple related entities in a single request. It is a powerful way for efficiently querying items with similar sort key prefixes.

It is important to note that if you do not know the complete value of the table or index's partition key attribute, the highly performant `Query`, `GetItem`, and `BatchGetItem` APIs cannot be used. In such cases, the `Scan` API becomes the alternative, but it should be used with caution due to its impact on time and read throughput. Let us learn more about it next.

Scan

The `Scan` API in DynamoDB, as the name implies, performs a comprehensive scan of the entire table, reading every piece of data within the table's responsibility. Unlike other operations, you do not need complete values for any of the attributes in the data, making the `Scan` API the most flexible but resource-intensive operation in DynamoDB.

Key points about the Scan API include the following:

- **Flexibility**: Scan allows filtering on every attribute in the data, supporting functions such as contains(), logical AND/OR operators, and more (4)

- **Costs and Limitations**:

  - **Time**: Executing a full table scan may require multiple paginated requests depending on the table's data volume.

  - **Read Throughput**: Like the Query API, each Scan API call can read a maximum of 1 MiB worth of items.

  - **Pagination**: If more items are present, DynamoDB provides LastEvaluatedKey as a continuation token. This token must be used in subsequent Scan requests until LastEvaluatedKey is null, indicating that DynamoDB has read all items in the table

To give an idea of how lengthy this process could be, performing a full table scan on a DynamoDB table of a size of 100 GiB would require an application to make at least 102.4k individual Scan API calls, each reading 1 MiB worth of items.

The Scan API in DynamoDB provides two options for conducting a full table scan: sequential scan and parallel scan:

- **Sequential scan**:

  - In a sequential scan, a single thread initiates a Scan request

  - The LastEvaluatedKey element from the response is used to issue the next Scan request, retrieving the next set of results along with the next LastEvaluatedKey element

  - This process continues until all items in the table are scanned

- **Parallel scan**:

  - Like the sequential scan, the parallel scan utilizes the LastEvaluatedKey element for subsequent requests

  - Instead of a single thread, the full table scan can be divided into logically exclusive segments

  - Multiple threads can be spawned, with each thread processing its individual segment concurrently

The advantage of a parallel scan over a sequential scan in DynamoDB lies in the time efficiency of reading through all items within the table, 1 MiB worth of items at a time. Both the sequential and parallel scans process the same amount of table data and consequently consume the same read throughput.

In terms of read throughput, the Scan operation is the costliest read operation. Conducting a full table scan in DynamoDB, where every item is traversed, consumes RCU equivalent to the size of all items in the table. While Scan is a potent operation, it should be avoided in latency-sensitive OLTP applications.

Avoiding full table scans in DynamoDB is recommended when a user is awaiting a response on the other end of the application. Scans do not scale efficiently, and as the table size grows, conducting a full table scan becomes increasingly expensive in terms of both latency and cost. Utilizing Scan for every customer request is highly inefficient.

The Scan API is suitable for scenarios where access is infrequent and there is no time constraint on completing the operation. Consider using it for tasks such as generating weekly or monthly reports. However, if the reporting use case involves complex analytics, the ExportToS3 feature of DynamoDB may offer a more cost-effective and performant solution. Some customers also employ the Scan operation for data validation checks in nightly background processes, which are rate-limited and have no SLAs on response times.

Next, let us learn about the ExecuteStatement and BatchExecuteStatement **PartiQL** APIs purely from the read standpoint, as they can also be used for performing writes to DynamoDB.

PartiQL read operations

PartiQL, a SQL-compatible query language, facilitates the transition from traditional RDBMS to NoSQL in DynamoDB. It allows you to interact with DynamoDB using SQL-like statements, including SELECT, INSERT, UPDATE, and DELETE. In terms of executing reads with PartiQL, two non-transactional APIs are supported: ExecuteStatement and BatchExecuteStatement. ExecuteTransaction, a transactional PartiQL API, also supports reads.

The foundation of PartiQL APIs includes SELECT, INSERT, UPDATE, and DELETE statements. For reads, SELECT statements retrieve data from a DynamoDB table. The outcome depends on the statement and the corresponding DynamoDB table key schema. For example, if the key schema includes a partition key as PK and a sort key as SK, the following applies:

- A SELECT statement such as SELECT * FROM mytable WHERE PK=x AND SK=y targets returning at most a single item, such as the GetItem operation
- A statement such as SELECT * FROM mytable WHERE PK=x could return multiple items, similar to a Query API
- If the statement lacks an equality clause on the PK, the SELECT statement can perform a full table scan

Unlike other DynamoDB APIs, a PartiQL API with a SELECT statement can act as a GetItem-like, Query-like, or Scan-like operation. The following code block illustrates these differences and their similarity to a non-PartiQL DynamoDB operation:

```
-- table schema
-- table name: Orders
-- partition key: PK | sort key: SK
```

```
-- similar to GetItem
SELECT * FROM "Orders" WHERE PK = 'o#12345' AND SK = '!'

-- similar to Query
SELECT * FROM "Orders" WHERE PK = 'o#12345'

-- similar to BatchGetItem
SELECT * FROM "Orders"
WHERE ((PK = 'o#12345' AND SK = 'p#1')
       OR (PK = 'o#999' AND SK = 'p#5')
       OR (PK = 'o#5567' AND SK = 'p#9'))

-- similar to Scan
SELECT * FROM "Orders" WHERE contains(non_key_attribute, '100')
```

This does not mean that the PartiQL APIs are server-side wrappers over the GetItem, Query, and Scan APIs but the behavior of the SELECT statement closely resembles these operations that we have already learned about in this chapter.

Unlike the BatchGetItem API, the BatchExecuteStatement API can only support up to 25 individual SELECT statements within a single network request at the time of publishing.

A notable distinction in the PartiQL read operations is found in the SELECT statement of the ExecuteStatement API. In contrast to the Query API, which can target data within exactly one item collection, the ExecuteStatement API offers a batch Query-like SELECT statement that can fetch data from multiple item collections. As a reminder, an item collection is a group of items within DynamoDB sharing the same value of the partition key attribute but differing in sort key values. For instance, the following SELECT statement, when part of an ExecuteStatement API, can retrieve multiple item collections as long as the overall data read by DynamoDB is under 1 MiB:

```
-- table schema
-- table name: Orders
-- partition key: PK | sort key: SK

-- parallel Query-like
SELECT * FROM "Orders"
WHERE ((PK = 'o#12345')
       OR (PK = 'o#999'
           AND SK BETWEEN 's#2020-08-11T12:00:00Z'
           AND 's#2021-08-30T12:00:00Z')
       OR (PK = 'o#5567' AND begins_with("SK", 'p#9')))
```

A recommended feature for all read APIs is the use of the `ReturnConsumedCapacity` parameter. By including this parameter in your DynamoDB requests, the response will provide information on the consumed RCU per request. This is particularly valuable during the development and testing phases, allowing you to observe how your read requests interact with the data model from a cost perspective. If non-key filters result in discarding a significant portion (30-40% or more) of the data read by DynamoDB, it could signal the need to reassess the data model for efficiency in handling your access patterns. While discarding a certain percentage of items due to non-key filters might be acceptable in some scenarios, excessive discards, especially with a large dataset, indicate inefficiency.

The `ReturnConsumedCapacity` parameter helps identify such inefficiencies, enabling you to ensure that the read throughput consumed aligns appropriately with the number of items returned in your read requests to DynamoDB.

Similar to the `ReturnConsumedCapacity` parameter, the read consistency level within DynamoDB can be configured for each of the DynamoDB read APIs.

It is worth noting that, among the read operations covered in this section, `BatchGetItem`, `GetItem`, and their corresponding `ExecuteStatement SELECT` operations cannot be performed on any secondary indexes (global or local). This limitation arises because the uniqueness of items within secondary indexes is ensured using the partition key and sort key of the base table. Therefore, single-item targeting operations cannot be used to retrieve data from these indexes. Instead, you can use the `Query` API and its corresponding `ExecuteStatement SELECT` operation on the secondary indexes, supplying a `LIMIT` parameter with a value of 1 to fetch the first item within the index that qualifies the given filters. You will learn more about this behavior of secondary indices in *Chapter 8, Secondary Indexes*.

Please note that we have not covered `ExecuteTransaction` and `TransactGetItems` from the read side, as these transaction APIs will be explored in the next dedicated section, *Transactions and atomicity*.

Transactions and atomicity

DynamoDB supports **ACID** transactions through dedicated multi-item APIs, providing ACID qualities for these operations. While single-item operations in DynamoDB are inherently ACID, the specific transaction APIs ensure these characteristics for multi-item operations. For those unfamiliar with the term ACID, it represents key properties in database transactions. Refer to (5) to read up more on these.

Note that DynamoDB's implementation of ACID transactions in a highly distributed system differs from traditional relational databases. Exploring these differences, highlighting key features, and diving into best practices—complementary to the information available on the AWS docs (6), which I highly recommend reviewing—will be our focus. Let us start by examining the primary differences.

Differences from traditional RDBMS transactions

In traditional RDBMS, transactions are typically opened and closed programmatically using specific directives. However, DynamoDB transactions operate differently. They consist of a group of read or write actions performed on one or more DynamoDB tables within a single HTTP request. Unlike traditional RDBMS, there are no explicit begin or close transaction directives.

For DynamoDB transactions, all put, update, and delete actions that need to be executed in an **all-or-nothing** fashion must be included in a single transaction write request. Similarly, all read operations that need to be performed on specific items in an all-or-nothing fashion must be part of a single transaction read request.

If any item involved in a read or write transaction request is part of another inflight transaction, the newer transaction request is immediately canceled with a `TransactionConflict` exception. This differs from traditional RDBMS transactions, where a newer transaction might wait in a queue until the inflight transaction is completed and then release the lock on the conflicting data object. In DynamoDB, the canceled transaction request is returned to the client, and it is the responsibility of the client's business logic to determine the next course of action. For instance, in a banking scenario where an inflight transaction is updating user A's account balance and creating a new bank transfer to user B's account, a new transaction attempting to interact with user A's account would be canceled with a transaction conflict. It is then up to the banking application's business logic to decide whether to retry the transaction or inform the user about the failed request.

From the preceding example and the difference in behavior from traditional RDBMS transactions, you may already have caught a hint of another difference between RDBMS transactions and DynamoDB transactions. Within a DynamoDB transaction request, it is not possible to read data from one action and write data to another item using a separate action based on the earlier read. The transaction must consist of either all write actions or all read actions. However, conditional write actions or separate condition check actions within a transaction are allowed. These condition check actions involve reading an item to validate a condition. In such cases, either all the writes and condition check actions pass, or none of them are successful.

To illustrate this difference, consider the same financial scenario but now the banking application wants to read user B's account info to use it in another write action for user A's bank transfer. In DynamoDB, these two actions cannot be part of the same transaction request. Instead, the recommended approach would be to perform a transaction read on user B's account info, use that information in a new write transaction to create the bank transfer, and include a condition check on user B's account info item within the same transaction. This condition check ensures that the transaction only succeeds if user B's account info has not changed since the application last read the account info item. The outcome is either the successful creation of the new bank transfer item (if user B's account info had not changed) or the failure of the transaction due to the condition check (if user B's account info did change). In case of failure, it is then up to the banking application's business logic to decide whether to retry the entire transaction or to inform user A about the failed bank transfer request.

Another small yet important difference from traditional RDBMS transactions is that each action within the same transaction request must operate on exactly one item, and no two actions within the same transaction can reference the same item.

Using our banking example, if a single transaction needs to perform actions such as debiting an amount from user A's account item, crediting user B's account with the same amount, and creating individual bank transfer line items in both users' wallets, then these must involve four distinct items. If, for some design-specific reason, the debit action and the line-item creation action need to be applied to the same exact item within DynamoDB (even though this may not be a scalable design), they must be part of a single `Update` action within the transaction request.

Now that you have explored the key differences between ACID transactions in DynamoDB and traditional RDBMS transactions, let us review some key highlights about transactions including some under-the-hood stuff for the advanced reader.

Reviewing key highlights of transactions

As covered in the preceding section, a write transaction request may include only write actions in addition to condition check actions. This means that in the same `TransactWriteItems` API, which is one of the write transaction APIs supported by DynamoDB, you may have the following:

- `Put` actions: New item writes, or in case of an item existing with the same keys, these would overwrite without any condition expression
- `Update` actions: Existing item updates, or in case of a non-existent item with the same keys, would create new items without condition expressions
- `Delete` actions: Existing item deletes, or in case of a non-existent item with the same keys, these would silently succeed without condition expressions
- `ConditionCheck` actions: Validation checks on individual items that must pass for the transaction to succeed

At the time of writing, a single `TransactWriteItems` request allows a maximum of 100 actions with any combination of `Put`, `Update`, `Delete`, and `ConditionCheck`. The same limitation applies to the PartiQL equivalent, `ExecuteTransaction`.

Returning to the financial example, the `TransactWriteItems` payload would be structured as follows to ensure the successful execution of actions such as debiting a certain amount from user A's account, crediting the same amount to user B's account, and creating transaction line items for both users as part of the bank transfer. All of these actions succeed only if user A's account has a sufficient balance.

```
 1 {
 2       "TransactItems": [
 3           {
 4               "Put": {
 5                   "TableName": "bank_accounts",
 6                   "Item": {
 7                       "pk": {"S":"user#A"},
 8                       "sk": {"S":"line_item#txnId_123"},
 9                       "receiver": {"S":"user#B"}, "value": {"N": "100"}
10                   }
11               }
12           },
13           {
14               "Put": {
15                   "TableName": "bank_accounts",
16                   "Item": {
17                       "pk": {"S":"user#B"},
18                       "sk": {"S":"line_item#txnId_123"},
19                       "sender": {"S":"user#A"}, "value": {"N": "100"}
20                   }
21               }
22           },
23           {
24               "Update": {
25                   "TableName": "bank_accounts",
26                   "Key": {
27                       "pk": {"S":"user#A"},
28                       "sk": {"S":"account_balance"}
29                   },
30                   "UpdateExpression": "SET balance = balance - :1a200",
31                   "ConditionExpression": "balance > :1a200",
32                   "ExpressionAttributeValues": {":1a200": {"N":"100"}}
33               }
34           },
35           {
36               "Update": {
37                   "TableName": "bank_accounts",
38                   "Key": {
39                       "pk": {"S":"user#B"},
40                       "sk": {"S":"account_balance"}
41                   },
42                   "UpdateExpression": "SET balance = balance + :1a201",
43                   "ExpressionAttributeValues": {":1a201": {"N":"100"}}
44               }
45           }
46       ]
47 }
```

Figure 6.1 – The TransactWriteItems payload for the bank transfer example

In terms of transaction reads, each action within a `TransactGetItems` request, one of the APIs supporting transactional reads, can encompass up to 100 individual `Get` actions. Each action targets precisely one item, and various actions within the same `TransactGetItems` request can focus on different DynamoDB table items, provided that the tables are in the same AWS account and region.

For the advanced reader, transactions within DynamoDB are implemented using a two-phase protocol. DynamoDB, being a highly distributed system, achieves the implementation of ACID-compliant multi-item transactions with predictable performance as a core design tenet. A detailed explanation of this implementation is provided in a dedicated 2023 whitepaper (7) authored by the DynamoDB team responsible for the feature.

Briefly, the two-phase commit protocol involves a `PREPARE` and a `COMMIT` phase:

- The `PREPARE` phase is when every action within a transaction read or write request is validated for errors, condition checks, throttling exceptions, transaction conflicts, among other sanity checks
- Once the `PREPARE` phase is successful, the `COMMIT` phase is executed when these vetted actions are committed in parallel by all backend nodes contributing to the transaction

If, for any reason, the `PREPARE` phase fails to pass all sanity checks, the `COMMIT` operation is never executed, and the transaction is rejected with the appropriate exception. Only when a `COMMIT` operation is executed for a transaction do the clients see the new image of data after the transaction. By design, DynamoDB does not have the concept of a dirty read.

A single actor, namely the **Transaction Coordinator** (**TC**), is responsible for orchestrating both phases of a transaction. All TCs rely on a common transaction ledger to maintain state information about ongoing transactions for a DynamoDB table. If a TC fails midway through a transaction, this is gracefully handled by monitoring the ledger for transactions in flight for more than a certain period. One of the healthier TCs is then tasked with picking up from where the failed TC left off in the execution of the transaction.

Due to the two-phase protocol of transactions within DynamoDB, the transaction APIs cost double the throughput to execute compared to the non-transactional DynamoDB APIs. For example, if a single `PutItem` request costs 1 WCU to write, the same payload in a `TransactWriteItems` request would cost 2 WCU. Similarly, the `TransactGetItems` request would cost double the RCU compared to the same read being requested using a strongly consistent `GetItem` operation.

Another notable feature of the transactional APIs is their support for a **Client Request Token** (**CRT**), which contributes to making your operations idempotent. Idempotency ensures that an operation performs a task exactly once, even if multiple duplicate requests enter the system. When using transaction APIs, you can include a CRT in your request, allowing DynamoDB to deduplicate any requests with the same payload that were duplicated due to network issues or application bugs. For DynamoDB to deduplicate two requests, their CRT values must be the same. If the request parameters change but the CRT remains constant, DynamoDB will raise an exception. The CRT is valid for a period of 10 minutes from its first use, and it must be regenerated for each application request. The good news is that if you use any of the AWS SDKs, they automatically generate and manage CRTs by default, eliminating the need for manual management.

That covers the essential information you need to start using transaction APIs in your applications with DynamoDB for OLTP access. The AWS docs on transactions (6) provide details on the database isolation levels supported by different DynamoDB operations, conflicts, and guidance on handling conflicts. As I do not intend to duplicate the docs, I recommend reading them to gain insights into these aspects of transactional APIs. Next, let us dive into some best practices, including situations when to use or avoid transactional APIs in your applications.

Using transactional APIs in your application

When learning about DynamoDB after being a tenured user of traditional RDBMS, whether for open source or commercial database technologies, your natural inclination would be to use ACID transactions in DynamoDB everywhere. In my experience working with customers transitioning to NoSQL and DynamoDB, even the most tenured chiefs and principal engineers tend to apply their wealth of RDBMS experience and use DynamoDB as a relational database.

When incorporating transactional APIs into your application, it is crucial to evaluate whether multi-item ACID-compliant operations are genuinely necessary and whether eventual consistency would impact your business. If your use case requires strong read-after-write consistency and all-or-nothing execution, transactional APIs are suitable. However, if your inclination toward these APIs stems from familiarity with ACID transactions in previous database technologies, a deeper examination of your application architecture may be needed.

For financial use cases or scenarios demanding multi-item operations with critical consistency requirements, transactional APIs make sense. On the other hand, for tasks such as adding comments in a social media application, typically involving single-item writes, opting for transactional APIs might be unnecessary.

In situations where bulk updates for tens or hundreds of items are required for infrequent access patterns, using transaction APIs may not be warranted unless there is a risk of significant impact on user experience or business. Non-transactional operations such as `BatchWriteItem`, `BatchGetItem`, and `BatchExecuteStatement` are more suitable for bulk operations, providing potential cost savings, improved average latencies, reduced complexity, and high availability. However, it is essential to consider potential downsides, such as preventing data reads during updates or informing users of changes with a last update timestamp.

When utilizing the transaction API for writing multiple items with a concern for all-or-nothing behavior, it is advisable to use transaction APIs for reading data as well. The reason behind this recommendation lies in the parallel execution of the COMMIT phase during a transaction write. DynamoDB, being a distributed system with storage nodes distributed across Availability Zones, processes parallel COMMIT phase requests independently. As a result, when the transaction is still in progress, a non-transactional read on any participating item may or may not reflect the latest committed data, as some nodes may have processed the COMMIT phase successfully while others may not have.

Only after the write transaction is marked as complete and all actions within it are successful will a subsequent transaction read be able to retrieve the desired data from the DynamoDB table. If a transaction read was made on an item that was part of an ongoing transaction write, the transaction read would be canceled by DynamoDB with a `TransactionConflict` error. Since non-transactional multi-item APIs such as `BatchGetItem`, `Query`, and `Scan` have a read-committed isolation level with a transaction write, these APIs can fetch the updated state of items even if the complete transaction has not been marked as complete. This is not a dirty read, as the transaction had completed its `PREPARE` phase. It is bound to complete successfully but may do so with a bit of latency. Consider this when designing your applications with the use of both transactional and non-transactional APIs.

> **Important note**
>
> Consider this as a reminder to delete any resources you may have created while going through this chapter in your AWS account.

That is all about transactions, atomicity, and how to think about transactions within DynamoDB. Let us wrap this chapter up by summarizing all that we learned.

Summary

In this chapter, we explored read consistency in DynamoDB, covering the various read operations and best practices associated with each. We started by examining the different consistency models offered by DynamoDB: eventually consistent reads and strongly consistent reads. Eventually consistent reads provide a cost-effective option with the trade-off of potential staleness, while strongly consistent reads ensure the most up-to-date data at a relatively higher cost.

The chapter then dove into specific read operations available in DynamoDB, such as `BatchGetItem`, `GetItem`, and `Query`, each serving distinct use cases. We covered the importance of specifying the right consistency model for your read operations, as well as balancing performance needs and data accuracy.

Next, we shifted our focus to transactions and atomicity within DynamoDB. Despite DynamoDB being a NoSQL database, it offers ACID-compliant multi-item transactions through dedicated APIs. We outlined the key differences between DynamoDB transactions and traditional RDBMS transactions, emphasizing the absence of explicit begin and close transaction directives. The two-phase protocol of DynamoDB transactions, involving a `PREPARE` phase for validation and a `COMMIT` phase for execution, was explored in detail.

Under the hood, transactions in DynamoDB are orchestrated by a TC, utilizing a common transaction ledger to maintain the transaction state. We highlighted that transactions incur double the throughput compared to non-transactional APIs due to their two-phase nature. The importance of considering when to use transactional APIs in applications was reviewed, with a focus on assessing whether the use case truly requires ACID-compliant operations and understanding potential cost implications.

In conclusion, this chapter provided a decent overview of read consistency, read operations, and transactions in DynamoDB, arming you with the knowledge to make informed decisions when designing and implementing read data access patterns in your applications.

In the upcoming chapter, we will delve into a widely used data-organizing strategy known as **vertical partitioning**. This approach offers scalability, availability, and cost efficiency when retrieving data for related entities from DynamoDB.

References

1. AWS Docs: Read Consistency – `https://docs.aws.amazon.com/amazondynamodb/latest/developerguide/HowItWorks.ReadConsistency.html`

2. DynamoDB SLA – `https://aws.amazon.com/dynamodb/sla/`

3. AWS Docs: DynamoDB API Reference – `https://docs.aws.amazon.com/amazondynamodb/latest/APIReference/API_Operations_Amazon_DynamoDB.html`

4. AWS Docs: DynamoDB Expression Syntax – `https://docs.aws.amazon.com/amazondynamodb/latest/APIReference/API_Operations_Amazon_DynamoDB.html`

5. Wikipedia: ACID properties – `https://en.wikipedia.org/wiki/ACID`

6. AWS Docs: DynamoDB Transaction APIs – `https://docs.aws.amazon.com/amazondynamodb/latest/developerguide/transaction-apis.html`

7. USENIX: Distributed Transactions at Scale – `https://www.usenix.org/conference/atc23/presentation/idziorek`

7
Vertical Partitioning

In the preceding chapters of this book, we went from grasping the fundamentals of DynamoDB to exploring interactions and adopting a NoSQL mindset. We dove into the nuances of data modeling and optimizing data models as per our access patterns, considering the application's read- or write-heavy requirements.

Our exploration also covered the misconception that NoSQL implies non-relational, clarifying that while data is always relational, its representation and organization in NoSQL systems differ. We examined the approaches of single-table design and multi-table design, emphasizing the mantra *Data that needs to be accessed together must be stored together*. Balancing performance and cost considerations is crucial in deciding between these approaches.

This chapter mostly focuses on single-table design – specifically, on the vital concept of data modeling known as **Vertical Partitioning**. At a high level, vertical partitioning involves breaking items down into smaller chunks of data and associating all relevant items with the same partition key value. This design pattern ensures scalability, availability, and cost-effectiveness.

From a data modeling perspective, vertical partitioning future-proofs applications, accommodating evolving access patterns and the integration of new features or data structures. Apart from single-table design concepts, vertical partitioning can provide users with the ability to break down large chunks of data and work around the 400 KB item size limit of DynamoDB.

Through practical examples, we will dive into the principles of vertical partitioning, discussing when to use it and how to implement it as per access patterns. The chapter will cover the construction of read and write operations, aligned with specific access patterns, based on data models implementing vertical partitioning.

To understand this design pattern, we will explore fundamental concepts, including **Item Collections**. Building upon prior knowledge, we will examine the advantages and potential downsides of vertical partitioning, considering key considerations for different scenarios.

By the end of this chapter, you will have a solid understanding of the vertical partitioning design principle, supported by practical examples. It is important to note that the learning of this chapter may or may not apply if you use DynamoDB solely as a simple key value lookup database.

In this chapter, we are going to cover the following main topics:

- Understanding item collections
- Breaking down data for vertical partitioning
- Future growth and expansion

Understanding item collections

It is likely that not all access patterns for an application would only deal with a single item all the time. Several access patterns for a modern application could require fetching multiple items related to each other in some manner. Like a simple key value acting in a singleton manner, items having the same partition key could have different sort key values, such that for each item, the partition key and sort key combination remains unique across the table data.

A group of items having the same partition key value but different sort key values is called an item collection. An example of an item collection would be maintaining the version history of all mutations performed for a user item or a game state object. If I were to allow my users to request a history of all mutations for their profile, I would store the profile data in such a way where the partition key for the item collection is `user_id` and the sort key could be an epoch timestamp or ISO 8601 format timestamp strings, such as `2024-01-13T14:37:07.999Z`. Such formats are naturally sorted in ascending order and hence an ordered read operation on this item collection would by design return a result set of all mutations made for a certain profile in chronological order.

The following figure illustrates this example use case:

user_profile_history

Primary key		Attributes (Item Collection)	
Partition key: user_id	Sort key: update_timestamp		
aman_dhingra	2024-01-13T14:37:07.999Z	new_state	entity_type
		{"First Name":{"S":"Aman"},"CityPreference":{"S":"London"}}	profile_update
	2024-04-15T11:35:57.557Z	new_state	entity_type
		{"First Name":{"S":"Aman"},"CityPreference":{"S":"Dublin"}}	profile_update
	2024-07-10T20:34:48.296Z	new_state	entity_type
		{"First Name":{"S":"Aman"},"CityPreference":{"S":"Dubai"}}	profile_update
lorenzo_ripani	2023-10-29T22:35:42.427Z	new_state	entity_type
		{"First Name":{"S":"Lorenzo"},"CityPreference":{"S":"Milan"}}	profile_update
	2024-10-18T06:11:32.318Z	new_state	entity_type
		{"First Name":{"S":"Lorenzo"},"CityPreference":{"S":"London"}}	profile_update

Figure 7.1 – Item collection example

If I were to read from such an item collection, I would use one of the multi-item operations supported by DynamoDB, the `Query` API. With the `Query` operation, I can only ever read from a single item collection, that is, read from a group of items with the same partition key value and different sort key values. By default, the `Query` operation would return items from an item collection sorted by the sort key values in ascending order. The sorting order for strings is also called the **lexicographical order** (1). Lexicographical order could be thought of as the computer science term for alphabetical order. The `Query` API not only supports simple ascending- and descending-ordered sorting functions for an item collection but also supports additional operators on the sort key values.

For example, for my user profile history use case, I also want my application to support queries that have a starting and ending date. Given `user_id aman_dhingra`, I want all profile state history between `2024-01-01T00:00:00.000Z` and `2024-03-31T23:59:59.999Z`, which refers to all state history for the first quarter of 2024. Using the `Query` API, this is possible with the highest efficiency, regardless of my profile potentially having thousands of profile updates before and after this period.

One of the most essential sort key functions supported by the `Query` operation is the `begins_with()` operator. You could design sort key values in such a way that using the `begins_with` operator acts like a traditional `GROUP BY` function. In terms of the user profile history example, access patterns such as *Get all versions for the month of January 2024* could be easily supported with `sort_key begins_with('2024-01')` as the `Query` condition. A common practice is to also use prefixes based on the entity type. For an e-commerce scenario, if warehouses, shopping carts, orders, and invoices are different entities modeled in a DynamoDB table, prefixes could be some part of the entity names themselves: `w#<warehouse_id>`, `c#<cart_id>`, `o#<order_id>`, and `i#<invoice_id>`. We will see these examples later in the chapter.

With the preceding user profile history example, it is important to note that given the partition and sort key design, it assumes that no user will update their profile twice at the same time, down to millisecond granularity. If your system needs to account for multiple writes at the same time in a similar scenario, you must consider one of two approaches:

- **Combine the sort key timestamp with a differentiating attribute value**: To ensure uniqueness within the same item collection despite multiple writes, consider combining the sort key timestamp with a distinct attribute value, such as the unique identifier `request_id`. By incorporating this differentiating attribute into the sort key, the combination of the partition key and sort key remains unique even when multiple writes are performed for the same item collection.

- **Use the same design, but in a Global Secondary Index (GSI)**: We will dig deeper into this topic in the upcoming chapter, *Secondary Indexes*. Notably, in a GSI, the partition key and sort key combination does not require uniqueness. Instead, uniqueness across GSI items is maintained by using the item's table partition key and table sort key combination, which must be unique. Hence, within a GSI, multiple items can coexist with identical GSI partition key and GSI sort key values under the same design.

Item collections provide not only a means to logically group related data but also contribute to minimizing the impact of network partitions. Storing all data for a specific item collection together within a DynamoDB table, it is probable that it resides on the same physical partition. In the event of an outage affecting one physical partition, the remaining data on other partitions could still be accessed successfully. This strategy effectively mitigates the impact of temporary network partitions that may arise during the lifetime of a DynamoDB table, reducing the potential blast radius.

In the example of the user profile history, the table was initially designed to exclusively store data for the `profile_update` entity. In this case, the table's partition key attribute name could be `user_id`, and the sort key attribute name is `update_timestamp`.

However, a significant advantage of adopting the single-table design and organizing data into item collections is the ability to store data for multiple entities within the same table. These entities can have various relationships, such as one-to-one (*1:1*), one-to-many (*1:m*), or many-to-many (*m:n*). Hence, the same partition key attribute could be associated with several other entities in distinct ways, and each entity may have multiple items linked to it within the partition key attribute.

Consider modeling data in DynamoDB for a popular song-playing application using the following example. One of the primary access patterns is retrieving song details for a given song ID, and among the displayed stats on your application screen, one key metric is the total downloads the song has garnered. While other statistics could include play counts – both personal and overall app plays – let us focus on listing total downloads for this example.

In this context, it is evident that data that needs to be accessed together, such as song details and the total download count being stored together, though not necessarily in the same item. The data model for this scenario would resemble the structure outlined in the following figure:

data

Primary key		Attributes		
Partition key: PK	Sort key: SK			
s#song1	A	type	title	artist
		details	How Much Is The Fish?	Scooter
	A#downloads	type	count	
		total_downloads	100	
	d#download-1	type	user	
		download	amdhing	
	d#download-2	type	user	
		download	ripani	

Figure 7.2 – Song-playing application example data model

Following the structure depicted in the previous figure, the table's partition key is `song_id`, associated with various entities that need to be accessed together, such as song details, total downloads, and others. Each of these linked entities may consist of one or more items related to the song.

The design dictates storing this data in the same item collection, encompassing details such as recording each download. In this schema, the song establishes a *1:1* relationship with `total_downloads` and (song) `details`, and a *1:m* relationship with downloads.

Ignoring the A# and d# prefixes for now, this organization facilitates making query requests on the table by providing only the full value of the song ID as the partition key attribute. The result would include all items within the item collection, delivered in pages of 1 MiB worth of items each.

Opting for multiple tables in this scenario, each dedicated to song details, song downloads, and download stats, could necessitate at least two requests to DynamoDB to read all the relevant information. This is particularly true if you do not need to list every song downloaded. However, this approach of using multiple tables in this scenario would incur additional network time and associated costs.

In the preceding example of the song-playing application, another important design consideration involves the use of generic names – **primary key** as **PK** and **sort key** as **SK** for the partition key attributes. This practice is termed the **overloading of key names**. Through key name overloading, it becomes possible to store data for multiple entities within an item collection. Without this practice, the item collection might be limited to holding data for at most two entities, with the partition key attribute containing an identifier for one entity and the sort key attribute containing an identifier for the other.

You might wonder why, in the song-playing application, I didn't store the total downloads information as a single attribute within the song details item. This approach would eliminate the need to store and retrieve separate items (`details` and `total_downloads`) and manage them independently, right? Well, the decision to break down the data in this manner was driven by a cost optimization strategy.

In the next section, let us look into how this translates to cost optimization within the context of vertical partitioning and single-table design.

Breaking down our data for vertical partitioning

While *Chapter 5*, *Moving From a Relational Mindset*, briefly touched on single-table design and design patterns for breaking down data to enhance efficiency and scalability, this section aims to further expand on those concepts within the context of vertical partitioning. Our learning begins by reviewing the advantages of breaking down data and how this approach can prove advantageous in many scenarios.

Advantages of breaking down data for vertical partitioning

Breaking down large JSON structures into smaller chunks, forming part of the same item collection or collections meant to be retrieved together, offers a significant advantage – highly efficient data retrieval. This approach minimizes the need for multiple network requests when accessing related data. However, it is crucial to note that this breakdown might involve denormalization and data duplication, potentially requiring multiple writes from the application.

In many cases, breaking down data may involve additional effort during the writing phase to achieve super-efficient reads. This trade-off may not typically be optimal for write-heavy, low-latency workloads, as the focus is on optimizing data retrieval efficiency.

Another key advantage of breaking down data, as discussed in *Chapter 5*, is its role as a cost optimization practice to reduce overall reading and writing expenses. Let us review one such case next.

Breaking down data as an update cost optimization strategy

In scenarios like an e-commerce product catalog, storing product inventory alongside extensive product details in a single DynamoDB item can lead to inefficient cost structures. For instance, updating inventory often results in paying for the entire large item, even when the change is minor. Consider a 10 KB item containing product descriptions and inventory. Each update to the inventory in this scenario would cost 10 **Write Capacity Units (WCUs)**.

Compare this to a vertically partitioned design where product descriptions and inventory are separate entities, associated with a product ID within the same item collection. With this design, assuming the inventory item is smaller than 1 KB, each inventory update costs only 1 WCU instead of 10 WCUs, representing a 10X optimization in write costs. While creating these distinct items initially incurs slightly higher write costs (11 WCUs in this example), it is a 10% increase for a 10X optimization in frequent update costs.

On the read side, accessing both product description and inventory items together through a query operation does not affect **Read Capacity Units (RCUs)**, since query charges are based on the size of items read, and there is no difference compared to the earlier design. This trade-off provides a substantial reduction in DynamoDB write throughput costs with no impact on read costs.

Reducing the number of GSIs to manage

An additional advantage of utilizing vertical partitioning to break down data is a potential reduction in the need for numerous GSIs in your DynamoDB table. As you vertically partition the data, each larger chunk may transform into several smaller chunks within distinct items, allowing them to be indexed in the same GSI. This contrasts with having the larger chunk indexed on separate attributes across different GSIs.

In the song-playing application example, if I aim to retrieve details about a song download using a specific download ID and simultaneously need the song name indexed, I can accomplish this with a single GSI. Both attributes can be indexed on the same GSI, benefiting from the vertically partitioned data model. Refer to the following figure illustrating the GSI model for a clearer understanding.

data   GSI: GSI1

| Primary key | | Attributes | | | | | |
Partition key: GSI1PK	Sort key: GSI1SK						
Scooter	How Much Is The Fish?	PK	SK	type	title	artist	
		s#song1	A	details	How Much Is The Fish?	Scooter	
	Ramp! (The Logical Song)	PK	SK	type	title	artist	
		s#song2	A	details	Ramp! (The Logical Song)	Scooter	
d#download-1	s#song1	PK	SK	type	title	artist	user
		s#song1	d#download-1	download	How Much Is The Fish?	Scooter	amdhing
d#download-2	s#song1	PK	SK	type	title	artist	user
		s#song1	d#download-2	download	How Much Is The Fish?	Scooter	ripani
d#download-3	s#song2	PK	SK	type	title	artist	user
		s#song2	d#download-1	download	Ramp! (The Logical Song)	Scooter	amdhing

Figure 7.3 – Song-playing application – GSI

Certainly, the preceding example effectively illustrates the importance of breaking down data into smaller chunks and the reduction in the number of GSIs required. In scenarios where all downloads for a song are listed in a single blob alongside the song details, the requirement for numerous GSIs may arise, particularly if you need to index download IDs or include additional attributes such as artist names and song names.

However, by breaking down the data into smaller chunks, with each download and song detail as individual items, you can efficiently index them on the same GSI. The use of overloaded GSI1PK and GSI1SK attributes allows for organized data, supporting various access patterns. This example is applicable to music technology scenarios but if you relate more to e-commerce, then let us review a shopping cart one.

In the shopping cart example, where each cart item is an individual DynamoDB item, you can create a GSI with the cart item's product ID as the partition key. This enables efficient lookups for all shopping carts or all users with a particular product ID added to their carts, facilitating the analysis of product sales forecasts. Refer to the following figures of the table and GSI models for a visual representation of this concept.

[Table] shopping_cart   Total 2   100/page   1   Actions ⌄

shopping_cart   GSI: GSI1

Primary key		Attributes					
Partition key: PK	Sort key: SK						
u#user_1	ACTIVE#product_1	product_id	entity	user_id	cart_id	GSI1PK	GSI1SK
		product_1	shopping product	user_1	cart_a	product_1	user_1
	ACTIVE#product_3	product_id	entity	user_id	cart_id	GSI1PK	GSI1SK
		product_3	shopping product	user_1	cart_a	product_3	user_1
u#user_2	INACTIVE#product_1	product_id	entity	user_id	cart_id	GSI1PK	GSI1SK
		product_1	shopping product	user_1	cart_z	product_1	user_2

Figure 7.4 – E-commerce – shopping cart example model of table

The table's partition key, PK, is user_id, and each item corresponds to a shopping product entity. The sort key of the table, SK, is a combination of the product's status within the shopping cart (assuming items can be active or saved for later) and the product ID. With this breakdown in design, creating a GSI with the partition key GSI1PK representing the product ID alone and the sort key GSI1SK representing the user ID with the product in their cart would enable retrieving all users with a specific product in their shopping carts. This information is valuable for sales and forecasting teams. The following figure shows such a GSI.

[Table] shopping_cart   Total 2   100/page   1   Actions ⌄

shopping_cart   GSI: GSI1

Primary key		Attributes					
Partition key: GSI1PK	Sort key: GSI1SK						
product_1	user_1	PK	SK	product_id	entity	user_id	cart_id
		u#user_1	ACTIVE#product_1	product_1	shopping product	user_1	cart_a
	user_2	PK	SK	product_id	entity	user_id	cart_id
		u#user_2	INACTIVE#product_1	product_1	shopping product	user_1	cart_z
product_3	user_1	PK	SK	product_id	entity	user_id	cart_id
		u#user_1	ACTIVE#product_3	product_3	shopping product	user_1	cart_a

Figure 7.5 – E-commerce – shopping cart example model of GSI

The advantage of having fewer GSIs isn't primarily about performance but rather about resource management. While GSIs, in general, have minimal management requirements, considerations such as throughput, autoscaling, and other aspects can be more effectively managed with vertical partitioning and the breakdown of large chunks of data. This approach streamlines the overall management of resources related to GSIs.

Working around the DynamoDB max item size limit

One of the most common use cases for vertically partitioning data, even outside a single-table design scenario, is to model large chunks of data. DynamoDB imposes an individual item size limit of 400 KiB, which is a strict constraint. Workloads involving data storage exceeding 400 KiB cannot be efficiently accommodated in DynamoDB. The solution lies in breaking down and vertically partitioning large chunks of data into several smaller chunks, each constituting items of less than 400 KiB, all part of the same item collection. Despite the segmentation, these items can still be retrieved together when necessary, leveraging DynamoDB's fully managed, serverless, and highly available database features.

In my experience working with numerous customers, many adopt this strategy in non-single-table design scenarios. They might have substantial data blobs collectively amounting to a few megabytes. However, these are intelligently broken down into consecutive items, each containing bytes within the 400 KiB limit. During runtime, applications utilizing such data read multiple items in the same request and assemble them into an artifact that meets the expectations of their end users. Mind you, this is not similar to making runtime JOINs on RDBMSs, since the items are organized to be retrieved together, without the application having to perform its own heavy lifting.

Implementing logical grouping of data within item collections

In addition to the advantages already listed in this section, vertically partitioning data enables us to create logical groupings within item collections using distinct sort key attribute value prefixes. Referring to the song-playing application data model in *Figure 7.2*, incorporating the A# prefix for the song detail and total download items facilitates the use of the `begins_with()` function in `Query` operations when retrieving both items in a single request.

Without applying logical grouping through sort key attribute value prefixes, maintaining the lexicographical order of the items becomes essential. However, adding the A# prefix logically groups the items exactly as needed, streamlining data retrieval. For individual item retrieval, providing the complete sort key attribute value, such as A for song detail or `A#total_downloads` for total downloads, allows flexibility.

You might have noticed the PK, SK, GSI1PK, and GSI1SK references for key attributes so far in this section, as well as the book. In the next section of this chapter, *Naming conventions*, we will learn about common naming conventions and their underlying rationale. Before that, let us briefly explore scenarios where breaking down items might not be practical and some considerations when vertically partitioning data in DynamoDB.

Key considerations of breaking down data for vertical partitioning

Although you would definitely benefit from vertically partitioning data in a lot of cases, there are some scenarios where you would not want to break down the data to deeper levels. You still must have *data that needs to be accessed together, must be stored together* in mind when practicing data modeling in DynamoDB, but in some specific scenarios, breaking down larger items into incredibly tiny chunks may prove to be inefficient compared to keeping them relatively grouped. Let us review some of these key considerations.

Optimizing write performance with infrequent updates

In scenarios with write-heavy workloads, it's essential to optimize for a highly performant and efficient write path, especially when the read path can be more flexible. Such scenarios often involve infrequent updates. Consider a financial services use case involving stock trading information, where the primary concern is quick writes and precise, targeted retrievals, with minimal updates.

For these use cases, it may not be practical to break down a single stock trading tick into multiple vertically partitioned items. Doing so could lead to inefficiencies, as it would increase the WCU needed for initial writes. Additionally, the application responsible for these writes might need to transform the data before writing it to DynamoDB, adding unnecessary complexity.

If the use case also demands real-time aggregation of granular data, such as maintaining a rolling 24-hour summary, this can be efficiently handled using asynchronous processing. DynamoDB Streams or Kinesis Data Streams can be leveraged for this purpose, enabling continuous updates without impacting the primary write operations.

Higher write costs for the initial write

In addition to infrequent update scenarios, another that may not require highly broken down structures would be tiny items. Again, with some focus on balancing the costs of breaking down items into smaller chunks, if you already had items where the overall item size is less than 1 KB, breaking down such data further into individual chunks of a few hundred bytes would prove to be more cost inefficient. Since DynamoDB charges WCU based on 1 KB roundups, having several 100-byte items instead of, say, a single 900-byte item would be 9X more inefficient in terms of write costs when you write those items for the first time.

Requiring additional transformation for analytics

Finally, when broken down data also needs to support certain analytical use cases, wherein you export the DynamoDB table data into Amazon S3 or implement a data pipeline that obtains change data for a table via one of the streaming options (DynamoDB Streams or Kinesis Data Streams for DynamoDB), the analytics system might also require certain additional transformations to stitch some of this broken down data back together to perform aggregations and other sorts of analytics. In most cases, this would not be a sole reason not to break down your data for vertical partitioning, but it is important to get an idea of the overall effort you and your team would be making when implementing the vertical partitioning design pattern for your multi-entity related data in DynamoDB.

Now, let us turn toward some naming conventions that are popular in the DynamoDB community that help with developing and implementing the vertical partitioning strategy.

Naming conventions

It is crucial to emphasize that DynamoDB does not enforce any naming conventions on your data stored in a table, except for specific reserved words (2) for attribute names. However, adopting certain naming conventions can enhance the flexibility of your data model. In the following section, we will explore some recommended naming conventions that can be employed to achieve a more adaptable and organized data model.

Naming conventions for attribute names

In DynamoDB, there are no strict formal naming conventions for attribute names. However, it is a best practice to use attribute names that are both **short and meaningful**. This is crucial because the size of an item in DynamoDB is determined by the combined size of its attribute names and values, directly affecting both throughput and storage costs.

There is a misconception that using single-character attribute names is the best approach. While it might save significant space and optimize costs, it is generally not advisable as it complicates the implementation of business logic and data access layers for development teams. As an engineer working on cutting-edge, high-performance applications, it is beneficial to avoid the overhead of maintaining a mental mapping between attribute names in DynamoDB and their meanings. This allows you to focus more on optimizing performance and functionality rather than deciphering cryptic names, which can be especially valuable when dealing with complex systems or when onboarding new team members.

Instead, choosing attribute names that are concise yet descriptive ensures efficiency without sacrificing clarity. For example, using `p_id` for a unique person identifier is preferable to using a longer name such as `person_identifier`. This principle of short but meaningful attribute names is not only applicable to vertical partitioning but is a general best practice across DynamoDB usage.

Naming conventions for attribute values

In the context of single-table design, a key convention is representing the entity type of the item in the partition key or sort key value's prefix. This practice not only facilitates grouping data in an item collection by entity type but also enhances the model's troubleshooting friendliness.

Consider a data model involving various entities with unique system identifiers, often in the form of UUIDs. Without entity type prefixes, navigating through multiple UUID attribute values could lead to longer development and troubleshooting times. However, by prefixing UUID attributes with a meaningful part of the entity they represent, the model becomes more understandable and efficient. For instance, using o#abcd for `order_id` makes more sense than just abcd, and p#1234 for `person_id` is more informative than 1234 alone. Adopting this convention aids in making sense of the data model and even in troubleshooting it.

The entity type prefixes not only aid in troubleshooting but also enable logical grouping of data within an item collection, as discussed earlier in this chapter. In the case of the song-playing application example, where multiple entities are linked to the song ID, such as monthly downloads and weekly plays, the use of prefixes allows organizing these entities into cohesive chunks of items that may need to be accessed together.

For instance, the monthly downloads can be logically grouped, where a single item represents the downloads for a specific month. Each month is then associated with a DynamoDB item. The overall data model of the song-playing application including the monthly downloads might look like the following figure:

	Primary key		Attributes		
Partition key: PK	Sort key: SK				
	A	type		title	artist
		details		How Much Is The Fish?	Scooter
	A#downloads	type		count	
		total_downloads		100	
	d#download-1	type		user	
		download		amdhing	
s#song1	d#download-2	type		user	
		download		ripani	
	m#2022-07	type		count	
		monthly_downloads		5	
	m#2022-08	type		count	
		monthly_downloads		95	

Figure 7.6 – Sort key attribute naming convention for logical grouping

In the preceding figure, there are three logical groupings of items that need to be retrieved together by the application. The first group includes song details and total downloads items, the second group consists of individual items recording a user download, and the third group is for granular, monthly downloads. Each group employs a prefix in the sort key values, which can be utilized in a GROUP BY scenario for retrieving the relevant data. DynamoDB Query operations can incorporate a corresponding begins_with clause along with these prefixes to retrieve the groups efficiently.

The convention of using the hash (#) sign as a delimiter for prefixes and the actual values is common, as most data models avoid using this symbol in their applications. However, if your system does use the hash symbol, it can be easily replaced by any other symbol not employed in your system, such as a hyphen (-) or a pipe (|).

Note that the naming conventions for attribute values are only relevant when the data type of the attribute is `String`. For others, there are no such conventions followed for attribute values. With that, we can now move on to understanding how vertically partitioning data can impact future growth, the expansion of application features, or the evolution of the data model.

Future growth and expansion

In the realm of NoSQL systems, a well-recognized and widely accepted trade-off involves sacrificing query flexibility to achieve predictable performance and high scalability. While the horizontal scaling of NoSQL databases can deliver exceptional performance at scale, they are not optimized to seamlessly support read queries with numerous optional and unknown filters. Furthermore, in the context of databases such as DynamoDB, the requirement to have knowledge of 60-70% of your application's access patterns upfront may prove to add an additional layer of complexity for users.

For tables implementing non-overloaded partition key and sort key attribute names instead of the generic PK and SK, evolving the table schema may be an involved process. This evolution may require creating a new table with the evolved key schema and migrating data from the current table, often involving varying degrees of downtime requirements. This highlights a significant consideration in managing the schema evolution process for tables in NoSQL databases.

Vertical partitioning of data, coupled with the overloading of key attribute names, effectively eliminates some of the challenges associated with evolving and expanding the data model. In a table where keys are defined as PK and SK, and similarly for GSIs as GSI1PK and GSI1SK, the data model can be easily evolved or extended to support additional features in the application.

For instance, in the song-playing application example, if there is a need to support listing popular playlists associated with a song in the future, this can be achieved by simply adding additional items to the existing item collection, representing the new entities. Similarly, if features associated with an entity become deprecated, items related to that entity can be identified and removed through a rate-limited background sweeper process script.

This extensibility also extends to GSIs. The overloaded and generic nature of GSI key attributes, such as GSI1PK and GSI1SK, allows for flexibility in changing and evolving the data within them without requiring substantial time and effort on the database side. The use of overloaded or generic attribute names streamlines this process.

Furthermore, vertical partitioning supports extensibility in scenarios where data might grow beyond the 400 KiB maximum item size limit imposed by DynamoDB. By breaking down large chunks into multiple smaller chunks of data, each not exceeding the maximum item size limit, vertical partitioning can accommodate any size of data in DynamoDB. While there are alternative options for managing **large objects (LOBs)** involving services such as Amazon S3 or other purpose-built data stores, using DynamoDB itself for relatively large data chunks becomes feasible with vertical partitioning.

> **Important note**
>
> Consider this as a reminder to delete any resources you may have created while going through this chapter in your AWS account.

With that, we are good to conclude the chapter on vertical partitioning. The amount of information and guidance provided in the chapter is sufficient to implement vertical partitioning in your projects involving DynamoDB.

Summary

In this chapter, we leaned into the concept of item collections within DynamoDB, understanding when and how to use them, along with their advantages and key considerations, illustrated through practical examples.

The focus then shifted to breaking down data for vertical partitioning, exploring the pros and essential considerations associated with this approach. The discussion highlighted how vertical partitioning not only optimizes performance but also addresses challenges in evolving and extending the data model.

We explored the practical implications of vertical partitioning, showcasing its role in future-proofing the data model and enhancing extensibility, aspects crucial for maintaining flexibility and scalability in DynamoDB, especially in comparison to traditional NoSQL systems.

I recommend checking out the AWS official blog article on vertical partitioning authored by Mike and our colleague (3). It's a valuable resource that complements the learnings shared in this chapter, providing practical insights into the application of vertical partitioning. Happy reading!

So far in the chapters of this book, we have used global secondary indexes in various places to support additional access patterns for our DynamoDB data. In the next chapter, we will dig deeper into secondary indexes in general, the supported secondary indexes, including GSIs, and how we might want to use GSIs to our advantage to allow expansion or evolve our applications backed by DynamoDB.

References

1. Wikipedia: Lexicographic order: `https://en.wikipedia.org/wiki/Lexicographic_order`

2. AWS Docs: DynamoDB Reserved words: `https://docs.aws.amazon.com/amazondynamodb/latest/developerguide/ReservedWords.html`

3. AWS Blogs – Vertical Partitioning in DynamoDB: `https://aws.amazon.com/blogs/database/use-vertical-partitioning-to-scale-data-efficiently-in-amazon-dynamodb/`

8

Secondary Indexes

Welcome to the eighth chapter of *The Definitive Guide*! By now, you've probably developed a good understanding of the principles and data patterns that can be used to model data in DynamoDB. These include considerations such as when to denormalize data to avoid making runtime compute-intensive joins, when to duplicate data to prebuild the data as your application would need it, and when to go for a multi-table strategy as opposed to a single-table one. There is obviously more to do with NoSQL and DynamoDB than just these concepts, but in a nutshell, they do represent some of the key aspects.

When DynamoDB was launched back in 2012 (1), it was quickly adopted by customers. This was also documented only a few months later by Amazon's CTO, Werner Vogels, in a blog post (2) where he publicly shared how the data stored by customers in DynamoDB was doubling every couple of months and the cumulative throughput provisioned on the tables by these customers was sufficient to perform hundreds of billions of requests per day! DynamoDB did not support any form of secondary indexes at launch, but even then, customers found it suitable for many use cases across industries, from media and entertainment to advertising, and software and internet to gaming.

Soon after DynamoDB's launch, customers already wanted it to do more, and a common request was about indexing data in alternate views. The situation customers found themselves in was that DynamoDB was proving to be great for them in terms of scalability, flexibility, and the fully managed serverless aspect, but their applications needed to access the same DynamoDB data in multiple ways that they couldn't with just the base table available. For one, they needed to sort data on attributes other than the sort key of the table. Additionally, they also wanted to access data efficiently with a completely different key schema than that of the base table.

Not long after gathering feedback from customers about what their ideal indexing experience would look like, the DynamoDB team put their heads down to build and launch **local secondary indexes (LSIs)** in early 2013 (2). LSIs helped a lot of customers who wanted to sort their DynamoDB data with an alternate key in addition to the way data was sorted in the base table while retaining the same partition key. Shortly after launching LSIs, the DynamoDB team also introduced **global secondary indexes (GSIs)** (3) to address the requirements of those customers who wanted their table data organized in a completely different key schema. This included having a different sort key, as well as a different partition key than what the base table supported.

Both LSIs and GSIs have their own strengths in terms of **consistency**, flexibility, scalability, and overall management. Both provide an alternate view of the base table data that allows supporting additional access patterns for applications with varying functional requirements.

In earlier chapters of this book, whenever we used secondary indexes, it was always because the base table's key schema was designed to support a set of access patterns different from those for which the secondary indexes were needed. The set of access patterns the base table was designed to support may be more foundational for the application, may be requested more frequently, or may require a certain kind of access that aligns more with the characteristics of the base table than with those of the secondary indexes.

Using the base table itself designed in a way that was different from what would be efficient for the other access patterns, we may need to go through each item in the base table using a full table scan, which by now we know would not be optimal for **online transaction processing** (OLTP) access. Hence, it is common to rely on secondary indexes for part of the access patterns.

To describe secondary indexes using an analogy, imagine you're in a huge library with shelves filled with thousands of books. Each book has a title, but finding a specific book solely by its title can be challenging, especially if there are numerous books with similar titles.

Now, picture that the library introduces a new system where they create additional indexes based on the genre, author's last name, and publication year. These indexes act like cheat sheets, allowing you to quickly locate books based on different criteria without having to scan every shelf.

In this analogy, the library's main catalog is like DynamoDB's base table, while the secondary indexes are like the additional cheat sheets. They help you find specific data in different ways, making it easier to navigate and access information efficiently without having to search through every record like going through the entire library's main catalog.

In this chapter, we will dive into secondary indexes within DynamoDB – something we have touched on briefly throughout the earlier chapters of the book, but now we will learn about them with all our focus. Secondary indexes make DynamoDB even more awesome, letting you find things in different ways without getting lost.

By the end of this chapter, you will learn how to think about secondary indexes in cleverly organizing data to meet your specific access pattern needs, how and when secondary indexes can be created for your table, bits on data modeling specific for secondary indexes, and building sparse indexes to optimize on costs.

In this chapter, we are going to cover the following main topics:

- Handling additional access patterns
- Local secondary indexes
- Global secondary indexes
- Comparing secondary indexes
- Best practices for modeling data with secondary indexes

Let us start by learning about the different scenarios where you may encounter access patterns that would not be optimal for the base table because of its key schema, and where you may need one or two secondary indexes.

Handling additional access patterns

Consider any of the applications you have built up until today, no matter how big or small. If you have spent some amount of time in the development cycle of the application, be it in any capacity – as a developer, an architect, or any other engineering or product role – you may know that after the scope of the application is defined and the application is launched, it is highly likely that you may come across users of your app wanting a certain functionality that was not part of the initial scope defined for the app.

In this case, you probably would have gone back to the drawing board, figured out what the new feature request was, whether it should be part of the same application, and if yes, what would be the approximate amount of dev hours it might take, and finally, whether that is worth prioritizing. After all this work, you might need to see what changes may be needed on the database level, if any, and how you may need to add additional columns in your relational database management system (RDBMS) tables to make certain database access work to support this new functionality.

This is particularly true when you have an additional access pattern that you need to support in your application, and for that, you may need some additional work on the database to accommodate the access patterns. Using secondary indexes in DynamoDB to support additional access patterns is not too different from this scenario.

In such a scenario, you may need to use one or more secondary indexes for your DynamoDB table even when the overall set of your application's access patterns is not a big one. Let us validate this statement using a few examples.

Example model – e-wallet

Consider an online wallet use case that utilizes DynamoDB to manage transactions executed by users. The following figure illustrates the data model for the primary access pattern of the application – listing all transactions made by a user for a particular wallet.

e-wallet

Primary key		Attributes		
Partition key: PK	Sort key: SK			
wallet#1	2023-07-27T14:06:20.993Z#TXN#A	type	ts	other attributes
		transaction	2023-07-27T14:06:20.993Z	...other data...
	2024-01-17T21:52:12.208Z#TXN#B	type	ts	other attributes
		transaction	2024-01-17T21:52:12.208Z	...other data...
wallet#2	2024-07-23T01:29:50.211Z#TXN#Z	type	ts	other attributes
		transaction	2024-07-23T01:29:50.211Z	...other data...

Figure 8.1 – Base table for e-wallet use case

As per the preceding figure, the partition key of the table is PK, and the sort key is SK. These are generic names, as we learned in the earlier chapters of this book, being a best practice for future extensibility.

The partition key and sort key were chosen to store walletID and transaction_timestamp#uuid because the primary access pattern for this use case is to retrieve all transactions made in a particular wallet in chronological order, with the latest being displayed first. The DynamoDB table in this case would serve such an access pattern tremendously well. However, moving ahead, would it be true that this will be the sole purpose of the e-wallet application? To store and retrieve transactions made based on walletID? If a user owned multiple e-wallets, would they not like a global view of all the transactions between their wallets or other users' wallets? If that is an access pattern, you'd need the transaction data indexed on userID in addition to the existing index on walletID, which is the base table.

So, there is a need for the same data to be indexed with a different key to efficiently answer other access patterns, such as providing the global user view across wallets. In this case, let us consider a GSI with its partition key as userID and the sort key as a combination of transaction_timestamp and a unique identifier, uuid. The following figure illustrates this new GSI:

e-wallet   GSI: GSI1

Partition key: GSI1PK	Primary key				Attributes
	Sort key: GSI1SK				
u#user_1	2023-07-27T14:06:20.993Z#TXN#A	PK	SK		type
		wallet#1	2023-07-27T14:06:20.993Z#TXN#A		transaction
	2024-01-17T21:52:12.208Z#TXN#B	PK	SK		type
		wallet#1	2024-01-17T21:52:12.208Z#TXN#B		transaction
	2024-07-23T01:29:50.211Z#TXN#Z	PK	SK		type
		wallet#2	2024-07-23T01:29:50.211Z#TXN#Z		transaction

Figure 8.2 – GSI for e-wallet use case

If we had not used a secondary index here, we would have needed to maintain a separate list of wallets corresponding to users, and then have the application make one read to this list to retrieve the different wallet IDs corresponding to a particular user, and then make several read calls for each wallet ID to retrieve the transactions. Finally, we would also need to sort the cross-wallet transactions by their timestamps in a single list. This might still be okay if the access pattern was exercised, say, once a day or once a week. But doing this for every user request more often than that would be highly inefficient and lead to poor user experience.

Example model – app authentication

Consider another use case that may be generic across industries – application authentication. Regardless of the industry, you are likely to have applications used by end users or employees themselves for certain parts of their work. An important aspect of these applications would be authentication and authorization. Let us consider the authentication part here, where users of the app have multiple methods of authentication.

For example, in a food delivery app, this might mean authenticating via the plain old email-password combination or using third-party authentication and identity providers such as Google, X (formerly Twitter), Facebook, GitHub, and others. We are not concerned about the **two-factor authentication** (**2FA**) bits here, which may be important for security, but a user may want to authenticate to the app using any of the authentication methods supported. As an application owner, you'd want to integrate as many commonly used authentication and identity providers as possible into the app.

For such a use case, you would imagine the primary access pattern would be for a given identity key, match the key with other authentication-related information associated with the key and, in turn, the user that owns the key. The schema that supports such an access pattern would look like the one in the following figure:

auth-info

Primary key Partition key: PK	Attributes			
key#1A4EE58B-2530	type	other_attributes	ts	user_id
	google_auth	...g_auth_info...	2004-03-13T12:19:28Z	am_too_cool
key#01A5D733-1149	type	other_attributes	ts	user_id
	meta_auth	...m_auth_info...	2018-10-09T12:19:33Z	am_too_cool
key#5D47F1CB-2DDB	type	other_attributes	ts	user_id
	x_auth	...x_auth_info...	2005-11-19T12:19:38Z	am_too_cool
key#aman@example.com	type	other_attributes	ts	user_id
	email_auth	...auth_info...	2016-03-01T12:19:43Z	am_too_cool

Figure 8.3 – Base table for app authentication use case

As you can see from the preceding figure, the table's partition key, PK, is a unique key identifier, and there isn't a sort key in the table schema because each authentication key identifier must uniquely point to a single user. Can this mean that a user can have multiple auth key identifiers? Yes, in this case.

The schema does solve the primary access pattern of the authentication use case, but what if the app owner has requirements to periodically go through every auth key identifier tied to a single user to validate certain security policies and allow the user to manage all their authentication methods in a single window in the app? For that, one way with the existing base table schema would be to scan the whole table, match up different auth key identifiers against users, and potentially do a lot of in-memory processing. This obviously wouldn't scale or be cost effective, as we know that memory is generally an expensive resource. As a solution to this problem, we can use a secondary index to organize the same authentication data in a different view. The following figure illustrates the GSI we build on the DynamoDB table to answer our alternate access patterns:

auth-info   GSI: GSI1

Primary key		Attributes					
Partition key: GSI1PK	Sort key: GSI1SK						
u#am_too_cool	email_auth	PK	type	other_attributes	ts		user_id
		key#aman@example.com	email_auth	...auth_info...	2016-03-01T12:19:43Z		am_too_cool
	google_auth	PK	type	other_attributes	ts		user_id
		key#1A4EE58B-2530	google_auth	...g_auth_info...	2004-03-13T12:19:28Z		am_too_cool
	meta_auth	PK	type	other_attributes	ts		user_id
		key#01A5D733-1149	meta_auth	...m_auth_info...	2018-10-09T12:19:33Z		am_too_cool
	x_auth	PK	type	other_attributes	ts		user_id
		key#5D47F1CB-2DDB	x_auth	...x_auth_info...	2005-11-19T12:19:38Z		am_too_cool

Figure 8.4 – GSI for app authentication use case

As an application owner, you may also want to ensure that one user has exactly one Google authentication key identifier, but they may have an email authentication key identifier in addition to the single Google authentication key identifier. In that case, you may want the schema of the preceding GSI to be the schema of the base table. This is because GSIs permit duplicate primary key values, meaning multiple items in the GSI with the same partition key and sort key values can exist. Uniqueness across GSI items is maintained by projecting the base table's partition key and sort key, which is unique. Switching the base table and GSI schemas would ensure exactly one item exists for the am_too_cool user for Google authentication, in the preceding figure. We'll dive more into the duplicate primary key concept in a GSI further in this chapter.

Example model – the Employee data model

If you recall in *Chapter 3, NoSQL Workbench for DynamoDB*, we touched on adding a GSI to the main Employee table to retrieve employees by their designation in the company. To revisit the same scenario, the following is the data model of the base table of the employee use case. This is before adding the GSI to support retrieval by designation:

Employee   GSI: Name   GSI: DirectReports

Primary key Partition key: LoginAlias			Attributes	
johns	FirstName	LastName	ManagerLoginAlias	Skills
	John	Stiles	NA	["executive management"]
marthar	FirstName	LastName	ManagerLoginAlias	Skills
	Martha	Rivera	johns	["software","management"]
mateoj	FirstName	LastName	ManagerLoginAlias	Skills
	Mateo	Jackson	marthar	["software"]
janed	FirstName	LastName	ManagerLoginAlias	Skills
	Jane	Doe	marthar	["software"]
diegor	FirstName	LastName	ManagerLoginAlias	Skills
	Diego	Ramirez	johns	["executive assistant"]
marym	FirstName	LastName	ManagerLoginAlias	Skills
	Mary	Major	johns	["operations"]
janer	FirstName	LastName	ManagerLoginAlias	Skills
	Jane	Roe	marthar	["software"]

Figure 8.5 – Base table for the Employee use case

The `Employee` data model is one of the sample ones available in NoSQL Workbench for DynamoDB, but if you cannot find it, it is also uploaded to the GitHub repository for this book (4).

Once again, in this scenario, if a secondary access pattern for my application involves retrieving employees by their designation, with the base table (i.e., the preceding figure), a full table scan would be needed to read through each item and filter to get the desired items. However, if the table had a GSI on an attribute that contained the employee's designation, then the retrieval would take the least amount of time while scaling incredibly well. The following figure illustrates this additional GSI on an attribute named `Designation`, which I also added to the table data as part of creating the GSI. `LoginAlias` can be the sort key of this new GSI:

Employee   GSI: Name   GSI: DirectReports   GSI: DesignationIndex

Primary key		Attributes			
Partition key: Designation	**Sort key: LoginAlias**				
CEO	johns	FirstName	LastName	ManagerLoginAlias	Skills
		John	Stiles	NA	["executive management"]
CTO	marthar	FirstName	LastName	ManagerLoginAlias	Skills
		Martha	Rivera	johns	["software","management"]
Developer	janed	FirstName	LastName	ManagerLoginAlias	Skills
		Jane	Doe	marthar	["software"]
	mateoj	FirstName	LastName	ManagerLoginAlias	Skills
		Mateo	Jackson	marthar	["software"]
Assistant	diegor	FirstName	LastName	ManagerLoginAlias	Skills
		Diego	Ramirez	johns	["executive assistant"]
Ops Manager	marym	FirstName	LastName	ManagerLoginAlias	Skills
		Mary	Major	johns	["operations"]
BI Analyst	janer	FirstName	LastName	ManagerLoginAlias	Skills
		Jane	Roe	marthar	["software"]

Figure 8.6 – GSI for retrieval by designation

Those were some examples of diverse use cases where leveraging secondary indexes was necessary to support access patterns that would otherwise require a full table scan on the base table, resulting in scalability issues in terms of latencies and cost.

While all the preceding examples focused on handling additional access patterns using GSIs, it is important to note that LSIs have their own strengths and are well suited for supporting alternate views. Let us dive into LSIs in detail next.

Local secondary indexes

As covered in the introduction to this chapter, LSIs were the first type of secondary indexes available for DynamoDB tables. These were introduced based on customer feedback to provide an alternative view of the same DynamoDB table data. However, the data itself was indexed by the same partition key as that of the base table. The key difference between LSIs and base tables themselves was that LSIs could use a different sort key attribute than that of the base table. In other words, LSIs indexed the base table data by the same partition key but sorted the same data by a different attribute than the sort key of the base table, providing an alternative sorting capability for the DynamoDB data.

If you need your e-wallet transaction data sorted by a creation timestamp as well as transaction value, you can leverage an LSI to sort the same transaction data by two different attributes but indexed by the same partition key (i.e., `walletID`). Now, you can provide the ability to sort transactions by the transaction value as well as the transaction creation timestamp.

Similarly, if you need your distributed task processing state data sorted by last modified time as well as by a priority score, so that you can retrieve tasks that were potentially not progressing as well as tasks that needed to be processed with a priority, then you could use an LSI on your DynamoDB table to provide an alternate sorting capability for the data.

Some of you might think that GSIs could also be used in these examples, where you could simply keep the partition key the same as the base table, so what are the strengths of using LSIs anyway? Well, let us review the key features of LSIs next to understand their strengths.

Feature highlights of LSIs

LSIs store data in the same location as the base table partitions, ensuring close integration with the base table. This unique aspect grants LSIs distinct advantages not necessarily applicable to GSIs. Let us now review the key highlights of LSIs, with additional details available in the AWS docs (5).

Strong consistency

The tight coupling of LSIs with base tables enables support for both strong and eventually consistent reads on the LSIs, mirroring the behavior of a DynamoDB table. Reading from an LSI incurs charges similar to reading from the base table. A write made to a DynamoDB base table is deemed successful only after it is committed to the base table and any other LSIs on the table. The tight coupling between the base table and the LSIs aids in keeping the write latency overhead to be negligible.

For further insights into the tight coupling and architecture within DynamoDB, refer to *Chapter 10, Request Routers, Storage Nodes, and Other Core Components.*

Throughput management

Every write made to the base table that includes a value of the sort key set on an LSI would be projected to the LSI, incurring write capacity charges. Similarly, if any item containing an LSI's sort key attribute is updated or deleted on the base table, regardless of which attributes are being updated, those changes would also be reflected in the LSI by DynamoDB, potentially consuming additional write capacity for the projection.

Because of the close integration between the base table and any LSIs, write and read throughput for an LSI is consumed from the base table.

Projections

With secondary indexes in general, you have the option to decide whether you want all your base table data projected onto an index or choose only a specific set of attributes to be projected. Projecting only a subset of attributes to an index can lead to efficiency in terms of DynamoDB costs, provided it aligns with your use case. This is because any attribute updated in the base table item, if projected onto any LSI, would also incur updates to that LSI, consuming additional write units.

The recommendation is to build LSIs with the required projection type and only project the subset of the item needed to support the access patterns the LSI is designed for.

The projection type for an LSI could be one of the following:

- KEYS_ONLY: Every LSI item only projects the base table partition key and sort key, as well as the sort key of the LSI itself. All remaining attributes in the corresponding base table item are not projected to the LSI.

- INCLUDE: In addition to the KEYS_ONLY projected attributes, you may also provide a list of attributes that will be projected into the LSI from the base table item.

- ALL: The LSI will project all attributes present in the base table item, into the LSI. This projection type will essentially duplicate the base table item onto the LSI, albeit with a different sort order.

In a mailbox scenario, where each item in the DynamoDB base table represents an email with various attributes such as subject line, body, timestamps, and sender-receiver information, supporting a list view in the mailbox application poses a specific challenge. Users may only be interested in certain attributes such as sender-receiver information, timestamps, and message subject for this view.

To address this, I could create an LSI on the table with an INCLUDE projection type, including only the attributes needed to support the list view. This projection type would exclude the message body attribute, resulting in a significantly smaller LSI compared to the base table. As a result, read throughput consumption for the Query operations used to render the list view would be reduced, as they would charge read units based on the reduced amount of data read due to the projection type.

When a user selects a particular email to view it in full, the full email, including the large message body, can be lazily loaded as a separate read operation to the base table. This approach optimizes performance and cost effectiveness by efficiently managing the data accessed for different views in the application.

Uniqueness characteristics

From the supported projection types for LSIs, it is evident that each projection type will include at least the base table keys projected to the LSI, namely the partition key and the sort key of the base table. These keys are crucial for guaranteeing uniqueness among LSI items. As a result, the combination of the LSI's partition key and sort key values may not be unique. This characteristic can be both advantageous and a consideration to keep in mind while designing the DynamoDB table and secondary indexes.

Item collections

If you have not covered earlier chapters of this book, an **item collection** refers to a group of items that may share the same partition key but have different sort key values. In a DynamoDB table with LSIs, the item collection size is limited to a total size of no more than 10 GB. This size constraint takes into account both the base table item size and the item size in any LSIs the table may have.

If a new write to an item collection would cause the item collection size to exceed the 10 GB limit, the write will be denied with an `ItemCollectionSizeLimitExceededException` exception. Writing to such item collections will not be successful until existing items in that item collection are deleted. However, other item collections in the same table with a size of less than 10 GB may continue to process writes successfully.

Index creation

LSIs can only be created or deleted along with their base table. DynamoDB does not support adding LSIs to an existing table, nor does it support deleting LSIs from a table without deleting the entire table, which would result in the loss of all table data. During DynamoDB table creation, you must specify the LSIs you want on the table and their projection types. This is the only time you can define these configurations for LSIs, including the projection types.

With these features of LSIs in mind, let us review the considerations for using LSIs next.

Considerations when using LSIs

LSIs are uniquely powerful when you need **strong read-after-write consistency** or the ability to sort the base table data differently. However, there are some aspects of LSIs that you should be aware of before you start using them. Let us review them now.

Index creation

As mentioned briefly in the *Feature highlights of LSIs* section, LSIs can only be created or deleted along with their DynamoDB table. This means that once the table is created, LSIs cannot be added to the table, and existing LSIs cannot be deleted while retaining the base table and any other secondary indexes that exist for the table.

If you already have a table with LSIs and data, and you no longer need the LSI, you have two options:

- **Migrate table data to a new table without the LSI**: This option may or may not require downtime for your application, depending on the migration process
- **Remove or replace the LSI sort key attribute from the data**: By removing the attribute defined as the sort key for the LSI or replacing it with a differently named attribute storing the same value, you can prevent base table items from being projected onto the index

Shared throughput

Since LSIs share read and write throughput with the base table, the maximum rate of read or write throughput that you can perform on any individual LSI or the base table is also reduced. At the time of publishing, each DynamoDB base table partition can handle 1,000 write units per second or 3,000 read units per second. With LSIs on a DynamoDB table, this number remains the same but includes the reads and writes made to any of the LSIs as well.

For writes, any writes made to the base table can consume write throughput for writing to the base table itself and for projecting the item to each of the LSIs. This operation, including writing to the base table and projecting the item to all LSIs, is bounded by the 1,000 write units per second limit. For example, in a table with three LSIs, if one write unit is consumed for writing to the base table, then an additional three write units will be consumed for projecting the item to all three LSIs, totaling four write units to the DynamoDB table partition. Consequently, instead of being able to write to a DynamoDB table partition at 1,000 write units per second without LSIs, the write rate will be limited to 250 write units per second when LSIs are present, as the same amount of write units will be required to project the items to each LSI.

Similarly, for reads, any reads made to the LSI will consume read units from the base table partition's allocation of 3,000 read units per second. This may limit the ability to read from the base table if reads are simultaneously being made to the LSI for the same table partition. A read made to the base table, however, will not consume additional read units from the LSIs.

To maximize write throughput on the base table, you can use GSIs instead of LSIs, as GSIs do not affect the base table's read or write throughput. However, it is important to note that GSIs do not support strong read-after-write consistency.

Shared storage

Once again, due to the tight coupling between the base table partitions and LSI data, item collection sizes are capped at a maximum of 10 GB in total size. This includes the size of items in the base table for that item collection, as well as the size of items in every LSI on the table for the same item collection, with some minimal overhead. Without LSIs, these item collections have virtually no limit on the maximum size. However, even having a single LSI on the table can introduce this limitation on the item collection size.

Encountering the item collection size limit for an item collection can lead to interruptions in write activity for that collection, which is not an ideal situation. In such cases, you may consider migrating to a new table without the LSI. Even if you remove the LSI sort key attribute from the table data, which should delete the item from the LSI, the presence of the LSI would still limit the item collection size on the base table.

Supported operations

Due to the unique characteristics of LSIs, single-item read operations are not supported on them. This means that you cannot issue a `GetItem` operation or a `BatchGetItem` operation on an LSI, as there may be multiple items in the LSI with the same combination of the partition key and LSI sort key values. However, you can perform `Query` and `Scan` operations on the LSIs.

Extended lookup on the base table

LSIs offer a unique capability to perform extended lookups on the base table data for attributes that may not have been projected onto the LSI. For example, in a mailbox scenario where the LSI was created with an `INCLUDE` projection type and a set of attributes, if additional attributes are introduced in the data after some time, these new attributes cannot be projected onto the existing LSI as LSI projection types cannot be modified once created.

In such cases, if a read operation is issued on the LSI and a non-projected attribute is requested, DynamoDB conducts an extended lookup on the base table data, consuming additional read throughput, to retrieve the attribute. This capability is unique to LSIs.

That covers everything about LSIs. Certain characteristics of LSIs may appear rigid, such as the requirement for LSIs to be created or deleted along with the table or the fact that the LSI's partition key always matches that of the base table. GSIs offer more flexibility in these aspects, but they involve trade-offs in other areas and may be better suited for different use cases. Next, let us dive into GSIs, highlighting their features and considerations.

Global secondary indexes

Contrary to LSIs, GSIs offer the flexibility to provide a different view of the DynamoDB table data, including the ability to use a different partition key from the base table. This means that GSIs can have entirely different partition keys and sort key attributes compared to the base table, offering significant flexibility. DynamoDB indexes the complete table data in a different view specified by your GSI key schema.

Think of a GSI as a **materialized view** of your DynamoDB base table data. When creating a GSI, you specify the partition key and sort key attributes, and DynamoDB handles the creation of this materialized view on demand.

To illustrate, in the e-wallet use case, if you needed to retrieve all transactions across users for a particular currency, this could be achieved by using the currency as the GSI's partition key attribute, with the sort key being a combination of `walletID` and `transactionID`. This functionality may not be possible to achieve using an LSI, but with a GSI, it is a fundamental capability.

While GSIs offer powerful features, there are trade-offs involved in materializing a completely different view of DynamoDB table data. Let's review the feature highlights of GSIs next to gain a deeper understanding of them.

Feature highlights of GSIs

Unlike LSIs, GSIs are not tightly coupled with the base table partitions and are maintained independently of the base table. You can think of GSIs as separate DynamoDB tables, each with its own infrastructure, throughput, and storage. However, DynamoDB handles the task of projecting data to this infrastructure according to your configured GSI key schema.

Since GSIs are **decoupled** from base table partitions, they scale independently. This provides a significant level of flexibility when managing GSIs, a characteristic that may not apply to LSIs. Let us dive into the features of GSIs to understand them better, with additional details available in the AWS docs (6).

Eventual consistency

Because GSIs are decoupled from base table partitions, the propagation of writes from the base table to the GSIs occurs asynchronously. When a write is made to the base table, it is considered successful once the write processing is complete for the base table. An HTTP 200 response is returned to the client, and the same write is then validated against all the GSI schemas. If the write needs to be propagated to one or more GSIs, it is sent to the GSI partitions.

Due to the asynchronous nature of write propagation, GSIs can only support **eventually consistent reads**. For some access patterns, this may not be suitable. Therefore, the recommendation is to model those access patterns to leverage the base table instead of the GSI.

Throughput and storage management

Since GSIs are hosted on their own infrastructure, their throughput and storage are measured independently of the base table. This means that GSIs have their own read and write consumption, separate from the base table, and they also have separate read and write limits. When a write is made to the base table, additional write units may be consumed to propagate these writes to the GSIs. However, these additional write units are consumed from the individual GSIs, unlike LSIs, where the additional write units are consumed from the base table's provisioned throughput. Both read and write throughput units consumed for GSIs are the same as those for DynamoDB tables or LSIs.

Regarding storage, GSIs have their own storage, so the same per-GB charges apply as they do for the base table or LSIs, depending on the amount of data stored in the GSIs.

GSIs adopt the same capacity modes and storage modes as their base table. For example, if the base table has its capacity mode set to on demand and its table class as Standard-Infrequent Access, then each GSI will also have the same capacity mode and table class. However, individual GSIs can have different autoscaling configurations than the base table, particularly for the provisioned capacity mode. For more information about capacity modes and table classes, please refer to *Chapter 9, Capacity Modes and Table Classes*.

Projections

Just like LSIs, GSIs also support three projection types: KEYS_ONLY, INCLUDE, and ALL. These projection types can help reduce the amount of data stored in a GSI by allowing you to specify only those attributes needed by your application to answer specific access patterns.

For example, if a base table item size is 100 KB, and for a GSI, you only need to retrieve attributes worth 10 KB, it makes sense to use the INCLUDE projection type to reduce the amount of data projected onto the secondary index. Using the ALL projection type in this scenario would lead to a 9X cost inefficiency, as the GSI would be charged for storing extra data that is not needed. Also, read and write operations on the GSI would be charged based on the 100 KB item sizes, rather than the 10 KB the application requires.

For an existing GSI item, if the base table sees an update to the attribute that is the GSI's partition key or sort key, then the projection of the item to the GSI causes two write operations – one operation is to delete the existing item and the other is to create a new item with the new partition key or sort key. This is because the partition key or sort key of secondary indexes, as well as the base table, cannot be updated in place. Instead, any change requires the deletion of the existing item and performing a new put operation. However, in the case of secondary indexes, DynamoDB automatically handles the delete and put operations for you, whereas with the base table, the application would need to perform and manage this change.

Uniqueness characteristics

Like LSIs, GSIs also always project the base table partition key and sort key attributes in whichever projection type is configured. The combination of these two attributes is used to guarantee uniqueness among items in a GSI, so the combination of the GSI partition key and sort key attribute values need not be unique.

The sort order for items in a GSI with the same values of the GSI partition key and sort key is indeterministic. Therefore, applications must consider this while implementing GSI designs.

Item collections

GSIs do not impose any limits on data size for item collections on a table or a GSI. Even when a table may have an LSI, any GSIs on it can have item collections without an upper limit on the data size. However, neither the LSIs nor the base table can have item collections of more than 10 GB of cumulative data size. By design, GSI item sizes are not considered in the 10 GB base table or LSI item collection size limits.

Online index creation and management

The decoupled nature of GSIs from their base tables allows incredible flexibility when it comes to index creation and management. A GSI can be created or deleted at any time during a DynamoDB table's lifetime.

When a GSI is created while the base table already has data in it, DynamoDB performs a backfill operation. During this process, DynamoDB reads through every item in the base table, validates it against the GSI key schema, and if it qualifies to be in the GSI, issues a write of that item to the GSI. The reading of the base table data to qualify GSI items is not charged, but the writes made to the GSI during this backfill process are. The backfill process does not impact the performance of the base table serving live application traffic. Only when the backfill process is complete will the GSI be available to read from. More information about the different GSI creation phases can be found in the AWS docs (7).

That was all about the features and strengths that GSIs bring to the table (see what I did there?). Next, let us review any considerations that you must know about before designing data models using GSIs.

Considerations when using GSIs

There are a few considerations to address when using GSIs to provide an alternative index on your DynamoDB table data. Let us review them now.

Throughput and storage management

Since GSIs do not share throughput or storage with the base table, they require separate management of throughput and storage resources. This also entails additional costs associated with throughput and storage. However, the costs of throughput and storage for both GSIs and LSIs may be similar, so there is no significant difference in terms of cost.

For example, consider two scenarios: one with a base table that has one LSI, and another with a base table that has one GSI. As long as the secondary indexes in both scenarios use the same projection type, the write units consumed for an item in both cases would be the same. The additional storage required for the secondary indexes in both scenarios would also be similar. Therefore, from a pricing perspective, there is little difference between LSIs and GSIs in terms of throughput and storage.

GSI backpressure

Due to the design choice of the GSI partition key and sort key, application traffic directed toward the base table can create a skew in write propagation onto the GSI, potentially exceeding one or more GSI partition's per-partition throughput limits of 1,000 write units per second. In such cases, the affected GSI partition can apply backpressure to the base table partition, leading to throttling of application requests that target the affected GSI partition.

Backfilling GSIs

When creating a new GSI on a base table with existing data, the duration of the backfill process, where the GSI is populated by reading through the base table data, depends on the provisioned **write capacity units (WCUs)** allocated to the GSI, particularly in the provisioned capacity mode. If the provisioned WCUs are set to a low value, the backfilling process and ultimately the GSI creation may take longer compared to when higher provisioned WCUs are allocated.

During the backfilling process, autoscaling configurations on the GSI do not take effect because the process is reflected through Amazon CloudWatch metrics such as `OnlineIndexConsumedWriteCapacity` and `OnlineIndexPercentageProgress`, which are not monitored by autoscaling. However, once the GSI creation is completed, all read and write throughput consumption for the GSI is tracked through metrics such as `ConsumedReadCapacityUnits` and `ConsumedWriteCapacityUnits`, which are indeed monitored by autoscaling to manage throughput.

Supported operations

Like LSIs, due to the unique characteristics of GSIs, single-item read operations are not supported on them. This means that you cannot issue a `GetItem` operation or a `BatchGetItem` operation on a GSI, as there may be multiple items in the GSI with the same combination of the GSI partition key and sort key values. However, you can perform `Query` and `Scan` operations on the GSIs, both with eventual consistency only.

Write propagation delay and eventual consistency

Since writes made to a base table propagate over to its associated GSIs asynchronously, the GSIs are eventually consistent with the base table state.

A write made to the base table is propagated to the table's GSIs typically within the same millisecond or next, however, aspects of traffic driven toward the table and the GSI can impact the propagation times, and they can end up being more than a couple of milliseconds.

Customers often ask about this: "When will a GSI be consistent with its base table after a write is made?" One of the popular answers that go around is *eventually*. It is important to assume that the GSI may not have the latest state of an item when an application attempts to read it, and the application must react appropriately.

Updates to GSI keys

Given that writes to the base table are propagated to the GSIs, an update to an item in the table may require an update to the GSI partition key or sort key, depending on the GSI's key design. Because the partition or sort keys of a GSI or table cannot be updated directly, DynamoDB handles this by first deleting the existing item in the GSI and then inserting a new item in the same GSI with the updated partition or sort key value. This process consumes additional write units compared to updates on the base table, so it is important to account for this additional throughput, especially when the table and GSI are in provisioned capacity mode.

Therefore, it is advisable to provision more write throughput on GSIs than on their corresponding base table to accommodate these additional write operations. As a ballpark, provision **1.5X** write throughput on a GSI compared to X of the base table. This number can vary, but the idea is to provision more on the GSI in the case of provisioned capacity mode. For the on-demand capacity mode, the throughput consumed on the GSI will scale automatically.

Comparing LSIs and GSIs can provide valuable insights into when to choose one over the other. Let us dive into this comparison in the next section.

Comparing secondary indexes

The following table can be treated as a cheat sheet for learning about LSIs and GSIs:

LSIs	GSIs
Can only be created at table creation time	Can be created/deleted at any time
Shares WCU/RCU with the table	WCU/RCU independent of the table
Item collection size limit <= 10 GB	No size limits
Hard limit = 5	Soft limit = 20
Support strong and eventual consistency for reads	Support eventual consistency for reads
The partition key is the same as the base table, and the sort key can be different	The partition key and sort key can be different from the base table

Table 8.1 – Comparison of LSIs and GSIs

The comparison between LSIs and GSIs highlights their respective strengths and trade-offs. LSIs offer strong consistency but are less flexible, whereas GSIs provide more flexibility at the cost of eventual consistency. In practice, GSIs are more commonly used due to their versatility, reflecting a trend towards eventual consistency for scalability and high availability in modern applications.

However, it is important to remember that regardless of the choice between LSIs and GSIs, DynamoDB guarantees a high level of uptime through its **service-level agreements** (**SLAs**). For single-region tables and indexes, the SLA is **99.99%** monthly uptime, while for multi-regional global tables, it is **99.999%** (8).

That covered all the operational aspects of secondary indexes, including features, considerations, and best practices. However, these aspects leave out some important elements related to data modeling with secondary indexes. Let's conclude this chapter by exploring some best practices for data modeling on secondary indexes.

Best practices for modeling data with secondary indexes

Concluding this chapter with data modeling best practices for secondary indexes is a great way to round out the discussion. These practices delve into how to effectively structure your data to optimize performance and meet your application's access patterns. Let us dive into them now.

Designing sparse indexes

We know by now that secondary indexes can be used to model the same data in an alternate view with a different key schema compared to the base table. This alternate key schema is based purely on the access patterns that the application needs to support.

One common type of access pattern revolves around *finding a needle in a haystack* – retrieving a tiny subset of items with a certain filter from a large dataset. Scanning the whole table to retrieve this tiny subset of items is highly inefficient, as most of the items do not qualify for this access pattern, yet you are paying for reading through them all. Consider an internet of things (IoT) use case of storing device telemetry in DynamoDB. Every item in the table represents statistics about a device at a particular timestamp. Each device may also be escalated to a supervisor or technician for manual analysis or servicing. The following figure illustrates such an IoT data model in DynamoDB:

device-logs GSI: escalated-devices				
Primary key		**Attributes**		
Partition key: DeviceID	**Sort key: Timestamp**			
d#12345	2024-07-10T01:06:58.613Z	Operator / Conor	Date / 2024-07-10	
	2024-08-25T01:45:31.783Z	Operator / Conor	Date / 2024-08-25	
	2025-01-27T23:09:12.299Z	Operator / Barbara	Date / 2025-01-27	
d#999	2023-12-07T13:24:38.375Z	Operator / Barbara	Date / 2023-12-07	
	2024-06-23T01:30:08.063Z	Operator / Conor	Date / 2024-06-23	EscalatedTo / Lorenzo

Figure 8.7 – IoT use case sample data model

If an administrator wants to find all devices in a table that were escalated to a particular supervisor, this would be like finding a needle in a haystack – most devices may not be escalated at any given time, but only a tiny subset would be. In this use case, creating a GSI on the `EscalatedTo` attribute results in a sparse index, as most items do not contain this attribute. As a reminder, only those items are projected into secondary indexes that have valid values in the partition key and sort key attributes for the secondary index; all others are skipped. With this characteristic in mind, the sparse GSI with only those items where the device was escalated to a particular supervisor would look like the following figure:

device-logs   GSI: escalated-devices

Primary key		Attributes		
Partition key: EscalatedTo	**Sort key: Timestamp**			
Lorenzo	2024-06-23T01:30:08.063Z	DeviceID	Operator	Date
		d#999	Conor	2024-06-23

Figure 8.8 – Sparse index on the EscalatedTo attribute

Sparse indexes are efficient because they only project the data required to answer an access pattern, leading to lower throughput and storage costs compared to projecting the complete table data in a secondary index.

Leveraging KEYS_ONLY indexes

Not all applications backed by DynamoDB – or any database, for that matter – have intense latency requirements, such as single-digit millisecond averages for the database or end-to-end latencies under 20-25 milliseconds. Engineering is all about trade-offs, and some organizations might be willing to accept end-to-end latencies of 50-100 milliseconds if it comes with significant optimization in the total cost of ownership of the project.

One approach that some customers follow for specific large projects with lenient latency requirements is to use GSIs with the KEYS_ONLY projection type. As a reminder, the KEYS_ONLY projection type only projects the partition and sort key attributes of the GSI and the table. This makes those GSIs significantly smaller in size than the base table or a GSI with the ALL projection type. Additionally, KEYS_ONLY projections are less likely to see updates due to non-key attribute updates on the item via the base table, hence optimizing write throughput costs as well. The trade-off of using KEYS_ONLY GSIs is that every lookup on the GSI will also require an additional lookup on the base table to retrieve the complete corresponding item.

Such an approach has proven beneficial for use cases involving large amounts of data storage or certain write-heavy workloads that do not mind having average latencies beyond single-digit milliseconds. The overall database lookup latencies, even when using KEYS_ONLY indexes, have been observed to land around 25-30 milliseconds, which could be acceptable for many workloads given the cost optimization this approach offers. Considering a network time of 5-10 milliseconds, the average end-to-end latencies with such an approach can still be under 40-50 milliseconds.

Write sharding

When designing GSI or base table data models, it is crucial to assume that there may not be more than 1,000 puts, updates, or deletes against a single partition key. This is because, at the time of publishing, a single underlying partition in DynamoDB can only support 1,000 writes and 3,000 reads per second per partition key. Even if these per-partition limits increase in the future, handling high numbers of writes is a key consideration for large-scale applications. Some applications may need to index data in a way that exceeds these per-partition write limits.

For example, if an application needs to group profile data by biological gender, there will be two primary values with the most profiles: `male` and `female`. Gender is a low-cardinality attribute. Indexing profile data by gender using a DynamoDB GSI, a natural option would be to use `gender` as the GSI's partition key. Under normal circumstances, this would limit the application to 1,000 writes per second. Any writes beyond this limit would apply backpressure and be throttled by DynamoDB. To prevent such bottlenecks and throttling, a best practice is to artificially shard the skewed, low-cardinality partition key to support higher throughput at scale.

Sharding in DynamoDB involves adding multiple keys that belong to the same group. Sharding is typically done using random or calculated suffixes. Random suffix sharding involves adding N as a suffix to the original key, where N is a number between *0* and the total number of required shards. N is determined by evaluating the peak traffic expected on a single partition key. In the gender GSI example, if the application expects 100,000 writes per second at peak per-partition key, the key can be sharded to have suffixes between 0 and 99. This way, the keys become `Female_0`, `Female_1`, `Female_2`, ... `Female_99`. At write time, the application adds a random number between 0 and 99 as a suffix to the gender attribute value, distributing writes evenly across the shards, cumulatively supporting 100,000 writes per second per gender. When the application needs to read all profile data for a particular gender, it would need to be aware of the total number of shards (100) and make 100 individual `Query` API calls to fetch all the profile items.

Sharding keys with calculated suffixes can add intelligence to the read process compared to random suffix sharding, provided the use case fits. With calculated suffixes, you can use a key or non-key attribute to compute the shard an item must be written to. For example, if a use case requires retrieving user profiles from DynamoDB and partitioning data using the user's date of birth, as in sending marketing emails with promotions, a GSI can be used with the partition key as the date of birth in **ISO8601** format (`YYYYMMDD`). To accommodate popular birthdays in September (9) and handle general scale, the partition key of the date can be appended with a modulo of the hash of the `userId` value by N, where N is the number of shards needed to distribute the data based on expected peak traffic per-partition key.

With this design, if the application wants to access a particular `userId` along with their DOB, it can reach the exact partition key where the user item is located using the same GSI. If the user ID is known, the application could also use another GSI or table purely indexed on the `userId` attribute. However, with calculated suffixes, it is possible to spread traffic on the GSI that needs to group users by an attribute such as DOB while supporting higher scales than 1,000 writes, updates, or deletes per second.

Indexing data with keys of low cardinality

If write-sharding is not feasible for your use case of indexing data with keys of low cardinality, you must expect that the per-partition per-second write and read limits will apply to your partition keys, and DynamoDB may throttle requests exceeding these limits. As a reminder, at the time of publishing, these per-partition limits are 1,000 WCU per second and 3,000 RCU per second.

For use cases where the data needs to be indexed on an attribute such as a Boolean value (`True` or `False`), a status value (`Active`, `Inactive`, `Completed`, etc.), or any other categorical value attribute, without artificial sharding, the writes and reads to such indexes and individual item collections would be constrained by throughput limits.

Another alternative to artificial sharding for indexing low-cardinality attributes such as a Boolean is to use **sparse indexes**. Instead of having a status attribute, you can have the `isTrue` and `isFalse` attributes, where the corresponding attributes only exist if the value is non-empty. In the case of a Boolean type, the `isActive` attribute only has a `Yes` or `Y` string value if the Boolean value is `True`. With such an attribute design, a GSI can be indexed only on `isTrue` or `isFalse`, so the index only contains a subset of the overall data. This allows the application to query it quickly, without encountering per-partition per-second throughput limits.

Designing table replicas for skewed read access

For use cases where the read rate far outweighs the write rate, the per-partition per-second throughput limit of 3,000 RCU might be insufficient compared to the expected read traffic on the application. In such cases, the per-partition limits can be worked around by creating a GSI with the same key schema as the base table, effectively acting as a read replica of the table.

Every write to the base table with such a read replica GSI will lead to equivalent writes on the GSI as well. However, the overall read throughput per partition per second for the data will essentially double by utilizing both the base table and the GSI with the same key schema. It is important to remember that the read replica in such a case will be eventually consistent with the base table, so the application must account for this eventual consistency.

Using an LSI as a read replica is not advisable because LSIs share the throughput of the base table. Therefore, having an LSI with the same key schema as the base table will not increase read throughput for item collections beyond the per-partition limits.

Those were some of the important aspects associated with modeling data using secondary indexes. The AWS docs (10) also have some information on best practices associated with using secondary indexes, which is recommended reading.

> **Important note**
> Consider this as a reminder to delete any resources you may have created while going through this chapter in your AWS account.

Summary

In this chapter, we delved into the world of secondary indexes in DynamoDB, exploring their significance in expanding the accessibility and efficiency of database queries. We began by understanding the evolution of secondary indexes, tracing their development from customer feedback to the introduction of both local and global secondary indexes.

LSIs emerged as a solution for scenarios where additional access patterns were required within the confines of the base table's key schema. Their characteristics, including shared throughput and strong read-after-write consistency, made them valuable for specific use cases.

On the other hand, GSIs offered a more flexible alternative, allowing for independent throughput and storage management, catering to a diverse range of access patterns. Despite the presence of LSIs, GSIs became essential for supporting eventual consistency and accommodating complex querying requirements.

Throughout the discussion, we compared the strengths and considerations of LSIs and GSIs, highlighting their distinct roles in addressing different use cases. We also explored best practices for modeling data with secondary indexes, emphasizing strategies such as sparse indexing, appropriate projection types, and write sharding to optimize performance and cost efficiency.

As we conclude this chapter, we set the stage for the next, where we will dive into **capacity modes** and **table classes** in DynamoDB. Understanding these concepts will be pivotal in making informed decisions about resource allocation and scalability for DynamoDB tables.

References

1. Launch announcement: DynamoDB 2012: `https://aws.amazon.com/about-aws/whats-new/2012/01/18/aws-announces-dynamodb/`

2. Launch announcement: Local secondary indexes: `https://aws.amazon.com/about-aws/whats-new/2013/04/18/amazon-dynamodb-announces-local-secondary-indexes/`

3. Launch announcement: Global secondary indexes: `https://aws.amazon.com/about-aws/whats-new/2013/12/12/announcing-amazon-dynamodb-global-secondary-indexes/`

4. GitHub repository: `https://github.com/PacktPublishing/Amazon-DynamoDB---The-Definitive-Guide/tree/main/Chapter03`

5. Local Secondary Indexes: `https://docs.aws.amazon.com/amazondynamodb/latest/developerguide/LSI.html`

6. Using Global Secondary Indexes in DynamoDB: `https://docs.aws.amazon.com/amazondynamodb/latest/developerguide/GSI.html`

7. Phases of index creation: `https://docs.aws.amazon.com/amazondynamodb/latest/developerguide/GSI.OnlineOps.html#GSI.OnlineOps.Creating.Phases`

8. DynamoDB SLA: `https://aws.amazon.com/dynamodb/sla/`

9. Most common birthdays globally: `https://www.worldatlas.com/society/the-most-common-birthdays-in-the-world.html`

10. Best practices for using secondary indexes in DynamoDB: `https://docs.aws.amazon.com/amazondynamodb/latest/developerguide/bp-indexes.html`

11. GitHub repository: `https://github.com/PacktPublishing/Amazon-DynamoDB---The-Definitive-Guide`

Part 3:
Table Management
and Internal Architecture

This part covers key aspects of managing DynamoDB tables and understanding the underlying architecture that powers them. It begins with an exploration of different capacity modes and table classes, providing strategies for configuring DynamoDB to match specific workloads and budget requirements. Following this, the part offers an in-depth look at DynamoDB's architecture, explaining how request routers, storage nodes, and other core components work. Understanding DynamoDB-specific distributed systems concepts will provide context about DynamoDB's scalability, high availability, and durability characteristics. It will potentially also help you develop and troubleshoot DynamoDB-backed applications.

This part contains the following chapters:

- *Chapter 9, Capacity Modes and Table Classes*
- *Chapter 10, Request Routers, Storage Nodes, and Other Core Components*

9
Capacity Modes
and Table Classes

With DynamoDB, users only need to think about having a good design along with the read and write throughput depending on the capacity mode of the table. Capacity modes determine how users are billed for the use of their DynamoDB tables, and they also affect how the table can scale.

To understand capacity modes, let us consider an analogy. Imagine you are running a restaurant. You have a limited number of tables and seats to accommodate your customers, and you must decide how many servers and chefs to employ based on the expected number of customers. If you overestimate the number of customers and hire too many servers and chefs, you will be wasting money. On the other hand, if you underestimate the number of customers and have too few servers and chefs, you will be providing poor service and losing customers. Capacity modes in DynamoDB work in a similar way, and they help you to avoid the problems of over- or under-provisioning.

There are two capacity modes offered by DynamoDB tables: **provisioned capacity mode** and **on-demand capacity mode**. In provisioned capacity mode, you specify the amount of read and write capacity you require for your table, and DynamoDB reserves that capacity for you. This mode is ideal for workloads with predictable traffic patterns, where you know in advance how much read and write capacity you need. In on-demand capacity mode, DynamoDB automatically scales the table's read and write capacity based on the traffic to the table. This mode is ideal for workloads with unpredictable traffic patterns, where the number of requests to the table can vary greatly over time. You may decide the capacity mode of a table while creating it, or switch between the modes throughout the lifetime of a table. The following figure shows both modes:

Read/write capacity settings Info

Capacity mode

> ○ On-demand
> Simplify billing by paying for the actual reads and writes your application performs.

> ◉ Provisioned
> Manage and optimize the price by allocating read/write capacity in advance.

Figure 9.1 – AWS Management Console – capacity modes

DynamoDB also provides users with different **table classes** to cater to their specific needs. The two primary table classes offered by DynamoDB are **Standard** and **Standard-Infrequent Access** (**Standard-IA**). These table classes allow users to optimize the cost-efficiency of their DynamoDB workloads while maintaining optimal performance and minimizing operational overhead.

In this chapter, you will learn about the two capacity modes offered by DynamoDB tables and the features that enable you to manage your table's capacity. We will start with provisioned capacity mode and explore how to calculate and provision the right amount of capacity for your workload. We will also discuss the factors that affect the performance of your table and the best practices for optimizing your table's performance.

Next, we will dive into on-demand capacity mode and how it can automatically scale your table's capacity to handle traffic spikes. We will discuss the benefits and limitations of this mode, and how you can monitor and troubleshoot your table's performance in this mode.

We will then cover **AWS Application Auto Scaling**, a service that enables you to automatically adjust the capacity of your table based on the traffic to the table. You will learn how to configure auto scaling and set up alarms to notify you when the capacity of your table reaches certain thresholds.

After that, we will discuss **capacity reservations**, a feature that enables you to reserve a specific amount of capacity for your table, regardless of the traffic to the table. You will learn how to create and manage capacity reservations, and how they can help you to optimize the cost of your table.

Finally, we will discuss table classes, which determine the cost of your table based on the capacity mode and the type of data stored in the table. You will learn about the different types of table classes and how to choose the right table class for your workload.

By the end of this chapter, you will have a thorough understanding of the different capacity modes offered by DynamoDB tables and the features that enable you to manage your table's capacity effectively. You will also be equipped with the knowledge and best practices for optimizing the performance and cost of your DynamoDB tables.

In this chapter, we are going to cover the following main topics:

- Diving into capacity modes
- Using auto scaling for provisioned mode
- Cost optimizing with capacity reservations
- Choosing the right capacity mode
- Learning about table classes

Diving into capacity modes

The capacity mode of a DynamoDB table (and its secondary indexes) is a configuration that decides how throughput is or can be managed for a particular table, including how you are charged for that throughput as well as the amount of operational overhead that may be required. Capacity mode is one of the required configurations when creating a new DynamoDB table, although it can be modified throughout the lifetime of a table.

Regardless of the capacity mode chosen for a table, DynamoDB table data may be supported by one or more underlying partitions. At the time of publishing, each partition can physically support about 1,000 writes and 3,000 reads per second, or about 10 GB of data. These partitions may split as the throughput or storage requirements grow for a table and are completely managed by DynamoDB, with users having no visibility or control over the partitions directly. More about these partitions and how it all works under the hood is discussed in *Chapter 10, Request Routers, Storage Nodes, and Other Core Components*.

For now, let us dive deeper into the two capacity modes – provisioned and on-demand. We'll understand how each mode operates, and how DynamoDB does throughput management and scaling in both the capacity modes, including the control that a user may have in order to operate on their DynamoDB table throughput, given the capacity mode.

Provisioned capacity mode

Provisioned capacity mode is a mode in which DynamoDB provides you with control over two knobs to manage throughput. One allows you to specify the desired number of reads per second that the table (or global secondary indexes) should support, while the other enables you to define the desired number of writes per second.

You have the flexibility to determine the overall throughput you want the table to handle, and DynamoDB takes care of provisioning and managing the underlying infrastructure to meet those requirements. It is now your responsibility to utilize the provisioned throughput on the table, as charges are based on the provisioned throughput rather than the actual throughput consumed. Any requests exceeding the provisioned capacity on a table may be subject to throttling by DynamoDB, which could impact end user experience. It is important to configure appropriate levels of provisioned capacity on tables, which could also mean having some safety buffer over absolute expected throughput consumption levels to handle unanticipated usage.

To better understand DynamoDB's provisioned capacity mode, let us consider an analogy: hosting a conference.

Imagine you are organizing a conference and need to allocate resources to accommodate all attendees. In this analogy, each attendee represents a request made to DynamoDB, and the resources needed to handle these requests are the read and write capacity units.

In provisioned capacity mode, you have full control over the resources allocated, just like being the event planner. You determine the number of attendees (requests) you expect and pre-book a specific number of seats and staff (read and write capacity units) accordingly. These resources are reserved for your conference (DynamoDB table) regardless of whether they are fully utilized or not.

If your conference attracts a consistent number of attendees, provisioned capacity mode ensures that you always have enough seats and staff to handle the expected demand. It provides predictable performance and low latency since the resources are dedicated solely to your event. However, if the number of attendees fluctuates significantly or is difficult to predict, you might end up with either underutilized resources or, worse, not enough seats and staff to accommodate all requests.

To address scalability, DynamoDB's provisioned capacity mode allows you to adjust the allocated resources based on the anticipated demand. You can scale the number of seats and staff up or down as needed to ensure a smooth conference experience.

Hopefully, the analogy helped you understand the provisioned capacity mode better. Next, let us review how the reads and writes are characterized.

Units of reads and writes

In provisioned capacity mode, DynamoDB measures reads and writes in terms of **Read Capacity Units (RCUs)** and **Write Capacity Units (WCUs)**, respectively.

A single RCU represents a **strongly consistent** read operation for data up to 4 KB in size. Alternatively, it can represent an **eventual consistent** read operation for data up to 8 KB in size. To clarify, if you perform a strongly consistent read on a single item that is 3.5 KB in size, it will consume 1 RCU. On the other hand, if you perform a strongly consistent read on a single item that is 4.5 KB in size, it will consume 2 RCUs. For deeper insights into the read consistency options mentioned, review *Chapter 6, Read Consistency, Operations, and Transactions*.

In terms of writes, a WCU corresponds to a write operation for an item up to 1 KB in size. If you make a write to DynamoDB for an item that is 200 bytes in size, it will be charged as 1 WCU. However, if the item size is 1.1 KB, it will be charged as 2 WCUs.

Based on this, at the time of writing, every physical partition that may support a DynamoDB table (or any of its global secondary indexes) could support a maximum of 1,000 WCUs and 3,000 RCUs, or about 10 GB of data storage.

If a table was configured with 10 WCUs and 10 RCUs, that means the table is able to support 10 writes per second for item sizes up to 1 KB and 10 reads per second for item sizes up to 8 KB each in the case of eventual consistent read requests and 4 KB each for strongly consistent read requests.

It is quite common for applications to experience occasional traffic surges resulting from specific patterns of end user behavior. These surges may occur during short-lived time periods throughout the day or week, or they may be triggered by unexpected third-party events or marketing campaigns. In such situations, the provisioned capacity allocated to the DynamoDB table supporting the application may prove inadequate. To address this challenge, DynamoDB offers a range of features designed to provide support in such scenarios. One such feature is burst capacity, which we will explore in detail shortly.

Burst capacity

Burst capacity is a feature of provisioned mode tables that empowers applications to gracefully handle unexpected surges in traffic and workload, going beyond the provisioned capacity configured at the time. It serves as a safety net, ensuring that your DynamoDB table can seamlessly accommodate sudden spikes in read and write requests without experiencing throttling issues.

Burst capacity operates by continuously monitoring the unused provisioned capacity of a table and enabling the table to surpass its allocated capacity by utilizing previously untapped resources. DynamoDB keeps track of the table's unutilized provisioned capacity from the last 5 minutes or 300 seconds, reserving it as additional capacity that can be utilized if the consumed throughput for reads or writes exceeds the provisioned limits. It is important to note that this unutilized capacity is monitored on a rolling basis for only 5 minutes, which means that a table may experience temporary bursts beyond its provisioned capacity but may encounter request throttling if the overconsumption persists for extended periods.

For example, consider the following figure that shows the provisioned and consumed write capacity of a DynamoDB table.

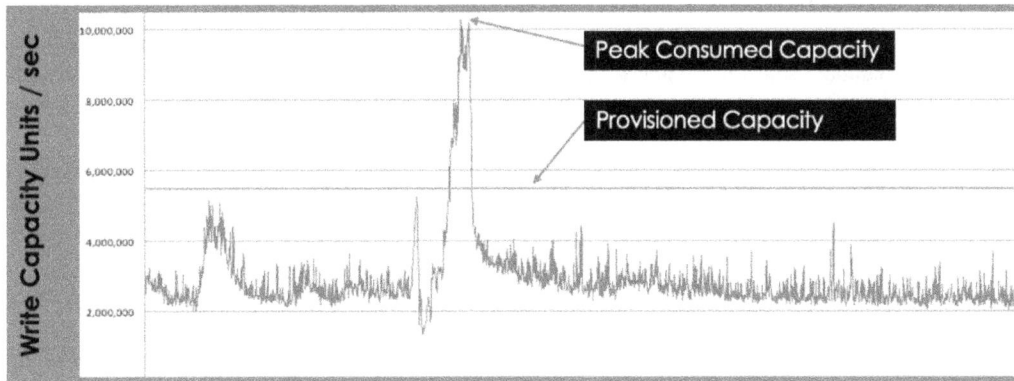

Figure 9.2 – Burst capacity

The table had a provisioned capacity of approximately 5.5 million WCUs based on the user's forecasts. Analyzing the graph presented, it is evident that the consumed WCUs remained well below the provisioned capacity for the majority of the duration. However, there was a brief period lasting just under a minute during which the application utilizing this table experienced an unexpected surge in traffic. Thanks to the burst capacity feature, the DynamoDB table could gracefully accommodate this sudden surge without any write request throttling, even though the consumed capacity reached approximately 10 million WCUs. This behavior contrasts with many other database systems that would typically experience performance degradation when the consumed throughput exceeds the preconfigured limits. DynamoDB, on the other hand, effectively handles these unexpected traffic surges through its burst capacity feature in provisioned mode tables.

It is crucial to emphasize that if the consumed capacity had exceeded the provisioned capacity for a longer duration than in the example we just mentioned, DynamoDB would have eventually exhausted its burst capacity. As a result, the table would have reached its limit and started enforcing throttling measures to restrict the excessive requests back to the provisioned capacity. While burst capacity provides temporary flexibility to handle unexpected spikes, it is not designed to sustain prolonged overconsumption beyond the provisioned limits.

Next, let's learn about another powerful feature of provisioned mode tables that helps imbalanced traffic across table partitions, called adaptive capacity.

Adaptive capacity

DynamoDB's **adaptive capacity** is a powerful feature that helps handle imbalanced traffic by automatically distributing throughput across partitions based on the workload. This ensures that heavily accessed partitions receive more capacity to handle the increased demand, while less utilized partitions receive less capacity, preventing hotspots and enabling efficient resource utilization. By dynamically adjusting the allocation of read and write capacity, adaptive capacity ensures that the system can effectively handle traffic spikes and maintain consistent performance even during periods of imbalanced workload. This capability allows DynamoDB to seamlessly adapt to changing traffic patterns and provide a reliable and responsive database solution for applications with varying workload demands.

Consider an example with the following figure. The table has four partitions where each partition sees a baseline of 50-WCU traffic. Ideally, if a table had 400-WCU provisioned capacity, 400 WCUs would be equally distributed across the four partitions to support traffic on each of the partitions equally. In real life, the traffic is almost never equally distributed across its partition keys and, hence, physical partitions.

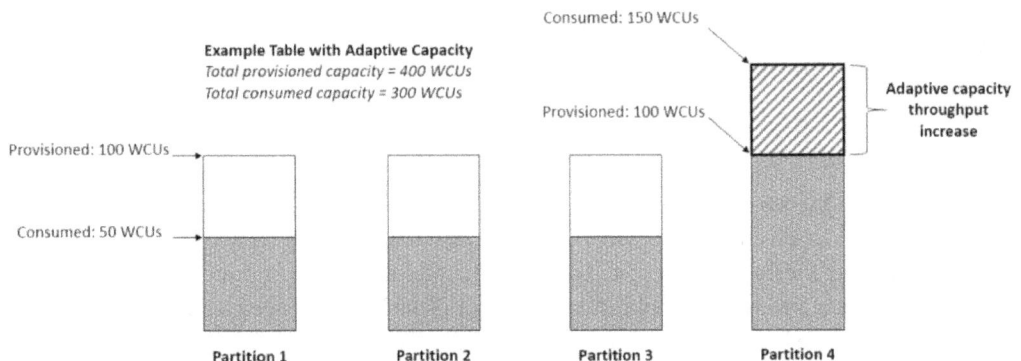

Figure 9.3 – DynamoDB – adaptive capacity

If there is a single partition out of the four with uneven traffic thrown at it, DynamoDB dynamically allows more throughput for the partition. It is important to remember that adaptive capacity only allows for imbalanced traffic across partitions until their individual partition limits are hit, which are 1,000 WCUs and 3,000 RCUs beyond which the excess requests would be throttled by DynamoDB, even when the overall table's provisioned capacity is underutilized.

Next, let us learn about how provisioned mode tables are metered and billed by DynamoDB.

Metering

In provisioned capacity mode, tables in DynamoDB are metered and billed hourly. Currently, at an arbitrary point within each hour, DynamoDB records the provisioned capacity of a table and charges for the entire hour based on that recorded provisioned capacity. This means that if a table was provisioned with 100 WCUs for most of the hour but briefly increased to 10,000 WCUs for a few seconds within that hour, DynamoDB may charge for the entire hour based on either 100 WCUs or 10,000 WCUs. To estimate costs and make projections, it is prudent to assume that the table will be metered at its highest provisioned capacity during the hour. This approach allows for worst-case cost estimation, providing a more accurate assessment of the potential charges.

For more information on provisioned capacity mode and its management, refer to the AWS docs (1). Next, let us learn about the other throughput mode: on-demand.

On-demand capacity mode

On-demand capacity mode is specifically designed to eliminate the need for capacity planning and forecasting in DynamoDB. With this mode, you can simply create a table and begin performing reads and writes without any prior configuration. The table automatically scales to accommodate your evolving requirements as they arise. In essence, you relinquish direct control over the specific knobs dictating the number of reads and writes supported by the table, allowing DynamoDB to handle the scaling dynamically. You have the flexibility to set a table's capacity mode as on-demand during its creation or modify the capacity mode to on-demand at any point during the table's lifetime. This mode provides a simplified approach, enabling you to focus on driving your application's functionality without the need for manual capacity management.

In on-demand capacity mode, you can scale-down to zero in a true serverless manner. This means that you are only billed based on the actual number of read and write requests made to the table. If there are periods when no reads or writes occur on the table, you will not be charged for any throughput during that time. This flexibility allows your table to remain inactive without incurring any additional costs for throughput. However, it is important to note that you may still be charged for the storage of data in the table, which is calculated on a gigabytes per month basis. This ensures that you only pay for the storage space used by your data within the table while offering the freedom to scale your throughput capacity dynamically and cost-effectively.

On-demand capacity mode tables share many of the scaling characteristics with provisioned capacity mode tables. Both modes utilize the same underlying infrastructure and store data in partitions using the same methodology. They also exhibit similar performance characteristics. The key distinctions lie in the scaling behavior and billing structure. On-demand capacity mode tables automatically scale based on the workload, dynamically adjusting to accommodate changing read and write demands. In contrast, provisioned capacity mode tables require manual provisioning and have predefined throughput limits. Furthermore, the billing for on-demand capacity mode is based on the actual number of read and write requests made, while the provisioned capacity mode is billed based on the provisioned throughput capacity. Despite these differences, both modes operate within the same foundational infrastructure and storage mechanisms, ensuring consistent performance and data management capabilities.

Continuing with the conference analogy, let us explore DynamoDB's on-demand capacity mode.

In on-demand capacity mode, imagine hosting a conference where you do not need to pre-book any seats or staff. Instead, the venue automatically adjusts the resources based on the number of attendees as they arrive. This flexible arrangement ensures that every attendee gets a seat, and the necessary staff are available to handle their requests.

Similarly, in DynamoDB's on-demand capacity mode, you do not need to provision any specific read and write capacity units. The system automatically adjusts the resources based on the incoming workload. Whether you have a sudden surge in attendees or a lull in activity, DynamoDB scales the capacity up or down to match the demand in real time.

This mode is particularly useful when the workload is unpredictable, as you do not have to spend time estimating and managing capacity. It offers high availability and responsiveness, ensuring that your requests are promptly processed regardless of workload fluctuations. Additionally, you pay only for the actual usage, allowing for cost optimization.

On-demand capacity mode is most suited for workloads with unpredictable usage patterns, highly variable request rates, and scale-to-zero requirements. Next, let us learn about how on-demand tables characterize reads and writes.

Units of reads and writes

Units for reads and writes for on-demand capacity mode tables are quite similar to those of provisioned mode tables; however, there are slight changes in the nomenclature.

In on-demand capacity mode, DynamoDB measures reads and writes in terms of **Read Request Units (RRUs)** and **Write Request Units (WRUs)**, respectively.

A single RRU represents a **strongly consistent read operation** for data up to 4 KB in size. Alternatively, it can represent an **eventual consistent read operation** for data up to 8 KB in size. To clarify, if you perform a strongly consistent read on a single item that is 3.5 KB in size, it will consume 1 RRU. On the other hand, if you perform a strongly consistent read on a single item that is 4.5 KB in size, it will consume 2 RRUs.

In terms of writes, a WRU corresponds to a write operation for an item up to 1 KB in size. If you make a write to DynamoDB for an item that is 200 bytes in size, it will be charged as 1 WRU. However, if the item size is 1.1 KB, it will be charged as 2 WRUs.

Similar to provisioned mode tables, if an on-demand mode table was configured with 10 WRUs and 10 RRUs, that means the table is able to support 10 writes per second for item sizes up to 1 KB and 10 reads per second for item sizes up to 8 KB each in the case of eventual consistent read requests and 4 KB each for strongly consistent read requests.

Fortunately, both provisioned mode tables and on-demand mode tables offer the same Amazon CloudWatch metric to track consumed throughput. The metric for tracking consumed reads is named `ConsumedReadCapacityUnits`, while the metric for writes is labeled `ConsumedWriteCapacityUnits`. This consistent naming convention simplifies monitoring and allows you to easily monitor and analyze the consumed throughput for both types of tables. Whether you choose provisioned mode or on-demand mode, you can rely on these CloudWatch metrics to gain insights into the actual read and write capacity being utilized by your DynamoDB tables.

Now that we have gained an understanding of read and write units, let's explore how on-demand mode tables automatically scale to accommodate increasing traffic demands while efficiently supporting scaling down to zero.

Scaling characteristics of on-demand tables

While on-demand capacity mode may give the impression that you can immediately start performing millions of reads and writes on a newly created table, that is not the case. On-demand tables follow a specific scaling pattern where they allow for doubling the previous peak throughput every few minutes. The previous peak refers to the maximum throughput in reads and writes that the table has experienced at any point in its lifetime.

When you create a new on-demand table, the initial previous peak is set to 4,000 WRUs and 12,000 RRUs per second. This establishes a baseline for the table's capacity. For more information on the reason behind these numbers and how DynamoDB works under the hood, read *Chapter 10, Request Routers, Storage Nodes, and Other Core Components*.

If you switch a provisioned mode table to on-demand, the on-demand capacity will support either the maximum provisioned capacity the table has ever had or a minimum of 4,000 WRUs and 12,000 RRUs, whichever is higher.

Tables in on-demand capacity mode can double their previous peak throughput every 5–30 minutes when the table's consumed throughput approaches or exceeds the previous peak levels. This triggers a scaling process where a new peak is established, doubling the capacity of the table. This scaling behavior is illustrated in the following figure:

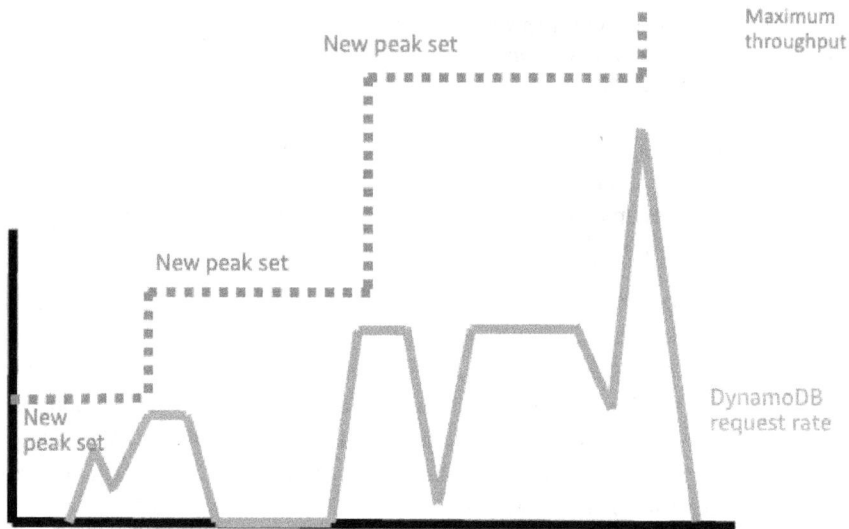

Figure 9.4 – On-demand mode tables – scaling behavior

As shown in the preceding figure, new peaks are established as the request traffic approaches the previous peak levels. It's important to note that the previous peak watermark is retained, irrespective of any periods of underutilization or complete lack of utilization. This means that even during times of low or no activity, the system retains the knowledge of the highest throughput previously observed. This ensures that the scaling mechanism can effectively respond to future increases in workload, providing the necessary capacity to handle peak traffic efficiently.

Metering

On-demand mode tables are the simplest to meter and bill for. You pay per number of requests throughout the month for on-demand tables. It does not matter how the requests are distributed across the month, the cumulative count of requests is solely metered and charged per the DynamoDB pricing rates (2).

With that, you should have a pretty good understanding of both of the capacity modes. Next, let us learn about an advanced feature that is supported by both capacity modes, and can help workloads with skewed data access.

Feature common to both capacity modes – isolating frequently accessed items

One important feature of DynamoDB tables, applicable to both capacity modes, is the capability to isolate frequently accessed items. This reactive feature aids in managing disproportionately high traffic that certain items may experience, resulting in imbalanced throughput for reads or writes. By isolating these heavily accessed items, more commonly called **hot keys**, DynamoDB ensures that the overall performance of the table is not impacted by the traffic imbalance, providing consistent and reliable throughput across the entire dataset. This feature allows DynamoDB to effectively handle scenarios where specific items receive a significantly higher workload, maintaining optimal performance and preventing bottlenecks.

The following figure illustrates the functionality of DynamoDB's workload isolation feature. Initially, a physical DynamoDB partition contains multiple items. However, two specific items on the partition experience significantly higher traffic compared to the other items, potentially leading to request throttling for the entire partition. Recognizing this imbalanced traffic pattern, DynamoDB isolates these hot keys by splitting the partition where they reside into two. Over time, as the high traffic persists, DynamoDB may continue isolating these items, resulting in a scenario where a single item resides on a physical partition capable of handling up to 1,000 writes (new inserts, updates, or deletes) and 3,000 reads:

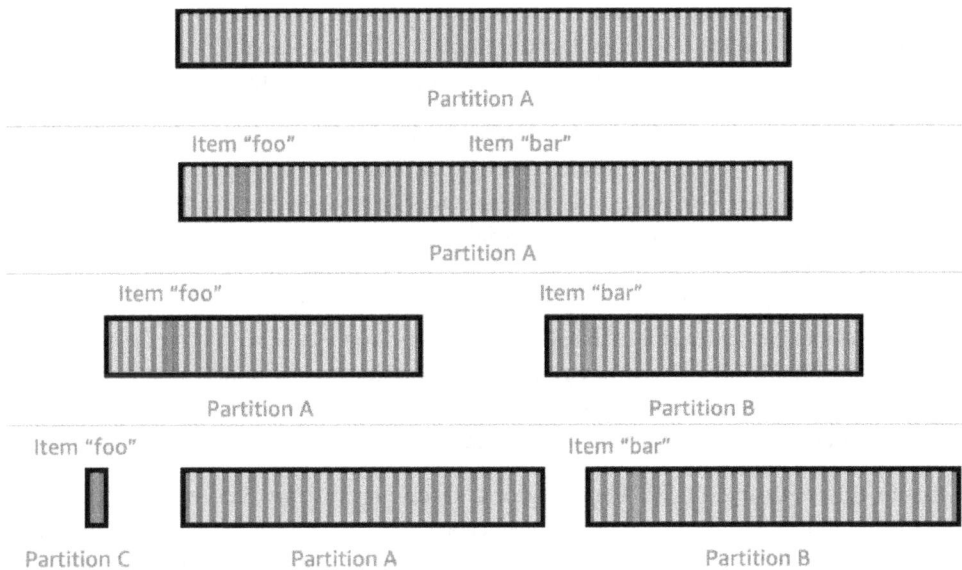

Figure 9.5 – Isolating frequently accessed items

This isolation process ensures that the heavily accessed items can receive the necessary throughput without affecting the performance of other items or the overall partition. For more information on this, read the AWS docs (3).

> **Important note**
>
> Isolating frequently accessed keys, or hot keys, is a reactive measure rather than a proactive one. Therefore, it should not be relied upon during the design phase. This approach is most effective in scenarios where a single item resides in its own partition. In such cases, the item may handle no more than 1,000 writes and 3,000 reads per second.

Now that we have explored both capacity modes, including their metering aspects and distinctive features, let us delve into auto scaling for provisioned mode tables. DynamoDB auto scaling enables the automation of adjusting the provisioned capacity of a table based on the observed traffic patterns.

Using auto scaling for provisioned mode

By utilizing auto scaling, the process of increasing or decreasing the provisioned capacity becomes automated, ensuring that the table can dynamically scale up or down to handle the varying workload. This feature eliminates the need for manual intervention, allowing the table to efficiently adapt to changes in traffic and optimize resource allocation. Auto scaling is a valuable tool in maintaining optimal performance and cost-efficiency for provisioned mode tables in DynamoDB.

Auto scaling for provisioned mode tables operates by continuously monitoring the consumed throughput of a table and comparing it to the provisioned throughput. This process triggers automatic scaling actions, eliminating the need for manual intervention in managing throughput. The key components involved in the functioning of auto scaling, alongside DynamoDB, include CloudWatch metrics and alarms. AWS Application Auto Scaling, an AWS service, handles the configuration of alarms that monitor the table's metrics. When triggered, these alarms make update calls to DynamoDB to modify the provisioned capacities of the table. Auto scaling operates independently for both reads and writes, encompassing the table itself and any global secondary indexes with auto scaling enabled. This flexibility allows for customized auto scaling configurations for reads and writes, as well as individual configurations for each global secondary index associated with the table.

When enabling auto scaling on a table or its global secondary indexes, you have the option to configure the minimum and maximum provisioned capacities that auto scaling can set for the table. These configurations allow cost optimization by limiting capacity beyond an anticipated limit and preventing under-provisioning. Alongside these values, a target utilization percentage is another important configuration. This setting determines the threshold at which the ratio between consumed and provisioned throughput triggers a scale-up or scale-down action. By defining the target utilization percentage, you can fine-tune the auto scaling behavior to align with your specific performance requirements and cost considerations.

The following figure shows the AWS console that allows for setting the auto scaling configuration values while creating a DynamoDB table:

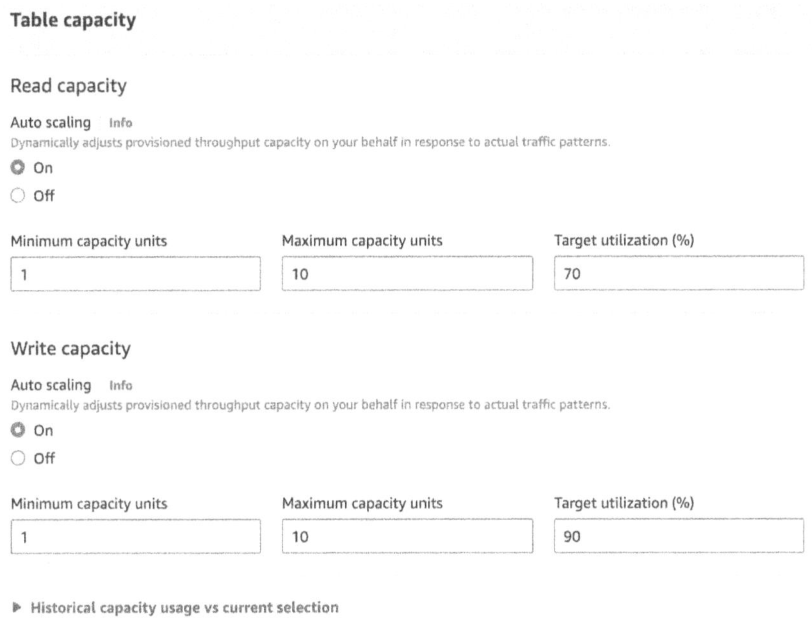

Table capacity

Read capacity

Auto scaling   Info
Dynamically adjusts provisioned throughput capacity on your behalf in response to actual traffic patterns.

⦿ On
◯ Off

Minimum capacity units	Maximum capacity units	Target utilization (%)
1	10	70

Write capacity

Auto scaling   Info
Dynamically adjusts provisioned throughput capacity on your behalf in response to actual traffic patterns.

⦿ On
◯ Off

Minimum capacity units	Maximum capacity units	Target utilization (%)
1	10	90

▶ Historical capacity usage vs current selection

Figure 9.6 – AWS Management Console – Auto scaling settings

Based on the depicted figure, when the consumed capacity of a table exceeds 70% of its provisioned capacity during a specific period, the auto scaling mechanism triggers a scale-up action. Conversely, if the consumed-to-provisioned ratio falls below the target threshold of 70%, as indicated in the figure, a scale-down action is initiated, given that the provisioned capacities remain within the defined minimum and maximum capacity unit values. Target utilization can be set to a lower value to increase the safety buffer between consumed and provisioned capacity depending on how quickly a workload's traffic is expected to see ramp-ups and -downs.

A good starting point for the target utilization percentage is 70%, which allows a buffer of 30% compared to the consumed capacity. However, it is recommended to fine-tune this value by observing the CloudWatch metrics for the consumed capacity on a table-by-table basis. This allows for better optimization and alignment with the specific workload requirements of each table.

Going into further detail, auto scaling closely monitors a table's consumed capacity units CloudWatch metric for a rolling 2-minute period (individually for reads and writes) to assess the need for a scale-up action. Conversely, a scale-down action is evaluated based on a rolling 15-minute period of the table's capacity units.

It is important to note that the propagation of CloudWatch metrics into metrics and the triggering of CloudWatch alarms can introduce an additional delay of approximately 3–4 minutes. In essence, the end-to-end process for a scale-up action, from the moment the table experiences traffic surpassing the target utilization to the triggering of the scale-up action on the DynamoDB table, may take around 4–6 minutes. This time frame accounts for factors such as the propagation of CloudWatch metrics, alarm triggering based on auto scaling settings, and the actual execution of the scale-up action.

The following figure demonstrates the functioning of auto scaling for a table's read and write capacities in provisioned mode. By dynamically adjusting the provisioned capacity based on actual workload, auto scaling optimizes costs compared to continuously provisioning for peak capacity, which would result in cost inefficiency:

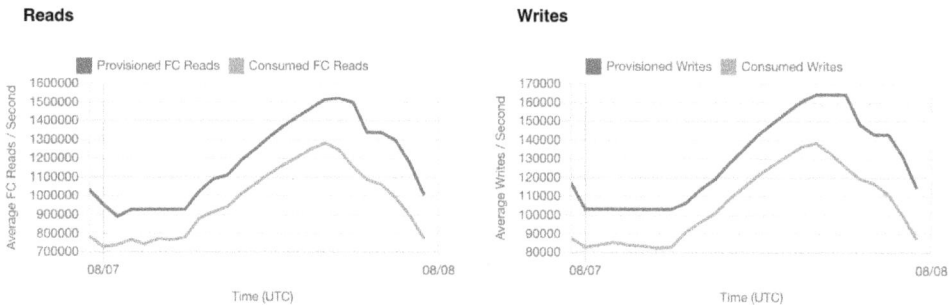

Figure 9.7 – Auto scaling in action for provisioned mode tables

Now that we have explored how auto scaling enables dynamic adjustment of a table's provisioned throughput for cost-efficiency, let us discuss another cost-saving option available for provisioned mode tables: upfront reserved capacity.

Cost optimizing with capacity reservations

DynamoDB reserved capacity is a cost-saving option that allows users to commit to a predefined amount of provisioned throughput for a fixed duration, typically one or three years. By reserving capacity upfront, users can benefit from significant cost savings compared to on-demand or public pricing of provisioned capacity modes.

Reserved capacity can be considered primarily as a billing feature, as its main impact is on the billing of an AWS account's regional provisioned capacity for a given month. It does not affect a provisioned mode table's performance, limits, or scaling behavior.

Reserved capacity is a valuable choice for minimizing DynamoDB costs in workloads characterized by consistent usage and predictable growth. Similar to the pricing of provisioned capacity, reserved capacity incurs charges regardless of utilization. This means that careful consideration should be given before purchasing long-term reserved capacity, ensuring that you anticipate using the majority or all the allocated capacity throughout the chosen term, whether it's a 1-year or 3-year duration.

Savings calculations

As per the Amazon DynamoDB pricing page (4), "By purchasing capacity up front, you can save **up to 54% (one-year term) or up to 77% (three-year term)** over the regular hourly rates."

Reserved capacity is charged in two phases:

- A one-time upfront fee
- An hourly fee for each hour during the term of purchase (1-year or 3-year)

For example, consider a 100-WCU reserved capacity purchased in the N.Virginia region for a 1-year term. The following calculations are based on the pricing page (4) for a reserved capacity of 100 WCUs:

- One-time upfront fee: $150
- Hourly fee (per 100 WCU): $0.0128
- Hours in a month: 730
- Months in a year: 12
- Total cost for 100 WCUs reserved for a year: 150 + (0.0128 x 730 x 12) = $262.128

This is the calculation if you were to use 100 WCUs of provisioned capacity for the same 1-year term without capacity reservations:

- One-time upfront fee: $0
- Hourly fee (per WCU): $0.00065
- Hourly fee (per 100 WCU): 0.00065 x 100 = $0.065
- Hours in a month: 730
- Months in a year: 12
- Total cost for 100 WCUs for a year: 0 + (0.065 x 730 x 12) = $569.4

Comparing the amortized costs for 100 WCUs for a year, reserved capacity is about 54% (*100 - (100 x (262.128/569.4)))* more cost-effective than non-reserved provisioned capacity.

Reserved capacity metering

Reserved capacity is acquired for a specific AWS account's region and can be utilized by all tables within that region. The capacity purchased is measured on an hourly basis, meaning that if, for example, 100 WCUs of reserved capacity are purchased, it equates to 100 WCUs per second for every hour. It is important to note that any additional capacity provisioned on provisioned mode tables in the region will be subject to the public pricing rates (4), separate from the reserved capacity. This implies that you are not limited to provisioning only 100 WCUs across all provisioned mode tables, but rather, you are billed as per reserved capacity pricing for the initial 100 WCUs that are provisioned, and any additional provisioned capacity on the tables post the 100 WCUs would be charged as per standard public pricing rates for provisioned capacity.

If your organization follows a multi-account strategy in AWS, where each workload has its own AWS account and all accounts are linked to a single payer account, the reserved capacity purchased by any account is shared among all the linked accounts. In this case, it is advisable to acquire reserved capacity from the payer's account. By doing so, any provisioned mode tables in the payer's account will utilize the purchased capacity first. Any remaining reserved capacity from the payer's account can then be utilized by provisioned mode tables in any of the linked accounts. This approach ensures optimal utilization of reserved capacity across the organization's accounts and helps maximize cost savings.

Tools to help identify the capacity to reserve

A general guideline for determining the amount of capacity to reserve for provisioned mode DynamoDB tables is to ensure that you utilize all or a significant portion of the purchased capacity in each hour throughout the term of purchase (1-year or 3-year).

To determine the amount of reserved capacity that can be purchased, a simple approach is to calculate the sum of the auto scaling minimums for all provisioned tables in the region. In the case of linked accounts, this sum should include all tables across all linked accounts in the region. Since provisioned mode tables typically have auto scaling enabled to accommodate fluctuations in throughput, setting appropriate auto scaling minimums and maximums helps manage capacity dynamically. By adding up these auto scaling minimums, you can identify the minimum amount of reserved capacity that would provide some level of cost-efficiency.

Another approach to determine the reserved capacity to purchase is by examining the CloudWatch metrics for provisioned read and write units across all provisioned mode tables. By analyzing these metrics, you can identify the minimum baseline of provisioned throughput that, if purchased, would be fully utilized. This approach allows you to gauge the actual usage patterns of your tables and make an informed decision on the appropriate amount of reserved capacity to purchase.

A more technical approach to determine the reserved capacity to purchase is by analyzing the **AWS Cost and Usage Reports** (**AWS CUR**) data for provisioned mode DynamoDB tables in the region. The CUR data is a feature provided by AWS Billing, which offers detailed information about the consumption of individual resources within AWS. To leverage this data, you can set up automation (5) to deliver the CUR data to your own S3 bucket. Subsequently, you can analyze this data using Athena queries or dedicated tools to identify the optimal amount of reserved capacity for maximum cost-efficiency.

AWS provides a blog (6) with detailed instructions on how to use Athena queries to identify the appropriate reserved capacity that should be purchased. Additionally, AWS has open-sourced a Python-based tool (7) that can analyze the CUR data, providing comprehensive recommendations on the reserved capacity that should be purchased and the potential cost-efficiency that can be achieved by following those recommendations.

It must be noted that reserved capacity is only available as a cost-efficiency measure for provisioned capacity mode tables. There is no such cost-efficiency support for on-demand capacity mode. Next, let us learn how to choose the right capacity mode for your workload.

Choosing the right capacity mode

Choosing the right capacity mode involves evaluating your specific requirements and considering factors such as expected traffic patterns, workload predictability, and cost considerations. By carefully assessing these factors, you can make an informed decision to determine whether provisioned capacity mode or on-demand capacity mode aligns better with your workload's needs.

Use provisioned mode when the workload has the following characteristics:

- Is steady and more or less predictable (could be cyclical)
- Has gradual ramp-ups and -downs
- Is driven by known and controllable events
- Can be monitored on an ongoing basis (for forecasting)
- Sees limited short-term bursts of traffic

Use on-demand mode when the workload shows the following behavior:

- Is highly unpredictable
- Has frequently idle periods of no traffic
- Is driven by unknown and uncontrollable events
- Need to "set it and forget it"
- Would like to pay per request count

As a rule of thumb, if you can consistently utilize at least 14.4% of your table's provisioned capacity in provisioned mode, it would be the most cost-efficient option for your table. This number is obtained by comparing the pricing of provisioned and on-demand modes. In cases where your average utilization can remain above this level, the traffic patterns align with the provisioned mode's characteristics, ensuring that a considerable portion of the provisioned capacity is always used. However, there are situations where the traffic on a table is unpredictable, making it challenging to effectively utilize provisioned mode with auto scaling. In such scenarios, on-demand capacity mode would be more suitable and cost-efficient for those tables. It provides the flexibility to scale dynamically without the need for upfront provisioning and can be beneficial for workloads with varying or unpredictable traffic patterns.

If we disregard the considerations of buffer capacity and traffic characteristics, and solely compare the costs between provisioned and on-demand modes for the same traffic over a period of time, provisioned mode with auto scaling is approximately seven times more cost-efficient in terms of throughput compared to on-demand mode. It is important to note that this cost-efficiency is based on the assumption of optimal utilization and effective scaling in provisioned mode. However, it is crucial to consider other factors such as workload patterns, traffic variability, and the ability to predict and manage capacity requirements when choosing the appropriate capacity mode for your DynamoDB tables.

See the AWS docs (8) for more information on evaluating the right capacity mode for your DynamoDB table. Next, let us learn about the different table classes supported by DynamoDB, how one differentiates from the other, and how to choose the appropriate table class for your DynamoDB table.

Learning about table classes

Tables within DynamoDB can have different capacity modes, and they can also be assigned different storage classes. There are two table classes available in DynamoDB: **Standard** and **Standard-Infrequent Access (Standard-IA)**.

The table class in DynamoDB primarily pertains to the billing aspect, as there are no differences in terms of performance, integration, or availability between the two classes. Both the Standard and Standard-IA table classes offer similar features, including single-digit millisecond average response times for data access. They utilize the same underlying infrastructure and exhibit the same behavior when it comes to scaling throughput or accommodating data size. Both table classes support seamless integration with other AWS services. The main distinguishing factor lies in the cost structure associated with throughput and data storage.

When comparing the costs of throughput (read and write) and storage between the Standard and Standard-IA table classes, it is observed that the storage costs for Standard-IA are 60% less than those of the Standard class. However, the throughput costs for Standard-IA are 25% higher than those of the Standard table class. This implies that the Standard table class is well-suited for workloads where the cost of throughput is proportional to the storage costs, and the table experiences many read and write requests relative to the amount of data stored. On the other hand, the Standard-IA class is ideal for workloads where a significant amount of data is stored, but only a small portion of the data is accessed at any given time.

In a banking system, historical transactions provide another example where the Standard-IA table class can be applicable. Storing all transaction data is crucial for a banking service provider, ensuring that users can access their past transactions promptly. While storing historical transactions on a colder storage tier may be an option, it can result in longer retrieval times, impacting user experience.

By utilizing the Standard-IA table class, the banking system can efficiently store and manage large volumes of historical transaction data. The lower storage costs offered by Standard-IA make it cost-effective for storing infrequently accessed transaction records, while still providing fast access when needed. This ensures a seamless user experience and allows the banking system to balance cost optimization with data availability.

Table class considerations

You can change the table class for a table up to 2 times within a 30-day trailing period. Switching table classes has no impact on the table's performance during or after the switch. It is safe to switch production tables without experiencing any degradation in performance. The duration of the switch can vary from a few moments to several minutes, depending on the table's scale. However, there is no performance impact during the switching process.

One important aspect to consider when using the Standard-IA table class is that reserved capacity purchases do not apply to provisioned mode tables with the Standard-IA class. If you have already purchased reserved capacity for 1-year or 3-year terms, it is advisable to let those terms complete before converting tables to Standard-IA for cost-efficiency purposes. If you have other provisioned mode tables in the Standard table class that could utilize the reserved capacity currently allocated to the table you wish to switch table classes for, then it may be a cost-effective decision to proceed with the class switch. In such cases, switching the table class can help optimize the rebalance of reserved capacity allocation and ensure efficient utilization of the reserved capacity overall.

Both table classes are supported by both capacity modes and global tables. A table's storage class is applied to all its global secondary indexes.

See the AWS docs (9) for more information on table class considerations. Next, let us learn how to choose the right table class for your DynamoDB table.

Choosing the right table class for your workload

When building a new application on DynamoDB or migrating an existing application to a DynamoDB-based solution, it is recommended to start with the Standard table class. Initially, the storage costs for the new DynamoDB table are usually minimal compared to the write operations (and possibly read operations) performed on the table. Once the migration or data bootstrapping is complete, you can evaluate the storage and throughput costs and determine whether switching the table's class to Standard-IA is appropriate.

When comparing the throughput and storage costs of a table in the Standard class, if storage is the main cost driver and exceeds 42% of the table's total costs, then the table would be most cost-effective in the Standard-IA table class. On the other hand, if the storage costs are less than or equal to 42% of the throughput costs, then the Standard table class would be the more cost-efficient option. By considering the proportion of storage costs to throughput costs, you can determine the most suitable table class for optimizing costs based on the specific characteristics and usage patterns of your table.

There are multiple methods and tools available to compare the storage and throughput costs of a Standard class table. One straightforward approach is to utilize table tags. By assigning a unique tag to each DynamoDB table representing that specific table, you can leverage the cost allocation tags feature (*10*) provided by AWS Billing. This feature allows you to analyze and compare the storage and throughput costs of individual tables. Additionally, AWS has released an open source tool (*11*) that automatically assigns tags representing the table name to DynamoDB tables, eliminating the need for manual tagging efforts.

Another effective approach for comparing storage and throughput costs of a Standard class table is by utilizing the CUR data. By grouping the costs based on resources, such as tables, you can gain insights into the cost breakdown for each DynamoDB table. Leveraging tools such as Athena, you can run queries on the CUR data to analyze the historical storage and throughput costs of your DynamoDB tables. This enables you to identify potential candidates for the Standard-IA table class based on the cost patterns and make informed decisions regarding table class optimization.

Those were some of the important aspects associated with table classes for DynamoDB. The AWS docs (*12*) also have some information on evaluating the right table class for your workloads, which is recommended reading.

> **Important note**
> Consider this as a reminder to delete any resources you may have created while going through this chapter in your AWS account.

Summary

In this chapter, we learned about the different capacity modes supported by DynamoDB and gained a deep understanding of their individual characteristics. We discovered the distinctions between provisioned capacity and on-demand capacity and when each mode was most suitable for specific workloads. The concept of auto scaling in provisioned mode was explained, allowing DynamoDB tables to automatically adjust their capacity based on your fluctuating traffic patterns.

We also explored the different table classes available for DynamoDB tables and learned how to choose the right table class based on the nature of the workload. The considerations for storage and throughput costs were discussed, and you discovered various tools and methods to analyze and compare these costs effectively.

Now, you should have a comprehensive understanding of capacity modes and table classes in DynamoDB. This knowledge can equip you with the insights and tools necessary to optimize your DynamoDB usage, making informed decisions about capacity provisioning and table class selection to meet the specific needs of your applications and workloads.

This also marks the end of *Part 2, Core Data Modeling*. The next part deals with under-the-hood information about how DynamoDB operates at the scale it does, and how it performs various functions in a distributed manner to provide blazing-fast data access and virtually unlimited scale while being highly available for its customers. The next chapter is all about the core components of the DynamoDB service, which interact with each other to serve every single data plane request, and cumulatively serve hundreds of millions of customer requests per second at peak.

References

1. Managing settings in provisioned capacity mode: `https://docs.aws.amazon.com/amazondynamodb/latest/developerguide/ProvisionedThroughput.html`

2. DynamoDB on-demand mode pricing: `https://aws.amazon.com/dynamodb/pricing/on-demand/`

3. AWS docs – Isolating frequently accessed items: `https://docs.aws.amazon.com/amazondynamodb/latest/developerguide/bp-partition-key-design.html#bp-partition-key-partitions-adaptive-split`

4. DynamoDB provisioned mode pricing: `https://aws.amazon.com/dynamodb/pricing/provisioned/`

5. AWS docs – Athena and CUR: `https://docs.aws.amazon.com/cur/latest/userguide/cur-query-athena.html`

6. AWS blog – Reserved capacity using Athena queries: `https://aws.amazon.com/blogs/database/calculate-amazon-dynamodb-reserved-capacity-recommendations-to-optimize-costs/`

7. GitHub – Python-based tool for reserved capacity recommendations: `https://github.com/awslabs/amazon-dynamodb-tools/tree/main/reco`

8. AWS docs – Cost Optimization, *Evaluate your table capacity mode*: `https://docs.aws.amazon.com/amazondynamodb/latest/developerguide/CostOptimization_TableCapacityMode.html`

9. AWS docs – *Considerations when choosing a table class*: `https://docs.aws.amazon.com/amazondynamodb/latest/developerguide/WorkingWithTables.tableclasses.html`

10. AWS docs – AWS Billing, *Cost allocation tags*: `https://docs.aws.amazon.com/awsaccountbilling/latest/aboutv2/cost-alloc-tags.html`

11. GitHub – Table Tagger tool: `https://github.com/awslabs/amazon-dynamodb-tools#eponymous-table-tagger-tool`

12. AWS docs – Cost Optimization, *Evaluate your table class selection*: `https://docs.aws.amazon.com/amazondynamodb/latest/developerguide/CostOptimization_TableClass.html`

10
Request Routers, Storage Nodes, and Other Core Components

Welcome to part three of this book, where we will explore the fundamental components that make **DynamoDB** an impressive database system. This part comprises a single chapter, in which you will gain insights into the essential components of DynamoDB and how they collaborate to efficiently handle database requests. We will also discuss the secure storage of data, ensuring its reliability. This section may be particularly interesting if you are curious about distributed systems and their workings.

Throughout this chapter, you will learn about each core component, highlighting their role within DynamoDB. You will discover how these components work together to deliver fast and dependable performance when reading or writing data. We will also dive into the methods that are employed to store and safeguard data, including authentication and fine-grained access control.

By the end of this chapter, you will have a clear understanding of how DynamoDB's components combine to create a powerful and scalable database system. Whether you are an experienced professional or simply intrigued by distributed systems, you will be provided with valuable insights, expanding your knowledge.

In this chapter, we will focus on the essential building blocks of the DynamoDB architecture, as outlined in the latest DynamoDB white paper (*1*) and explained in an informative *Under the Hood* talk from the 2018 re:Invent on YouTube (*2*). By reviewing the significance of these components and their roles, you will gain valuable insights into how DynamoDB consistently delivers predictable performance at any scale.

This chapter will cover several key components, such as **Request Routers (RRs)**, **Storage Nodes**, **Auto Admin**, and **Metadata** services. These components collaborate to handle various tasks within DynamoDB, such as creating new items or creating new tables themselves. Think of DynamoDB as a large distributed system, where different parts work together to accomplish a single task for the end user. You will learn about each component and how they interact.

The following figure illustrates a distributed system, where components such as **Request Router**, **Storage Node**, **Auto Admin**, and **Metadata** form DynamoDB. This system is accessed by end users through a straightforward HTTPS endpoint:

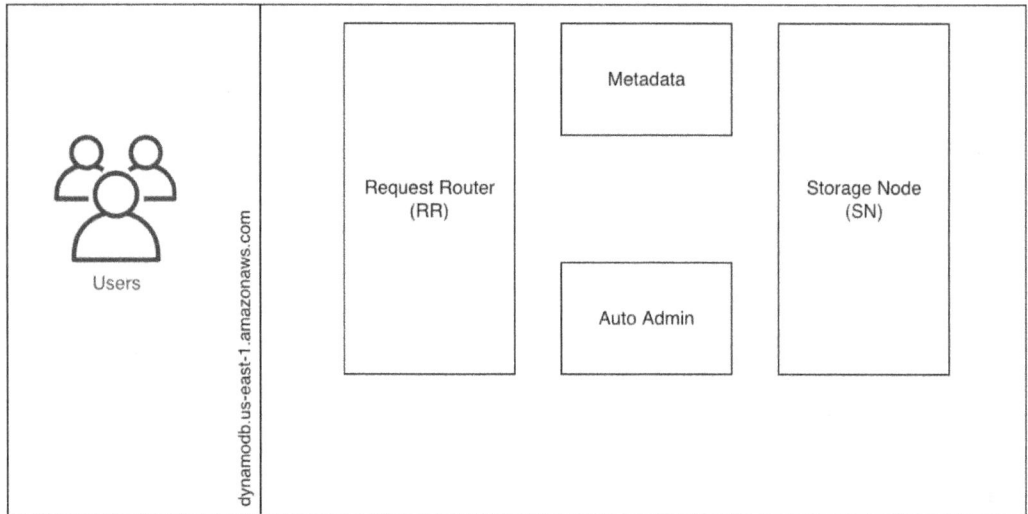

Figure 10.1 – DynamoDB from a distributed systems view

We will also explore the complete journey of a read and write request within DynamoDB, providing you with a comprehensive understanding of how the system functions. This exploration will connect the core components of the database system and help you grasp how DynamoDB achieves its capabilities, both from a user's and an engineer's perspective.

Throughout this chapter, we will closely examine each component, understanding its functions, interactions, and the mechanisms that enable DynamoDB to achieve its impressive capabilities. By the end, you will have a deep understanding of DynamoDB's architecture and the knowledge to maximize its potential for your projects.

Specifically, we will cover the following topics:

- RR
- Storage needs
- Auto Admin and Metadata
- The journey of a DynamoDB request

RR

The RR serves as the frontend in the DynamoDB system architecture. These are stateless hosts responsible for routing requests to other service components based on the nature of the requests. Being stateless means that there is no specific allocation of RR hosts for different customers; any RR can process requests for any customer. RRs are designed to handle all types of DynamoDB requests from end users, including both data plane and control plane requests.

In the context of DynamoDB, data plane requests directly interact with table data. For instance, **create, read, update, and delete** (**CRUD**) operations that act on the data itself are considered data plane requests. On the other hand, control plane requests involve managing DynamoDB data in various ways. Examples of control plane requests in DynamoDB include creating, deleting, or updating a table, modifying a table's provisioned capacity, or enabling/disabling a DynamoDB Stream for a table.

In an Amazon Web Services (AWS) region for DynamoDB, there can potentially be tens of thousands of RR hosts actively serving customer requests. When dealing with tables with high throughput rates, it is common to have multiple or even many of these hosts dedicated to handling requests for a single table. The horizontal scalability and statelessness of DynamoDB enable it to efficiently manage millions of requests per second for a table. As DynamoDB-based applications continue to thrive, this capability will become even more critical.

A notable example of DynamoDB's scalability is observed with customers such as Zoom Video, who rely on DynamoDB to manage their meeting metadata. They experienced a remarkable surge in traffic, growing from serving 10 million customers per day to a staggering 300 million customers per day during the 2020 pandemic. Such scalability was made possible by DynamoDB's ability to scale horizontally, starting with its frontend fleet of RRs. You can read more about Zoom Video and other DynamoDB customers on the DynamoDB landing page (3).

Situated in front of the RR service is a set of highly available load balancers. These load balancers play a crucial role in efficiently distributing client connections across the RR hosts, ensuring optimal resource utilization. They employ intelligent algorithms to direct requests to the most suitable RR host while considering factors such as current load and available capacity. In some cases, load balancers may even employ routing strategies that prioritize hosts with the fewest active connections, further optimizing the distribution process.

Due to the stateless nature of RRs, the DynamoDB service can handle a virtually infinite number of client connections. This distinguishes DynamoDB from other database systems that have limitations on the maximum number of client connections they can support. As the number of connections to those systems increases, performance degradation often occurs. However, with DynamoDB's stateless architecture, it can efficiently manage and process a vast number of client connections without suffering from any performance degradation.

Next, we'll dive into the roles and responsibilities of RRs in ensuring consistent performance at scale for DynamoDB's customers.

Role and responsibilities of RRs

The responsibilities of RRs extend beyond mere request routing. They also encompass a range of essential processing tasks before forwarding requests to their destination. These tasks include authenticating and authorizing incoming requests, thoroughly validating request parameters, publishing latency and other CloudWatch metrics for monitoring and analysis, enforcing rate limits, and intelligent throttling mechanisms to ensure fair and efficient resource allocation.

Authentication and authorization

When a request is sent to a DynamoDB service endpoint and received by an RR host, it undergoes authentication and authorization checks. Authentication verifies the identity of the requester by confirming if they are who they claim to be. For example, if there is an Identity and Access Management (IAM) user named Aman associated with an AWS account when a DynamoDB request is made for reading data within that account, the authentication check ensures that the request is made using the IAM role Aman. Typically, authentication is performed by matching the sign-in credentials with a trusted principal (such as an IAM user, federated user, IAM role, or application) associated with the AWS account.

Authentication operates by the client sending a signed request to the AWS service, and internally, AWS IAM processes the request to obtain authentication information. For more information about the signing of requests, see the AWS docs (6).

On the other hand, authorization validates whether the requester is permitted to perform the requested action. It answers the question, *"Are you allowed to perform the action you just requested?"* The permissions of the IAM entity that are used to sign the request are checked against the request parameters to determine if the request should proceed. Both authentication and authorization are performed by comparing the request information with AWS IAM through service-to-service API calls. These validations must be executed quickly to ensure efficient processing of each request. DynamoDB, specifically for data plane requests, aims to maintain single-digit millisecond average latencies. To achieve this, RRs may securely and temporarily cache certain static authentication and authorization information.

An important point to note is that unless a request passes these authentication and authorization checks, among other validation checks, the request is not passed on to the storage nodes, which are responsible for serving the table data. This helps in the efficient use of the storage nodes so that they only serve genuine data plane requests.

Accessing metadata

Since the actual customer table data is stored on storage nodes in the form of one or more partitions, the RR must direct data access requests to the authoritative partition and its corresponding storage node. In simpler terms, a table's data can be divided across multiple physical partitions, and any storage node can host a specific partition for a table. In a multitenant system such as DynamoDB, there can be thousands of storage nodes at any given time.

To ensure accurate routing of data access requests, the RR relies on the metadata system. The metadata system stores and manages mapping information that associates partitions with their respective storage nodes, along with other relevant routing details. When handling a customer's CRUD operation request, the RR interacts with the metadata system to obtain the necessary routing information. This information includes details about all the partitions in a table, the key range of each partition, and the storage nodes responsible for hosting each partition. Metadata access may also be required when handling control plane requests.

The following figure illustrates a typical GetItem operation being served by DynamoDB:

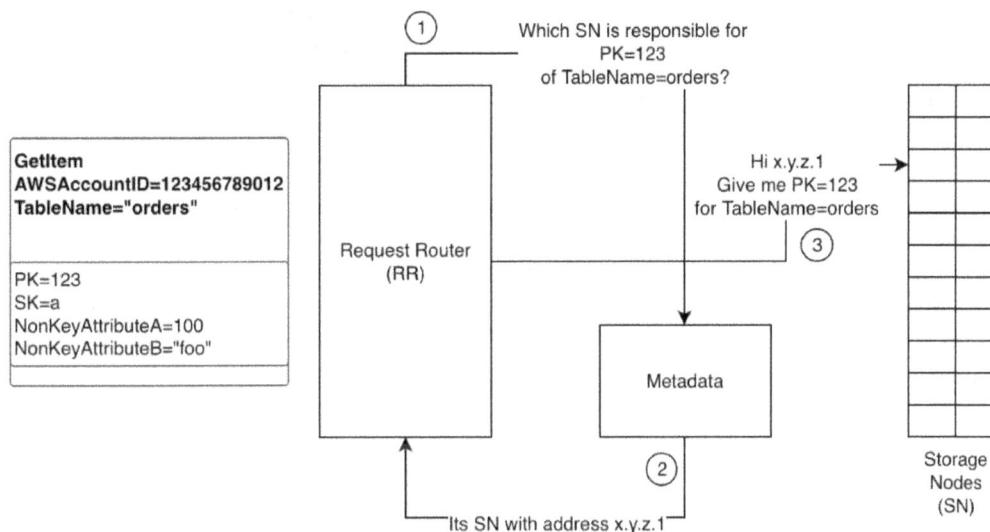

Figure 10.2 – RR interaction with the metadata system

After completing authentication, authorization, and other validations, the RR proceeds to query the metadata system for details about the storage node tasked with hosting data associated with a specific DynamoDB key, such as PK=123. The metadata system furnishes the RR with this routing information, enabling the RR to retrieve the data from the designated storage node and subsequently deliver it back to the client.

Without the routing information retrieved from the metadata system, the RR would have no knowledge of where to direct a specific customer operation on a table's data.

Since the inception of DynamoDB, a portion of the routing metadata has been relocated to a subsystem within the metadata system known as MemDS. MemDS is a distributed in-memory data store specifically designed for DynamoDB. Later in this chapter, we will delve deeper into the metadata system, including MemDS, in the *Auto Admin and Metadata* section. Next, we will learn about how RRs perform rate limiting for user requests.

Rate limiting

RRs in DynamoDB perform rate limiting for both control plane and data plane requests based on different criteria. In provisioned mode, data plane rate limiting is applied at the overall table level to ensure that requests for a single table stay within its provisioned capacity. Additionally, RRs enforce limits on the overall read or write requests to prevent exceeding the throughput limits set for the AWS account in the specific region. For on-demand capacity mode tables, rate limiting may be enforced on a per-table basis, applicable to each on-demand table.

Due to the stateless nature of RRs, achieving strict rate limiting across the numerous RRs can be challenging. To address this, DynamoDB utilizes a central subsystem called **Global Admission Control** (**GAC**) to assist RRs in implementing appropriate rate limiting strategies.

Rate limiting within DynamoDB operates through the use of token buckets (7). The GAC service is responsible for admission control on an overall table level. Each RR manages its local token bucket instead of communicating with the GAC service for every user request. Periodically, typically every few seconds, the RR synchronizes with GAC to renew its tokens. GAC dynamically evaluates a temporary state, depending on incoming client requests. Within their local systems, RRs handle a set of tokens. Upon receiving an application request, the router deducts tokens from its allocated bucket. If the tokens are depleted either due to usage or expiration, the RR seeks additional tokens from GAC. Utilizing information from the client, GAC gauges the global token usage and allocates tokens for the subsequent time interval, ensuring that varying workloads focusing on specific items can maximize their partition capacity. The GAC service may respond to an RR with a negative token number, which would make the RR throttle requests for the table until communication with GAC results in a positive number of tokens for the DynamoDB table.

By utilizing token buckets and the coordination of GAC, RRs in DynamoDB effectively perform rate limiting to maintain optimal performance and prevent excessive usage of resources, allowing the system to handle a wide range of workloads efficiently. Next, we will learn about how RRs help with the observability of DynamoDB tables.

Publishing CloudWatch metrics

In the DynamoDB architecture, the RR plays a critical role as both the first and last component in the request journey. While load balancers may be present in front of the RR, the traceability of a request begins when it is initially received by one of the RR hosts. Consequently, RRs are well-positioned to measure server-side latencies for all types of DynamoDB operations, including both data plane and control plane requests.

Like other AWS services, DynamoDB publishes request metrics, including latency metrics, in Amazon CloudWatch. CloudWatch provides essential capabilities such as aggregation, dashboarding, alarming, and more, enabling comprehensive observability across AWS services. When processing a customer request, the RRs publish metrics for latencies, throttles, server-side errors, and other response types to CloudWatch.

The following figure illustrates a CloudWatch metric specific to DynamoDB tables, focusing on `PutItem` latency. This metric, named `SuccessfulRequestLatency`, provides insights into the server-side latencies of data access requests. As per the following figure, `PutItem` request latencies consistently fall within the range of 2 to 3 milliseconds over the observed period of several days:

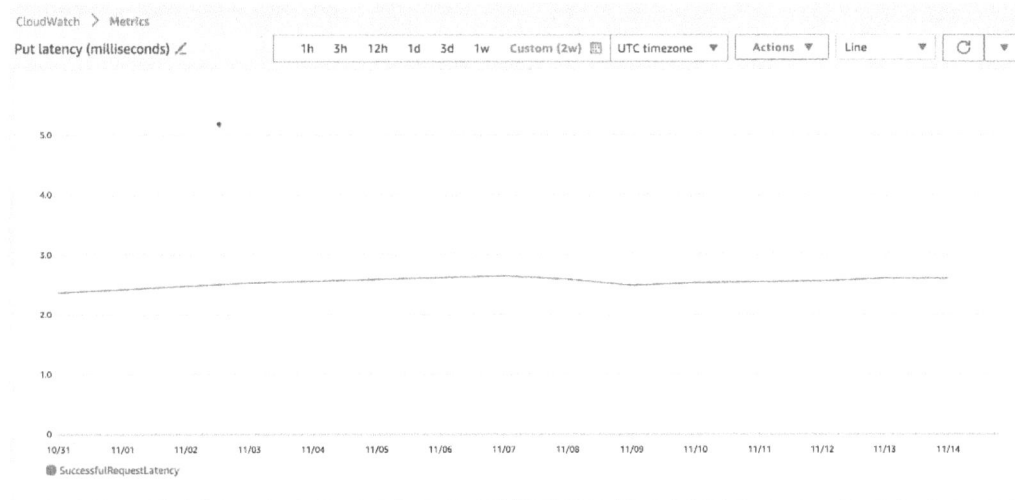

Figure 10.3 – Sample SuccessfulRequestLatency CloudWatch metric for PutItem

Both the RR and CloudWatch metrics systems incorporate intelligent and, at times, asynchronous mechanisms to ensure that DynamoDB request processing remains unaffected by transient or prolonged unavailability of downstream systems. These mechanisms are implementation details and hence it may not be necessary to dive deeper into these.

It is worth noting that the DynamoDB architecture is designed to maintain its availability even when one or more dependencies experience periods of unavailability. Such ability of software systems to continue to operate normally without the need to make changes during the failure or unavailability of dependencies is called **static stability** (8). In DynamoDB, static stability is ensured through several design choices, such as avoiding circular dependencies with other core AWS services and maintaining the capability to serve customer requests, regardless of external conditions.

Local lazy caching

Despite being stateless, RRs in DynamoDB leverage highly available systems, such as the metadata system, to retrieve essential information for processing incoming requests. However, RRs also employ short-lived caches to locally store various types of data, aiming to avoid unnecessary external calls when serving requests with the same target.

For example, in the case of a DynamoDB PUT request, where two write requests are intended to be routed to the same storage node and customer partition, it is logical to cache the routing information after processing the first request. This caching applies only to metadata and does not involve caching real customer data. Additionally, certain data, such as the encryption key for a DynamoDB table, may also be lazily cached for short durations to reduce the need for costly external calls to corresponding service components, thus improving overall server-side latency.

In situations where a cache miss occurs, RRs have no choice but to make additional external calls to process the request. While most of the cached information remains internal to the service, it is important to note that if a DynamoDB table is encrypted at rest using a **Key Management Service (KMS)** managed key, each RR locally caches this key for a fixed duration of 5 minutes. These caches are loaded lazily, meaning the KMS key is fetched and cached by an RR host only when processing the first request for the specific table.

By employing caching mechanisms, RRs aim to optimize performance by reducing the need for repeated external calls for certain types of data, ultimately contributing to efficient request processing within DynamoDB.

Now that we have familiarized ourselves with the various tasks that are performed by the RR component in processing a DynamoDB request, let's dive into storage nodes, which serve as the service component responsible for hosting the actual customer data.

Storage nodes

As the name implies, the storage node fleet is responsible for storing the actual user data in DynamoDB. Unlike RRs, storage nodes are stateful instances that store data for one or more tables belonging to one or more AWS accounts. This means that a single storage node can host partitions for multiple DynamoDB tables owned by different AWS accounts simultaneously. Each partition hosted by a storage node has its own security measures, including encryption of data at rest using either AWS-owned or customer-owned encryption keys. All read and write requests issued by users of DynamoDB would ultimately be processed by the storage nodes.

Partition data within a storage node is stored in a B-tree structure along with write-ahead logs, referred to as replication logs. Both these are critical in terms of serving read and write requests.

When a request is forwarded by the RR to a storage node, the storage node verifies whether it hosts the partition associated with the request. Since the storage node possesses up-to-date information about all the partitions it hosts, there may be cases where the partition metadata on the RRs becomes outdated, and the intended partition is no longer hosted by that specific storage node. In such situations, the storage node informs the RR that the requested partition is not hosted, and the RR takes the appropriate action accordingly.

Next, we'll review the roles and responsibilities of storage nodes.

Roles and responsibilities of storage nodes

At a high level, the main responsibilities of the storage nodes are to handle read and write requests for the table partitions they are responsible for. However, they also perform additional tasks to ensure that DynamoDB operates as a high-performance enterprise-level database system. These tasks include managing encryption of table data at rest, monitoring the health of their partitions (including exchanging heartbeats with other relevant storage nodes), implementing partition-level rate limiting, archiving replication logs, and responding to control plane actions for various management tasks.

Let's dive into the responsibilities of storage nodes, starting with managing and maintaining a replication group of its partitions.

Replication group and quorum

Each storage node in DynamoDB hosts a partition, which represents one of three copies of the same partition data forming a replication group. This replication group is designed to ensure the high availability and durability of the data, protecting against potential network partitioning (4) that can occur in a distributed system. Each copy in the replication group is called a **replica**. Consequently, a single storage node can host multiple replicas of partition data for various DynamoDB tables. To enhance availability, each replica is hosted on a storage node running in a distinct Availability Zone.

The following figure illustrates how partition data is typically replicated across Availability Zones:

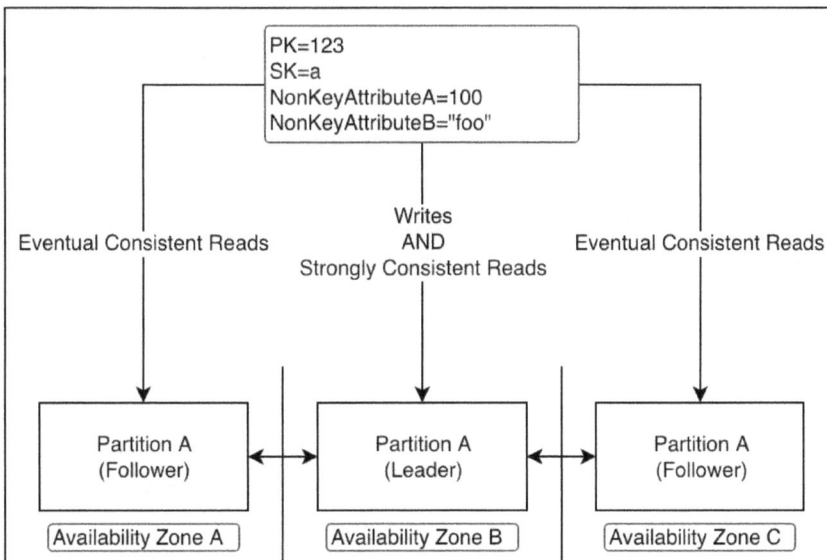

Figure 10.4 – Partitions, replicas, and Availability Zones

As we can see, every item within partition A, such as **PK=123**, is replicated across Availability Zones **A** through **C** in a replication group.

In the replication group, each replica maintains communication by exchanging heartbeats with the other replicas. Among the three replicas, one is elected as the leader, while the remaining two act as followers. The leader election process employs the consensus algorithm named **Paxos** (5), where all three replicas participate.

The leader replica assumes the responsibility of accepting all write and strongly consistent read requests. Upon receiving a write request, the leader replica first persists the write to its local storage and then propagates the write to both the other replicas within the replication group. The write is considered durable and successful only when two out of three replicas have persisted. For instance, in a replication group of three, once the leader replica has persisted the write and receives an acknowledgment of persistence from one of the two follower replicas, the write is considered successful. The leader then responds to the RR, which subsequently returns a successful response to the user. The third replica eventually acknowledges the persistence as well, although the leader does not wait for it since the quorum has already been established.

Encryption at rest

The encryption at rest feature in DynamoDB offers two options for key management: AWS-owned keys or customer-owned keys through AWS KMS. When AWS-owned keys are used, DynamoDB takes care of managing the encryption keys on behalf of the user, simplifying the key management process. On the other hand, customer-owned keys provide users with greater control over their encryption keys, including key rotation and management.

Irrespective of the key type chosen, DynamoDB ensures that each partition's data is encrypted with a unique data key. These data keys are encrypted using the designated encryption key, adding an extra layer of security. The encrypted data keys are stored alongside the data, allowing DynamoDB to retrieve and decrypt the data when requested by authorized users or applications. The responsibility for encrypting the data at rest lies with the storage nodes. When a storage node receives a request for a table's data, it decrypts the table's data key using the user's chosen encryption key and temporarily uses the decrypted key to serve the request. The decrypted key may be cached for a short period to serve multiple requests and maintain efficient latencies. This approach, known as envelope encryption, provides an additional safeguard for data security. More detailed information about envelope encryption can be found in the AWS documentation (9).

The following figure illustrates the envelope encryption technique that DynamoDB uses to encrypt DynamoDB table data at rest:

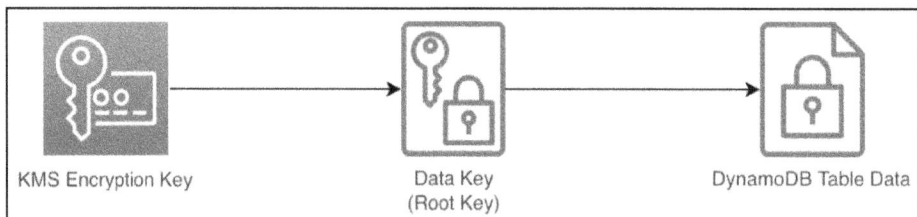

Figure 10.5 – Envelope encryption in Amazon DynamoDB

As we can see, the encryption type of your choice is used to encrypt and decrypt a data key that DynamoDB manages under the hood to encrypt your table's data at rest. Without access to the KMS encryption key, data could neither be read nor written to the DynamoDB table.

Next, we'll learn about the log archival process within storage nodes. This archival process serves multiple purposes, including ensuring data correctness checks and providing backup features for DynamoDB users, both on-demand and continuous.

Log archival

At a high level, DynamoDB generates replication logs for every mutation that occurs on a table. These logs serve various purposes, such as running live data correctness checks, ensuring consistency across replicas, maintaining state within the replication group, and supporting backup and restore features.

Storage nodes play a crucial role in the log archival process. They run agents that are responsible for capturing and archiving replication logs and creating periodic snapshots of user tables. Replication logs contain information about the changes that have been made to data in a DynamoDB table partition. These logs are archived to Amazon S3, preserving a historical record of modifications. In addition to replication logs, periodic snapshots are created to capture the state of the data at specific points in time.

Archiving replication logs and creating snapshots serve multiple purposes, including data correctness checks and backup functionality. DynamoDB supports both on-demand backups and continuous backups:

- With on-demand backups, users can create point-in-time copies of their tables
- Continuous backups, on the other hand, automatically capture data changes and create snapshots at regular intervals.

These backup features empower users to restore their data to a specific point in time, safeguarding against accidental data loss or corruption.

For more detailed information about log archival and backup functionality in DynamoDB, please refer to *Chapter 11, Backup, Restore, and More*. Next, we'll explore some of the failure modes that storage nodes are designed to handle and assist with.

Failure modes

As a critical component of DynamoDB's infrastructure, storage nodes play a key role in ensuring the availability and reliability of data.

One of the failure modes that storage nodes address is the failure of individual storage devices or disks. DynamoDB mitigates this risk by replicating data across multiple storage nodes and Availability Zones. If a storage node experiences a disk failure, the replicated copies of data on other storage nodes can still serve read and write requests, ensuring the system's overall availability.

Storage nodes also handle the failure of an entire storage node. Since each partition is replicated across multiple storage nodes, if one storage node becomes unavailable, the other replicas can seamlessly take over the responsibility of serving requests for the affected partitions. In the case where the failed storage node was hosting a leader replica of a table partition, the remaining replicas undergo leader election, and a new leader is chosen. This ensures that data remains accessible even in the event of storage node failures.

Within expansive services such as DynamoDB, incidents such as memory and disk malfunctions are frequently encountered. Should a node experience a failure, the replication groups residing on that node are reduced to merely two copies. Rectifying a storage replica can be time-consuming, often spanning several minutes, as it requires duplicating both the B-tree and write-ahead logs. To maintain durability during this time, the leader of a replication group can add a log replica.

A log replica contains only the replication log for a table partition and can accept writes to its local storage. However, a log replica cannot assume leadership of a replication group or serve read requests. Incorporating a log replica requires only a brief duration, given that the system merely has to transfer the recent write-ahead logs from a healthy replica to the new one, excluding the B-tree. This swift integration of log replicas provides prompt recovery of affected replication groups, upholding high durability for the latest writes.

By addressing various failure modes and employing replication strategies, storage nodes contribute significantly to the availability, durability, and reliability of data in DynamoDB.

Now that we've gained a better understanding of the storage nodes and their responsibilities, let's explore two other core components within DynamoDB that facilitate interaction and management between the RRs and storage nodes.

Auto Admin and Metadata

As DynamoDB continues to grow in scale and complexity as a distributed database, the manual administration of the database becomes impractical and inefficient. The range of administrative tasks required, such as maintaining healthy quorums for partitions, launching new nodes, and decommissioning underperforming nodes, would typically necessitate a team of administrators. However, given the immense scale at which DynamoDB operates, relying solely on human administrators is not feasible.

For instance, during AWS's re:Invent 2022, it was revealed that DynamoDB handles approximately 10 trillion requests daily. To effectively manage this scale, automation is essential. This is where Auto Admin comes into play, taking on the responsibility of performing these administrative tasks for DynamoDB. Auto Admin ensures that the database runs smoothly by automating various critical administrative operations, allowing for efficient and hassle-free management of DynamoDB at its massive scale.

Auto Admin

Auto Admin is a vital service within DynamoDB that acts as an automated database administrator. It serves as the central hub for all resource creation, update, and data definition requests within the system. When you issue a `CreateTable` request to DynamoDB, it is first received by the RR service. After performing basic processing tasks such as authentication, authorization, and rate limiting, the RR passes the `CreateTable` request to one of the queues processed by the Auto Admin. The Auto Admin service then takes the necessary actions to process the `CreateTable` request. These actions may include organizing and determining which storage nodes to place table partitions on. Similarly, all other control plane operations related to DynamoDB tables are handled by the Auto Admin service.

On a broader scale, Auto Admin's main responsibility is to ensure the health and stability of the DynamoDB fleet. It continuously monitors the health of partitions, replicas, and core components of DynamoDB. If any replicas are identified as slow, unresponsive, or hosted on faulty hardware, Auto Admin takes immediate action by replacing them with healthy replicas. For example, if an unhealthy storage node is detected, Auto Admin triggers a recovery process to replace the affected replicas hosted on that node, thereby restoring the system to a stable state. These actions are performed by the Auto Admin service continuously. By automating these tasks, Auto Admin plays the role of being the central nervous system in maintaining the availability, reliability, and performance of DynamoDB.

Next, we'll learn about the metadata service within DynamoDB, which is responsible for serving metadata information about every table within DynamoDB.

Metadata

At any given moment, DynamoDB can have a vast number of storage nodes responsible for hosting customer table data, and this number continues to grow. A single DynamoDB table can be partitioned into multiple partitions, which are distributed across these numerous storage nodes. It is crucial to have access to the mapping information between customer tables and the storage nodes where their partitions reside to efficiently handle any requests within DynamoDB. This is where the metadata service plays a pivotal role as it serves as the host and manager of all the essential mapping information within the system. Understanding the metadata service becomes valuable in comprehending the internal workings of DynamoDB.

According to the USENIX paper on the latest state of DynamoDB (*1*), the metadata in DynamoDB is stored in two different data stores. Initially, all types of metadata were served directly from DynamoDB tables themselves (yes, from its inception!). Specific system-only DynamoDB tables were dedicated to hosting and serving various types of essential metadata for DynamoDB's functionality for its customers.

When an RR receives a request for a table it has not encountered before, it accesses the metadata service to load all the necessary table routing information into its local cache. Subsequent requests for the same table would then be served directly from the RR's local cache. Since the metadata for partition routing changes infrequently, these local caches on the RRs typically achieve high cache hit rates. The following figure shows how the DynamoDB system looks with metadata served off system DynamoDB tables:

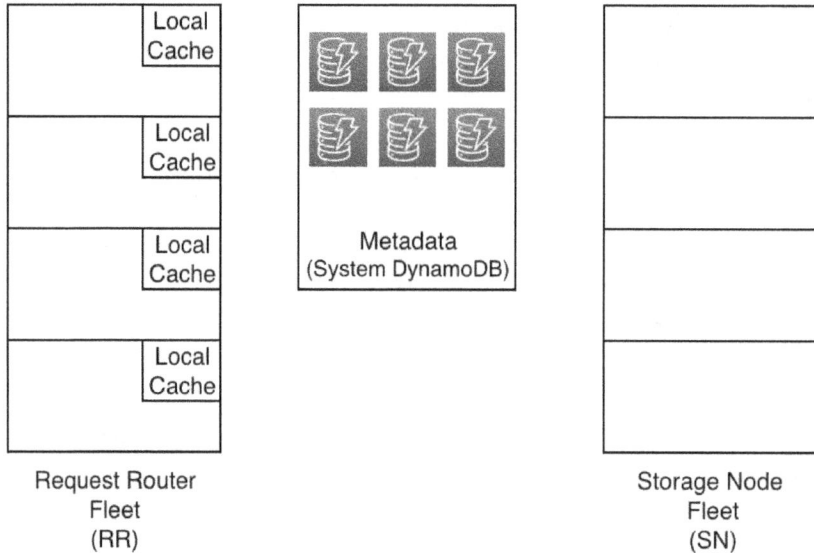

Figure 10.6 – Partial DynamoDB architecture with system DynamoDB tables in metadata

In the preceding figure, each RR within the fleet is configured with a local cache responsible for lazily storing metadata related to customer tables and partitions. If an RR's cache lacks the required metadata, it initiates a request to the metadata system. In the earlier setup, this metadata system consisted of several DynamoDB tables designated as internal, strategically avoiding circular dependencies. The information retrieved from these system DynamoDB tables equips the RR with the routing details needed to reach the authoritative Storage Node host responsible for the specific customer item(s).

However, there were certain challenges with this approach. For instance, during deployments or when new routers were added to the fleet, the RRs would often have cold caches. This would result in a sudden influx of requests to the metadata service, potentially degrading its performance and destabilizing the system. This phenomenon is referred to as **bimodal behavior**, where the system's reaction to certain workloads can vary, making it difficult to predict the specific outcome.

Another concern was that the RRs would always retrieve all partition map metadata, even if they only needed the metadata for a single partition of the customer table.

To address the challenges of bimodal behavior, cold cache thundering herds, and inefficiency in loading table metadata into local RR caches, DynamoDB implemented a highly available and distributed in-memory data store called **MemDS**. This data store stored all partition metadata in-memory and ensured high availability by replicating the data across multiple nodes. The following figure shows how the DynamoDB architecture provides a high-level introduction to MemDS:

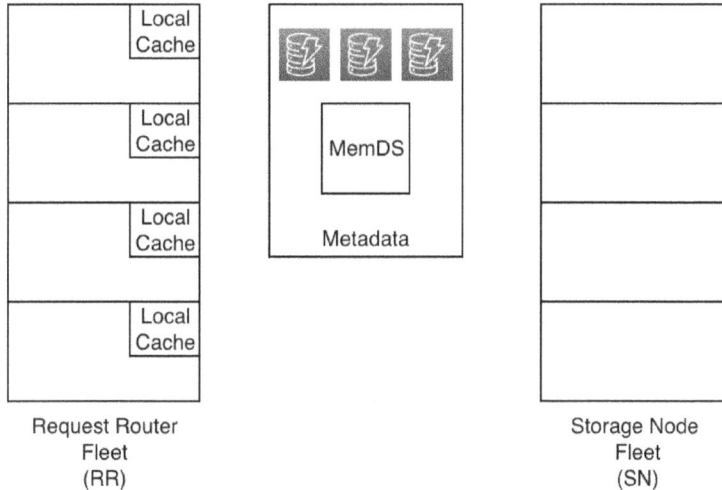

Figure 10.7 – Partial DynamoDB architecture after the introduction of MemDS

With the introduction of MemDS, the partition map information (which maps partitions to storage nodes) was moved to this new data store, while some other types of metadata continued to be served using the existing setup with system DynamoDB tables, which are not expected to see data access surges. This change improved the overall robustness of the metadata service.

Furthermore, improvements were made to the behavior of RRs when loading partition map metadata. After the introduction of MemDS, RRs would only fetch metadata for the specific table partition required to serve a customer request, instead of loading metadata for all partitions of the table. This optimized approach reduced unnecessary data retrieval and improved efficiency.

To address the issue of bimodal behavior and mitigate cold cache thundering herds, RRs were modified to fire requests to the MemDS data store, even when the partition map information was available in the local cache of the router. This ensured steady and predictable traffic toward the MemDS fleet, allowing for reliable scaling monitors. Further, requests to fetch partition routing information were hedged to more than one MemDS host, to prioritize high availability and fast performance.

Although these changes resulted in significant amounts of redundant traffic toward the MemDS data store at DynamoDB's scale, the trade-off was deemed worthwhile to prioritize system stability. Overall, the introduction of MemDS and the adjustments made to the request processing protocol successfully eliminated scalability bottlenecks and mitigated potential outages, leading to a more efficient and stable DynamoDB system.

Now that you've gained a good understanding of the pivotal role that is played by the Auto Admin and metadata systems in making DynamoDB highly available, scalable, and capable of delivering predictable performance, let's consolidate this knowledge about DynamoDB's core components to comprehend the journeys of read and write operations within the system.

The journey of a DynamoDB request

Throughout this chapter, we've explored various core components of DynamoDB as a distributed database. Now, let's delve into an overview of the read and write paths, consolidating the information we have learned so far.

The journey of a read request (GetItem)

The following figure illustrates the journey of a read request within DynamoDB:

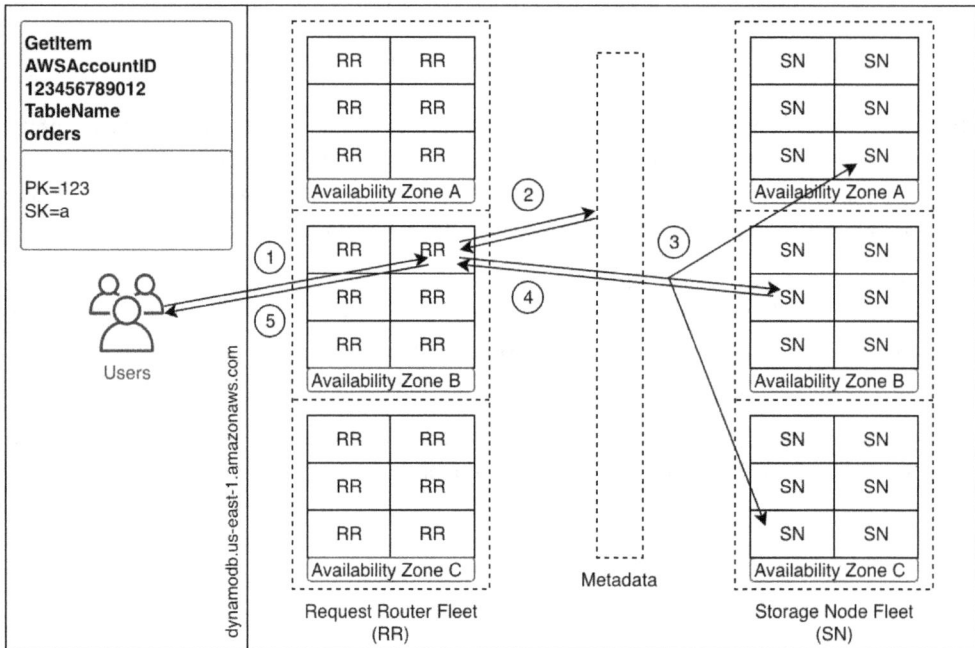

Figure 10.8 – DynamoDB read request journey

Let's explain the flow of the previous figure:

1. A `GetItem` request is sent to a regional DynamoDB endpoint. Through load balancing by the service, it reaches an RR host.

2. The RR performs various validations, including authenticating and authorizing the caller. It verifies that the caller entity (such as an IAM user or role) has the necessary permissions to perform the requested operation on the specified table. Additionally, the RR validates the request parameters, such as schema compliance, table name checks, and rate limiting. At this stage, the RR may interact with the metadata system to retrieve customer and table metadata or utilize locally cached information. It may also use its internal hash function output to obtain metadata about the partition responsible for the requested item.

3. After successful validation, the RR accesses the MemDS data store within the metadata system. It retrieves the partition map information for the specific partition responsible for hosting the data associated with the provided partition key in the request parameters. The response from MemDS includes the storage nodes hosting data of the particular partition. Given that three copies of the same partition are stored across three different storage nodes, the routing of a read request depends on the specified read consistency in the request parameters. In the case of a strongly consistent read request, it must be served exclusively by the leader storage node. On the other hand, for an eventual consistent read request, it can be directed to any of the three storage nodes hosting the partition data.

4. The RR contacts the appropriate storage node to retrieve the data for the specified partition and sort key. The storage node responds with the requested item.

5. The item that is returned by the storage node is then relayed back to the client by the RR. Additionally, the RR publishes data points to CloudWatch metrics, capturing information such as the consumed read throughput, encountered errors, and more.

The journey of a write request (PutItem)

The following figure illustrates the journey of a write request within DynamoDB:

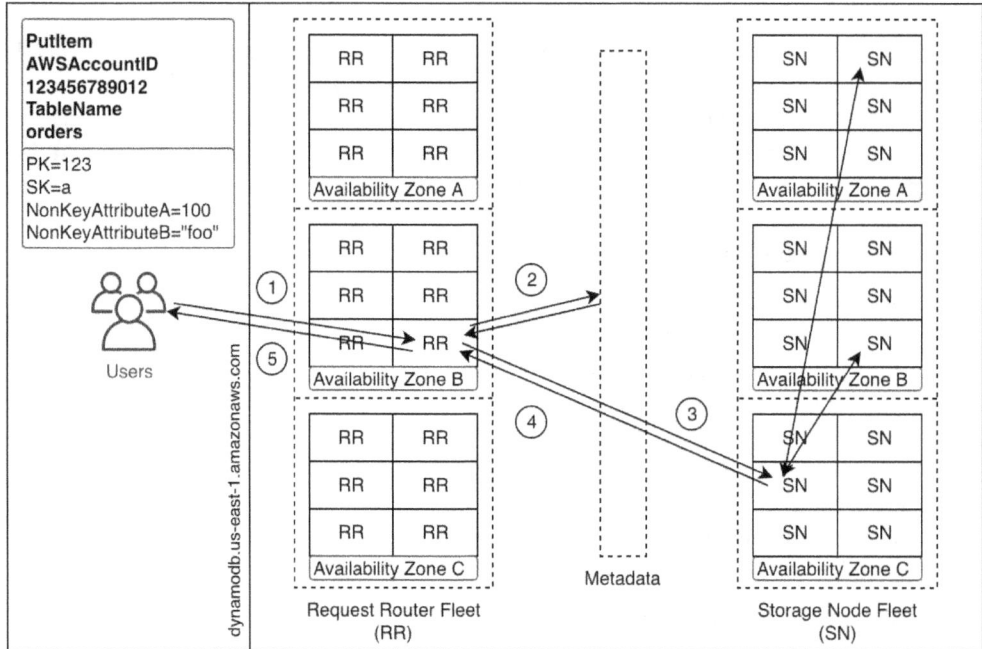

Figure 10.9 – DynamoDB write request journey

Let's explain the flow of the previous figure:

1. A PutItem request is sent to a regional DynamoDB endpoint. Similar to the GetItem request, it goes through load balancing to reach one of the RR hosts.

2. The RR performs authentication and authorization checks, validating the IAM entity (IAM user or role) that was used to sign the request. It also performs rate limiting, schema validation, and table name checks. The RR then applies its internal hashing algorithm to the partition key, generating a hash that determines the responsible partition for hosting the item. The RR retrieves this partition information either from the metadata system or its local cache. Once the partition information is obtained, the RR contacts the MemDS data store to acquire the partition map information for the storage nodes hosting the specific partition data.

3. It's important to note that each partition data can be hosted across three different storage nodes, with one serving as the leader and the others as followers in the replication group. The request is forwarded to the leader storage node, which performs local rate limiting and additional validations. The leader then executes the write operation on its local copy. Subsequently, the leader sends the write intent to both follower nodes.

4. On receiving acknowledgment of write intent from at least one follower, the leader returns a successful response to the RR.

5. Finally, the RR passes the successful response back to the client, indicating the successful completion of the `PutItem` operation.

> **Important note**
>
> Consider this as a reminder to delete any resources you may have created in your AWS account while going through this chapter.

Summary

In this chapter, we explored the various core components of DynamoDB from a distributed systems architecture perspective. We began by understanding the roles and responsibilities of RRs, including tasks such as accessing metadata, publishing CloudWatch metrics, and performing authentication and authorization checks while serving DynamoDB requests.

Our focus then shifted to storage nodes, which are responsible for hosting and managing customer data. Regarding storage nodes, we learned about establishing durability through replication groups, encryption techniques, archival processes, data correctness checks, and their involvement in different failure modes to ensure self-healing capabilities.

Additionally, we delved into the Auto Admin service, which supervises the self-healing of DynamoDB. We explored its capabilities, such as health checks and heartbeat monitoring, along with its role in serving control plane operations to ensure the efficient functioning of DynamoDB as a service.

Furthermore, we studied the metadata service within DynamoDB and its evolution over time. We learned about the balance between system stability and cost efficiency, as well as the importance of designing large-scale distributed systems to enable predictable scaling and prevent scenarios where large and small fleets interact, potentially leading to bimodal behavior.

Overall, this chapter covered several distributed computing concepts that have been adopted by DynamoDB to provide high availability and predictable performance for data access. We also gained insights into the journey of customer read and write requests and how the core components of DynamoDB communicate with each other to enable robust data access capabilities.

In the next chapter, we will explore how the service authenticates user requests and discuss various security use cases, including fine-grained access control.

References

1. USENIX 2022: Amazon DynamoDB Whitepaper: `https://www.usenix.org/conference/atc22/presentation/elhemali`

2. YouTube: Amazon DynamoDB Under the Hood: `https://youtu.be/yvBR71D0nAQ`

3. Amazon DynamoDB: Customers: `https://aws.amazon.com/dynamodb/customers/`

4. Wikipedia: Network Partition: `https://en.wikipedia.org/wiki/Network_partition`

5. Wikipedia: Paxos Algorithm: `https://en.wikipedia.org/wiki/Paxos_(computer_science)`

6. AWS Docs: Signing AWS API requests: `https://docs.aws.amazon.com/IAM/latest/UserGuide/reference_aws-signing.html`

7. Wikipedia: Token Bucket: `https://en.wikipedia.org/wiki/Token_bucket`

8. Static Stability: `https://docs.aws.amazon.com/whitepapers/latest/aws-fault-isolation-boundaries/static-stability.html`

9. AWS Docs: Envelope Encryption: `https://docs.aws.amazon.com/kms/latest/developerguide/concepts.html#enveloping`

10. GitHub repository: `https://github.com/PacktPublishing/Amazon-DynamoDB---The-Definitive-Guide`

Part 4:
Advanced Data Management and Caching

This part explores the advanced features of DynamoDB. It begins with a comprehensive coverage of backup and restore functionalities, including on-demand and continuous backups, **Point-in-Time Recovery (PITR)**, and native, no-code data import/export options. The different streaming options supported by DynamoDB and **Time to Live (TTL)** are discussed, highlighting their applications and best practices. The part then dives into designing and implementing multi-region architectures using global tables. Lastly, it examines **DynamoDB Accelerator (DAX)** and other caching alternatives, focusing on reducing latency and optimizing performance. This part equips readers to leverage advanced features for enhanced data management.

Part 4 has the following chapters:

- *Chapter 11, Backup, Restore, and More*

- *Chapter 12, Streams and TTL*

- *Chapter 13, Global Tables*

- *Chapter 14, DynamoDB Accelerator (DAX) and Caching with DynamoDB*

11
Backup, Restore, and More

As modern applications grow in complexity, they require more than just basic data access functionality. Supplementary features around backup and restore, cross-region local data reads and writes, streaming change data capture, in-memory caching, and complex analytics among others become essential for delivering scalable and performant applications. In the final set of chapters in this book, we will explore these additional features and functionalities of **Amazon DynamoDB**.

Throughout the next six chapters, you will learn about the features and functionalities of **DynamoDB** that extend beyond standard data **create, read, update, and delete** (**CRUD**). You will understand the overall ecosystem of the service and how these fully managed native functionalities can eliminate the undifferentiated heavy lifting that may be needed to perform these tasks manually. By the end of this part, you will be able to take full advantage of the power and capabilities of DynamoDB and develop applications that deliver a better user experience.

From backups and restores to in-memory caching, DynamoDB has many powerful features that can enhance the performance and scalability of your applications. With cross-region local data access and change data capture, you can build resilient and fault-tolerant systems that can operate across multiple regions. With complex analytics, you can extract meaningful insights from your data and use them to optimize your application's performance. These chapters will provide you with the knowledge you need to take advantage of these features and functionalities and develop applications that can deliver the performance and reliability your users demand.

Backups and restores are essential functions for any managed database system. The ability to predictably take backups of data and restore them quickly and efficiently is crucial for ensuring the integrity and availability of your data. In this chapter, we will explore the backup and restore functionalities of DynamoDB, along with additional data ingestion and export features available.

We will begin by examining the fundamental concepts of backup and restore in DynamoDB, including how backups are taken and how they can be used to restore data. We will then dive into two key features of DynamoDB backup and restore: **on-demand backup/restore** and **continuous backups** or **point-in-time recovery (PITR)**. Continuous backups automatically create incremental backups of your DynamoDB tables, allowing you to restore your data to any point in time during the retention period. PITR provides even more fine-grained control over on-demand backups, allowing you to restore your data to a specific point in time down to the second.

In addition to backups and restores, we will also explore the native export and import features of DynamoDB. These features allow you to move data in and out of DynamoDB easily and efficiently, using a variety of formats, including JSON and CSV. We will examine the various options available for exporting and importing data.

Finally, we will conclude with some best practices for working with DynamoDB backups and restores. We will discuss strategies for optimizing backup and restore performance and recommendations for securing and protecting your backups. By the end of this chapter, you will have a solid understanding of the backup and restore functionalities of DynamoDB and how to use them effectively in your applications. You will also have the know-how to use the native export and import features and understand where they could fit in your overall data architecture.

In this chapter, we are going to cover the following main topics:

- Backup fundamentals
- Restores
- Export to S3
- Import from S3

Understanding backup fundamentals

The backup and restore functionalities in DynamoDB operate independently of actual table reads and writes. When a backup is requested or performed, it does not impact the application's access to the table in terms of latency or throughput. Similarly, when a restore is executed against a DynamoDB backup, it is performed in isolation from the source DynamoDB table. This isolation is due to DynamoDB's use of **write-ahead logs**, which record every write operation performed on the table. These logs are used to perform backup and restore functions and ensure durability and continuous data correctness checks. The following figure is an illustration of a typical write-ahead log, presented for learning purposes and not as an exact representation of the logs used within DynamoDB:

PUT	UPDATE	PUT	DELETE
LSN: 001	**LSN**: 002	**LSN**: 003	**LSN**: 004
Key: Foo	**Key**: Abc	**Key**: Bar	**Key**: Xyz
....	....	....	....

Figure 11.1 – Example write-ahead log structure

As shown in *Figure 11.1*, at a high level, each write-ahead log entry contains details about a mutation, such as a PUT, UPDATE, or DELETE operation. It is accompanied by the partition key of the corresponding item. Additionally, the log entries include **log sequence numbers** (**LSNs**) to indicate the order of events in relation to one another. Each log entry contains enough information to enable the replay of mutations in the event of failures or to support advanced backup functionalities.

As per the 2022 DynamoDB whitepaper (2) presented at the USENIX conference, the **log archival agent** is an internal service component responsible for archiving write-ahead logs accumulated on storage nodes into Amazon S3 buckets owned by the DynamoDB service. The log archival agent runs on every storage node and uploads the write-ahead logs to S3 to leverage S3's 11 nines (99.999999999%) of durability. Each storage node containing data replicated for a single DynamoDB partition is accessed by the log archival agent to ensure that exactly one of the three agents is archiving the logs. The other two replicas that are not actively uploading logs to S3 track the progress of the replica that is performing the uploads, and they run data correctness checks and validations of the uploaded copies in S3 against their local copies of write-ahead logs.

In addition to the write-ahead logs, DynamoDB's storage nodes perform periodic snapshots of partition data and upload them to S3. This combination of periodic snapshots and write-ahead logs supports the different backup and restore functionalities within DynamoDB. At any given time, one of the three replicas of a DynamoDB table partition's log archival agent is uploading logs to S3, and when one of these partitions experiences high traffic due to customer requests, a different replica's agent may take on the task of uploading logs to S3. During restores, the periodic snapshots and write-ahead logs are used directly from S3, ensuring that there is no impact on the performance of the actual table partitions. To learn more about write-ahead logs and snapshots in DynamoDB, readers can refer to *Chapter 10, Request Routers, Storage Nodes, and Other Core Components*.

Now that we understand the basics of how data may be backed up by DynamoDB to perform restores, let us review the different backup restore features supported by DynamoDB. At a high level, these are as follows:

- On-demand backups
- Continuous backups

Over the next sub-sections, we will dive into each of these backup functionalities offered by DynamoDB, starting with on-demand backups followed by continuous backups.

On-demand backups

On-demand backups provide DynamoDB users with the flexibility to create backups as and when they need them. This option is ideal for use cases where backups need to be created at a frequency and these backups may be retained for longer durations. One example use case for using on-demand backups for DynamoDB is in a compliance-driven environment where data retention policies require the retention of backups for extended periods. For instance, in the healthcare industry, regulations such as HIPAA (Health Insurance Portability and Accountability) mandate that backup data should be kept for up to six years. In this scenario, an organization could use on-demand backups to create a backup of their DynamoDB data and configure **AWS Backup** to rotate backups using automation to ensure that the backup data is kept up to date and does not consume unnecessary storage space.

When an on-demand backup is created, the backup contains the full description of the table, including its schema, provisioned capacity settings, and any indexes. Additionally, the configuration for the backup itself, including the backup's unique identifier, the timestamp when the backup was created, and any applicable tags, is recorded in a backup object within DynamoDB. These configuration details make it easier to manage and organize backups, especially for large or complex tables.

One use case where you may not want to use on-demand backups is when you need to support the deletion of individual user records across your system due to certain geographical data privacy regulations. Deleting a single user record from within a backup may require restoring the entire backup to a new DynamoDB table, deleting the individual user record, and then creating a backup of the newly restored table. This can be a cumbersome process, especially if you have millions of items in the backup, just to delete a single item. In such scenarios, it may be more appropriate to use other features, such as native export to your own S3 buckets. This allows you to manipulate the data in a way that makes it inaccessible, such as deleting specific records, rather than relying on the backup alone.

On-demand backups in DynamoDB can be managed by either of two services, DynamoDB itself and **AWS Backup**. Let us review the features of each in terms of on-demand backups.

DynamoDB for on-demand backups

On-demand backups in DynamoDB can be managed natively by the service or by using AWS Backup. When backups are managed natively by DynamoDB, they are created and stored within the same AWS account as the DynamoDB table. This is different from when backups are managed by AWS Backup, where certain enhanced features such as cross-region copy and cross-account copy are also supported.

One advantage of using on-demand backups in DynamoDB is the instantaneous creation of backups regardless of table size. Unlike other databases, creating a backup in DynamoDB does not involve reading the data in the table. This means that backup creation is not impacted by table size, allowing users to create backups with minimal impact on application performance. Since write-ahead logs and periodic snapshots are used to serve on-demand backups, they are accumulated and archived on an ongoing basis. When a user requests an on-demand backup creation for a DynamoDB table, the service can quickly complete that request by simply recording the time at which the backup was created.

On the other hand, a significant advantage of on-demand backups is that multiple restores requested for a DynamoDB table can utilize relevant parts of the same archived write-ahead logs. This makes the overall restore process efficient, as it only needs to access the parts of the backup required to restore specific data. This contrasts with backup systems of other prominent databases where the entire backup must be restored before retrieving specific data, leading to longer restore times and increased costs.

AWS Backup for on-demand backups

AWS Backup provides features such as backup vaults and plans that allow for the centralized management of backups across AWS services, including DynamoDB. This can help to ensure compliance with backup policies across an organization and simplify the management of backups at scale. By default, DynamoDB is the service used for the management of backups in any AWS account. However, you may switch to using AWS Backup to manage your DynamoDB backups by turning on advanced features for AWS Backup from the DynamoDB management console's **Backups** page, as shown here:

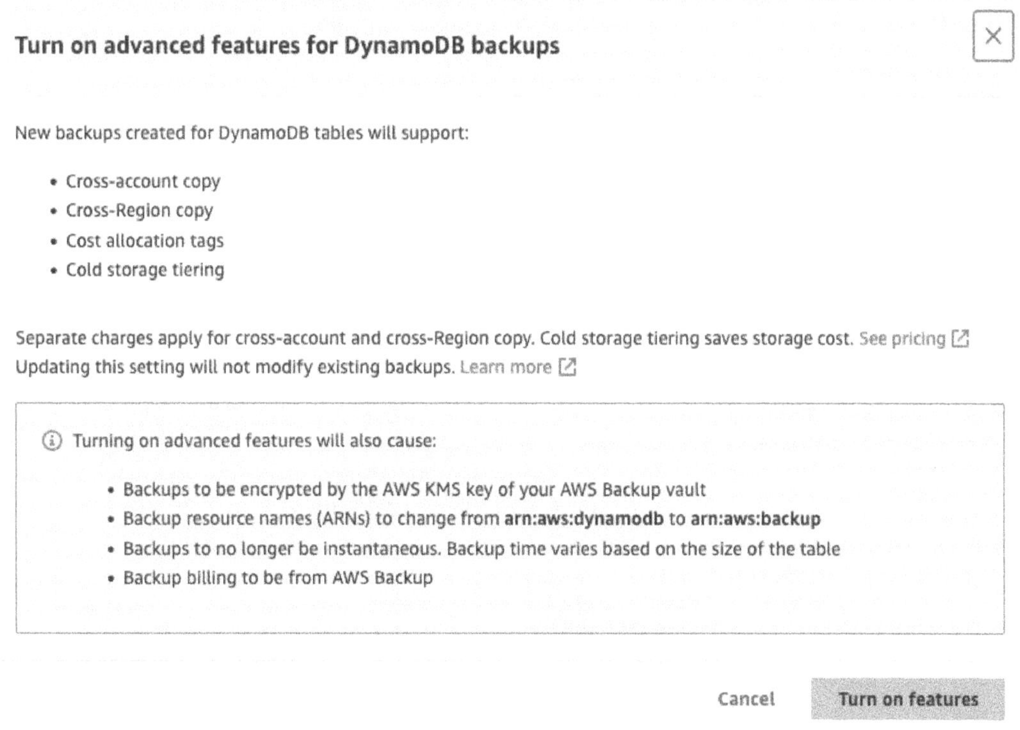

Turn on advanced features for DynamoDB backups   ⊠

New backups created for DynamoDB tables will support:

- Cross-account copy
- Cross-Region copy
- Cost allocation tags
- Cold storage tiering

Separate charges apply for cross-account and cross-Region copy. Cold storage tiering saves storage cost. See pricing ↗
Updating this setting will not modify existing backups. Learn more ↗

ⓘ Turning on advanced features will also cause:

- Backups to be encrypted by the AWS KMS key of your AWS Backup vault
- Backup resource names (ARNs) to change from **arn:aws:dynamodb** to **arn:aws:backup**
- Backups to no longer be instantaneous. Backup time varies based on the size of the table
- Backup billing to be from AWS Backup

Cancel   **Turn on features**

Figure 11.2 – AWS Backup enhanced features

On selecting **Turn on** in the DynamoDB management console, as shown in the preceding figure, all the previous on-demand backups that may have been retained in the AWS account would continue to be managed by DynamoDB alone, however, any new backups created would be managed and billed by AWS Backup as per their documented pricing (2).

By leveraging AWS Backup, an organization can create **backup vaults**, which are centralized repositories for backups that can be accessed across multiple AWS accounts and regions. Backup vaults also allow an organization to apply backup retention policies, such as how long to keep backups and automate the deletion of older backups when they are no longer required. This provides the organization with a scalable and efficient solution for managing their backup data while also ensuring compliance with data retention policies.

In addition to the standard same-region backup functionality, AWS Backup also allows enhanced features such as cross-account and cross-region copying of backups, as well as cold storage transitioning of backups and cost allocation tags on backups. To copy backups across accounts, the source and destination AWS accounts must be part of the same **AWS organization**. AWS Organizations is a service for managing accounts that allows you to bring together multiple AWS accounts into a single organization, which you have control over and can manage centrally. With AWS Organizations, you gain access to features such as account management and consolidated billing, empowering you to effectively address the financial, security, and compliance requirements of your organization. To learn more about copying DynamoDB backups across regions and accounts, see the AWS documentation (3).

Additionally, AWS Backup can also be used to transition backups to colder storage for cost efficiency. Colder storage options such as Amazon S3 Glacier or Amazon S3 Glacier Deep Archive offer lower storage costs, but higher retrieval times compared to standard storage. By creating an on-demand backup using AWS Backup, an organization can configure the backup to transition to colder storage after a certain period to save on storage costs. This ensures that the organization has a cost-effective backup solution that can be stored for extended periods as required by compliance regulations.

In summary, on-demand backups in DynamoDB offer the advantage of instant backup creation, irrespective of table size, without impacting application or database performance. This feature is particularly useful for scenarios where backups need to be created frequently or on an ad-hoc basis. By using AWS Backup to manage on-demand DynamoDB backups, users can benefit from enhanced features such as backup vaults and plans, allowing for automatic backup rotation and cold storage transition for cost efficiency. There are no downsides to using AWS Backup instead of DynamoDB directly to manage on-demand backups as they both use the same underlying building blocks in the DynamoDB system architecture.

The next section, *Continuous backups*, will explore another backup/restore feature in DynamoDB.

Continuous backups (PITR)

Continuous backups in DynamoDB are a feature that continuously backs up the data in your DynamoDB table, enabling the ability to restore the table data to any point in the past up to 35 days, or the point in time when continuous backups were last enabled on the table, whichever is more recent. This ability to restore to any point in time can potentially protect the table data from any accidental corruption. Continuous backups are also referred to as PITR.

Suppose you accidentally pointed a test application to a DynamoDB table in production, and the application started making unintended writes and deletes to the production table data. If you have a rough idea of when this data corruption began and the table had continuous backups enabled before that time, you could restore the table to a previous point in time before the corruption occurred. This would create a new DynamoDB table with the same data state that existed at the specified point in time, and you could use this restored table to fix the data corruption in the original live DynamoDB table.

The PITR feature of DynamoDB allows you to restore your table data to any second within a rolling 35-day window. This feature can be useful in scenarios where a table or its data is corrupted, accidentally deleted, or modified, or when there is a requirement to track and restore data changes over time. Using PITR, you can identify the time at which the table data was correct and restore the table to that point. This can prevent data loss and minimize downtime, as well as provide an easy way to recover from user errors, software bugs, or security breaches.

PITR also protects DynamoDB tables against accidental deletes by providing an automatically available system backup. In case a PITR-enabled DynamoDB table is deleted for whatever reason, DynamoDB automatically creates a system backup of the table with the latest known state of the data and makes it available to the user. Furthermore, AWS provides support for **deletion protection** for DynamoDB tables, which adds another layer of security by requiring the feature to be disabled before any table deletions can occur. This feature complements continuous backups and provides another layer of protection against accidental table deletions. Deletion protection is a checkbox on the DynamoDB management console or a Boolean parameter in the table management APIs.

I have personally assisted numerous customers who inadvertently deleted their production DynamoDB tables and sought immediate recovery assistance. A good number of them had PITR enabled, allowing a swift recovery once the PITR restore process was completed. However, a subset of customers had not implemented measures such as deletion protection and PITR, rendering them unable to recover their data.

Enabling PITR allows DynamoDB to use the same write-ahead logs and periodic snapshots that it archives in S3 buckets. Every write-ahead log entry is timestamped, so when a PITR is requested for a table with PITR enabled, DynamoDB applies entries from the archived write-ahead logs up until the requested timestamp to quickly create a new DynamoDB table with the state of data at that point in time.

Enabling PITR does not incur any performance overhead on a DynamoDB table when the service is backing up data or performing restores. For tables in production environments, it is recommended to have PITR enabled to prevent accidental corruption or deletion.

Next, let us learn about security in regard to backup and restore functionality within DynamoDB.

Securing backups (and restores)

Security is paramount for databases, and DynamoDB is no exception. When it comes to backups, security involves encrypting backup data and preventing unauthorized access during restoration.

For encryption, DynamoDB backups are automatically encrypted at rest using the **Key Management Service** (**KMS**) key associated with the DynamoDB table. However, when using AWS Backup for DynamoDB, you have the flexibility to configure a different KMS key for backup encryption.

To prevent unintended access to DynamoDB backups, IAM policies play a crucial role. Nearly every action in AWS can be restricted using IAM policies, as IAM entities are necessary for initiating actions. To secure on-demand backups, explicitly deny the `RestoreBackupFromBackup` and `DeleteBackup` actions for every IAM role or IAM user in the AWS account. See the following IAM policy for reference:

```
{
    "Version": "2012-10-17",
    "Statement": [{
        "Effect": "Deny",
        "Action": [
            "dynamodb:DeleteBackup",
            "dynamodb:RestoreTableFromBackup"
        ],
        "Resource": "arn:aws:dynamodb:*:123456789012:table/*"
    }]
}
```

For AWS Backup-managed DynamoDB backups, focus on the `StartAwsBackupJob` and `RestoreTableFromAwsBackup` actions. Additionally, employ AWS Backup Vault Lock in a **write-once-read-many** (**WORM**) scenario to safeguard backups against intentional or accidental deletion, ensuring a robust compliance posture.

Lastly, restrict IAM entities from the `RestoreTableToPointInTime` action to prevent them from restoring from continuous backups managed solely by DynamoDB and not AWS Backup.

The AWS documentation (4) has sample IAM policies to ensure the security of your DynamoDB backups is at the highest level. In the next section, we will learn about how restores work for both the backup features supported natively by DynamoDB.

Learning about restores

Restoring table data from backups reliably and efficiently in a predictable manner is crucial in ensuring business continuity and minimizing downtime. In the event of data corruption, accidental deletion, or other disasters, having reliable backups and a predictable restore process can help organizations recover their data quickly and without any loss of data. This ensures that operations can resume promptly and that there is no loss of revenue or reputation. Furthermore, efficient and reliable backups can help organizations meet compliance requirements, such as retaining data for a specific period.

Whether it is on-demand backups or continuous backups, DynamoDB uses similar service components to support the restoration of tables from backups. The write-ahead logs and periodic snapshots are crucial in supporting these native features. When a restore is requested, DynamoDB retrieves the archived logs and snapshots and uses the relevant information from those artifacts to create a new restored table. In the case of restoring an on-demand backup, the restored table reflects the state of the table data accurately to the point when the on-demand backup was created. On the other hand, when restoring a PITR-enabled table, the restored table data reflects the state of the time to which the restore was requested.

> **Important note**
>
> On-demand backups (via DynamoDB or AWS Backup) or continuous backups are always restored to a new DynamoDB table. The original tables are never impacted or modified as part of a restore.

The performance of restoring backups depends on several factors, including data distribution, backup size, and service-level parallelism thresholds. The skew of data across partition keys in a table is a major contributing factor to restore times. Tables with a highly cardinal partition key design and evenly balanced data distribution may have faster restores than tables with a low cardinal partition key. While DynamoDB does its best to address data skew during restores, addressing the skew may add latency to the overall restore process. Restores can be performed with or without secondary indexes, with restores without secondary indexes typically being more performant. The AWS documentation (5) indicates that about 95% of restores globally complete in under an hour. It's a best practice to regularly exercise restores of production environment backups to establish how restore times affect an application's overall recovery time objective.

By default, when restoring a table, configurations such as billing mode, read/write capacity, and table data encryption settings are copied from the source table at the time of the backup creation request or at the time the PITR is requested. However, these could be modified when requesting a restore. On the other hand, autoscaling settings, time to live (TTL) configurations, resource tags, and stream settings must always be set on a restored table and are not copied from backups. These characteristics of restores are valid for both backup/restore supported by DynamoDB, as well as backup/restore supported by AWS Backup. To learn more about restores, including how to issue restores via the AWS CLI or the AWS Management Console, see the AWS documentation in (6) and (7).

DynamoDB's native backup and restore features, including the enhanced capabilities offered by AWS Backup, provide fully managed solutions that do not grant direct access to the backed-up data. While this is an effective way to safeguard against data loss, some use cases may require access to the backed-up data for analytical purposes. In such scenarios, DynamoDB provides an additional native feature called **Export to S3**, which allows users to perform full data dumps of DynamoDB table data into their own S3 bucket. In this new section, we will delve into the Export to S3 feature, exploring its potential use cases and how it can facilitate efficient analytical access to DynamoDB data.

Export to S3

Export to S3 is a powerful feature in DynamoDB that enables users to export their table data to their own S3 buckets. While native backup and restore features supported by DynamoDB are fully managed, this feature offers additional control over the data dumps for analytical purposes. The Export to S3 feature is designed to be a simple and efficient method for DynamoDB users to export table data without the need for complex data pipelines or the risk of incurring high costs. At the time of writing, DynamoDB supports natively exporting table data in one of DynamoDB JSON or Amazon Ion (8) formats, with additional formats hopefully supported in the future.

The Export to S3 functionality offers two kinds of exports of your table data:

- Full export
- Incremental export

The full export, as its name implies, always performs a full data dump of a DynamoDB table to a specified S3 bucket and directory. Even if your table contains a billion items, the full export option would dump the billion items into several objects in S3.

On the other hand, the incremental export option is designed to export only those items that have changed in a certain period. As such, an incremental export requires a start and end timestamp, and the output in S3 is the delta of the table between those two timestamps.

Choosing between the two depends on the use case. Some scenarios may benefit from either option, while others might find value in utilizing the full table export once, followed by incremental exports in a daily batch job setup. Collectively, these export options provide the ability to copy DynamoDB table data into S3 to perform complex analytics, reporting, and AI/machine learning (ML) workflows.

To use the Export to S3 feature, it is mandatory to have PITR enabled. Enabling PITR or continuous backups allows DynamoDB to use archived write-ahead logs and periodic snapshots to perform the export. One of the significant advantages of the native Export to S3 feature is that it does not cause any performance overhead on DynamoDB tables. The export process is performed asynchronously in the background using the archived logs and snapshots, and it does not affect the performance of the table. This feature enables users to export table data without worrying about any impact on their application's performance.

Next, let us review some use cases where the Export to S3 feature could be and is used by customers of all shapes and sizes in their respective industries.

Use cases for Export to S3

One use case where the DynamoDB native Export to S3 feature can be leveraged is when data scientists need to perform complex analytical querying or train ML models using the DynamoDB table data. With this feature, data scientists can export the data from DynamoDB tables to S3 buckets in a format of their choice, which can then be loaded into their analytical tools, such as Amazon Redshift, Amazon Athena, or Apache Spark. They can then perform complex analytical queries or train ML models on the exported data without impacting the performance of the production DynamoDB tables.

Another use case would be a company that operates an e-commerce website and may use DynamoDB to store customer orders, user profiles, and product catalogs. The company's data scientists can use the native Export to S3 feature to extract data from DynamoDB tables and store them in S3 buckets. They can then use this data to perform complex analytics to improve customer experience, such as identifying popular products, predicting customer behavior, or detecting fraudulent activities. Additionally, they can use business intelligence and visualization tools, such as Tableau or Amazon QuickSight, to create interactive dashboards or reports to share their findings with other stakeholders. The following figure illustrates a similar pipeline that can be set up using this native Export to S3 feature.

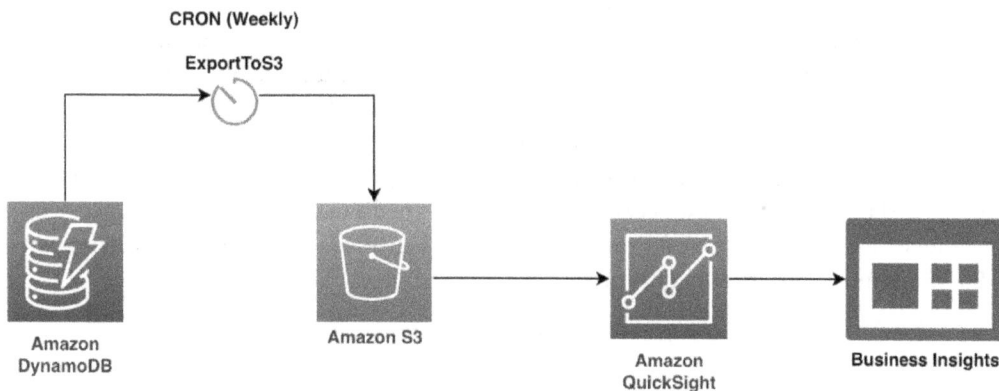

Figure 11.3 – Visualization pipeline using ExportToS3

Note that according to the preceding architecture, a new export will be created each time based on the cron setting. Therefore, the previously exported data may need to be cleaned up if there is no other use for the data. This is how some customers choose to use the export feature. By employing both the full export and incremental export features, the high-level architecture might resemble the following figure:

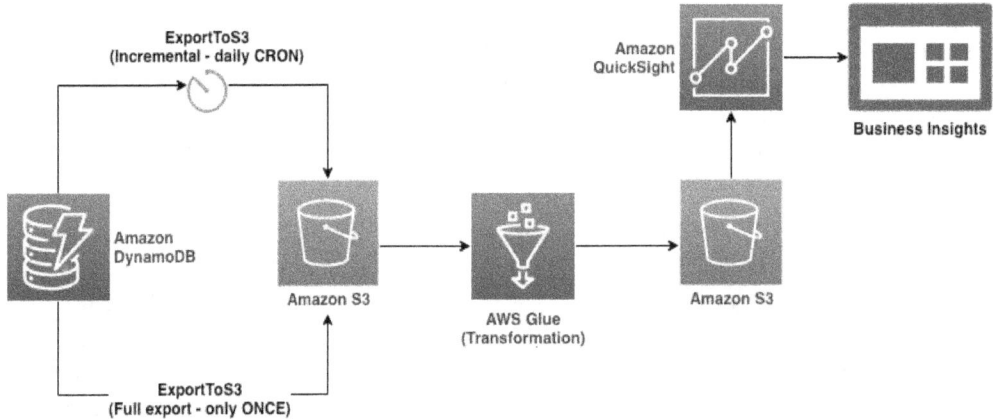

Figure 11.4 – Visualization pipeline using ExportToS3 – full and incremental export

As per the architecture in the previous figure, an initial one-time full export of DynamoDB table data is executed, followed by a scheduled daily or weekly cron job for the incremental export, capturing only the changes observed in the table data. Due to the differing structures between incremental and full exports, AWS Glue is employed to transform and merge the data into a unified, up-to-date copy within S3. This consolidated data is then utilized by Amazon QuickSight visualizations to generate insights for the business teams. Customers with large amounts of data in their DynamoDB tables tend to go this route since only a portion of their table data would change, and hence the incremental export would be much more cost-effective than performing a full export daily or weekly.

Now, let us look at a step-by-step guide on requesting an export of a DynamoDB table into S3.

Exporting table data into S3

Here's a step-by-step guide to exporting a DynamoDB table to S3 via the AWS Management Console:

1. Open the AWS Management Console and navigate to the **DynamoDB** service dashboard. From there, select the table you wish to export to S3 and click on the **Exports and streams** tab located in the top-right area of the screen. Then, hit the **Export to S3** button:

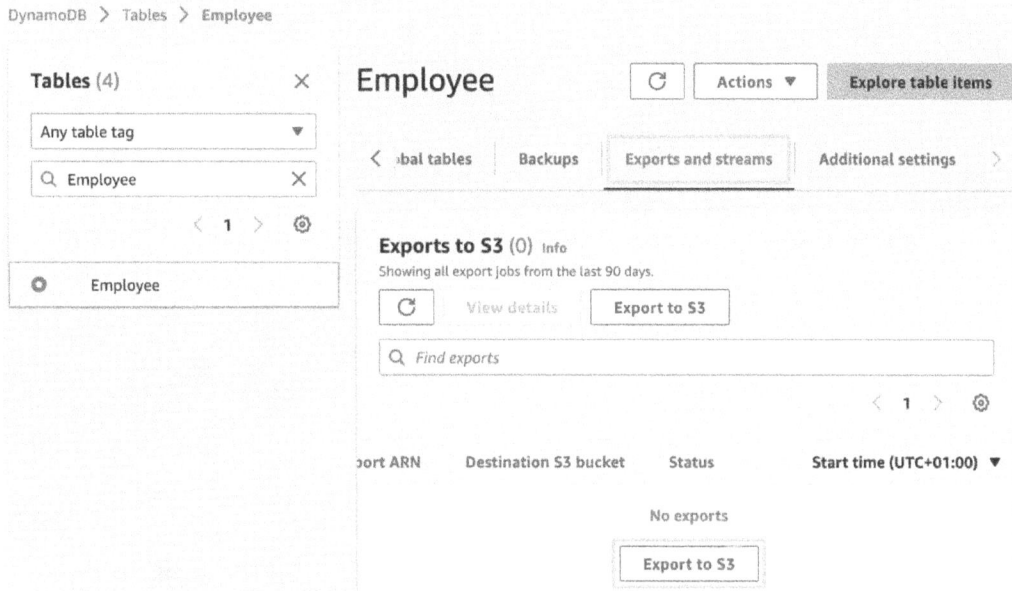

Figure 11.5 – AWS Management Console – the Exports and streams tab

2. In the Export to S3 wizard, the first step is to ensure that PITR is enabled on the DynamoDB table. PITR is mandatory for exporting to S3. After that, select the S3 bucket to which you want to export the data. You can either browse to the S3 location or enter the path of your S3 directory with the prefix. The S3 bucket can be located in the same or a different region or AWS account. In addition to specifying the S3 bucket path, you must also indicate whether the bucket is owned by the current account or a different AWS account. In either case, the user who requests the export must have permission to access and write to the S3 bucket:

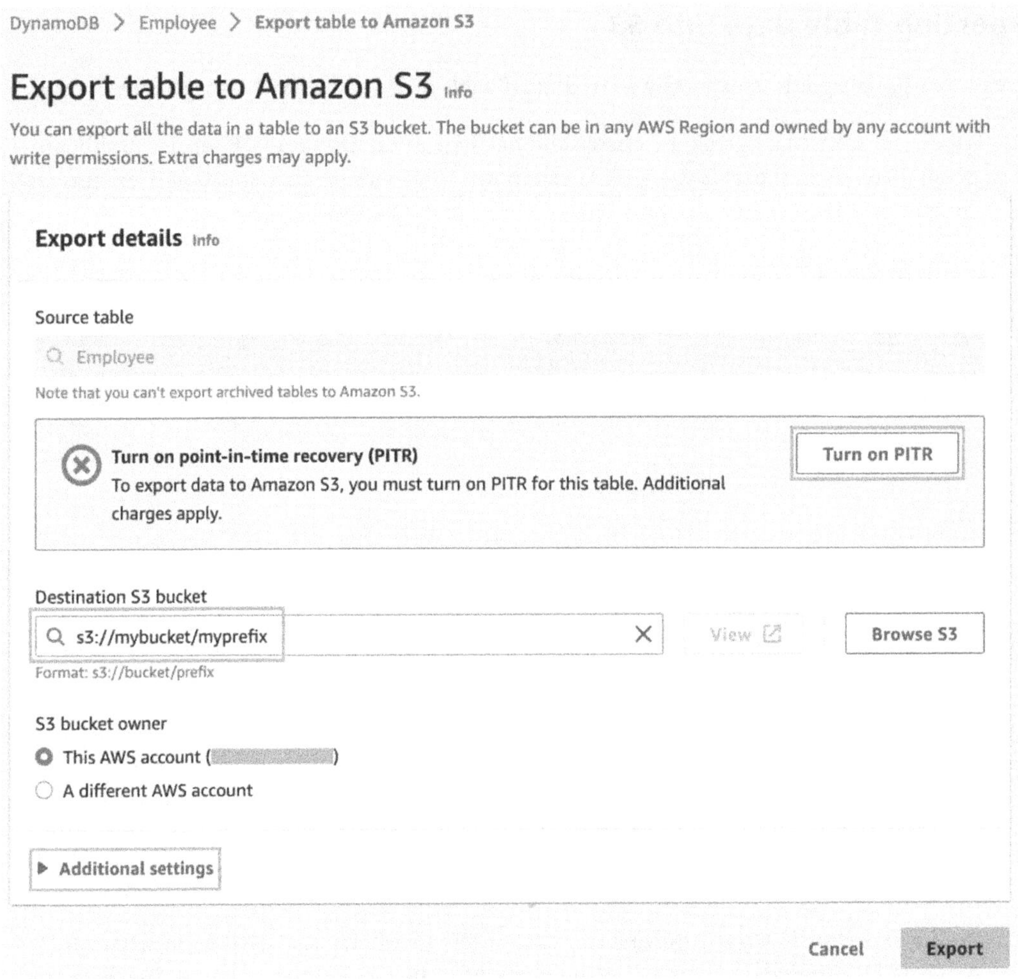

DynamoDB > Employee > **Export table to Amazon S3**

Export table to Amazon S3 Info

You can export all the data in a table to an S3 bucket. The bucket can be in any AWS Region and owned by any account with write permissions. Extra charges may apply.

Export details Info

Source table

Q Employee

Note that you can't export archived tables to Amazon S3.

> (✕) **Turn on point-in-time recovery (PITR)**   Turn on PITR
>   To export data to Amazon S3, you must turn on PITR for this table. Additional
>   charges apply.

Destination S3 bucket

Q s3://mybucket/myprefix   ✕   View ☐   Browse S3

Format: s3://bucket/prefix

S3 bucket owner

◉ This AWS account (▓▓▓▓▓▓▓▓▓▓)
○ A different AWS account

▶ Additional settings

Cancel   **Export**

Figure 11.6 – AWS Management Console – the Export to S3 wizard

3.   Next, navigate to the **Export settings** section as shown in the following figure, where you can specify your preferred export settings. You can choose the **Full export** option, exporting the entire table data, or the **Incremental export** option, which only exports data that has changed within specific time ranges. For the full export, you can pinpoint the specific point in time or opt for the latest version of the data. Both export options allow you to select the data format – either **DynamoDB JSON** or **Amazon Ion**. Additionally, you can configure encryption-at-rest settings for the exported files in the target S3 bucket. For now, choose the **Full export** option. Once you've chosen your configurations, click the **Export** button to initiate the process:

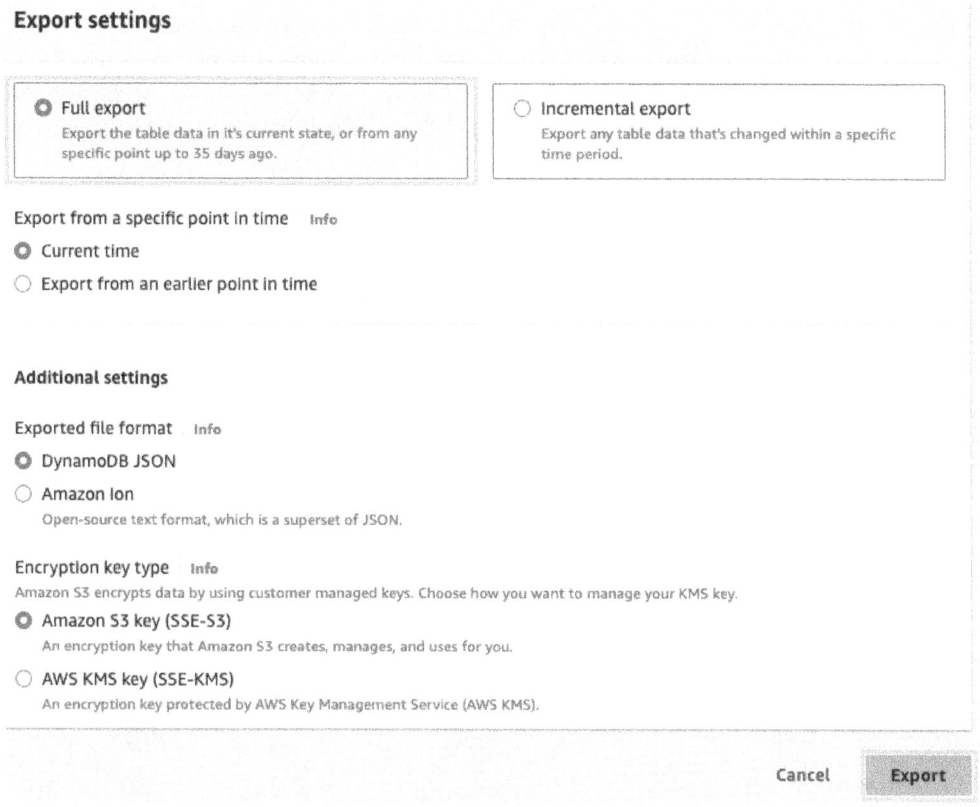

Export settings

● Full export	○ Incremental export
Export the table data in it's current state, or from any specific point up to 35 days ago.	Export any table data that's changed within a specific time period.

Export from a specific point in time   Info

● Current time

○ Export from an earlier point in time

Additional settings

Exported file format   Info

● DynamoDB JSON

○ Amazon Ion

Open-source text format, which is a superset of JSON.

Encryption key type   Info

Amazon S3 encrypts data by using customer managed keys. Choose how you want to manage your KMS key.

● Amazon S3 key (SSE-S3)

An encryption key that Amazon S3 creates, manages, and uses for you.

○ AWS KMS key (SSE-KMS)

An encryption key protected by AWS Key Management Service (AWS KMS).

Cancel   **Export**

Figure 11.7 – AWS Management Console – Export to S3 – Export settings

4. After initiating the export process, you should see the **Exporting** status on the console. Within a few minutes, depending on the table data size, the export should complete, with a **Completed** status. After the export completes, we will review the output files using the AWS CLI.

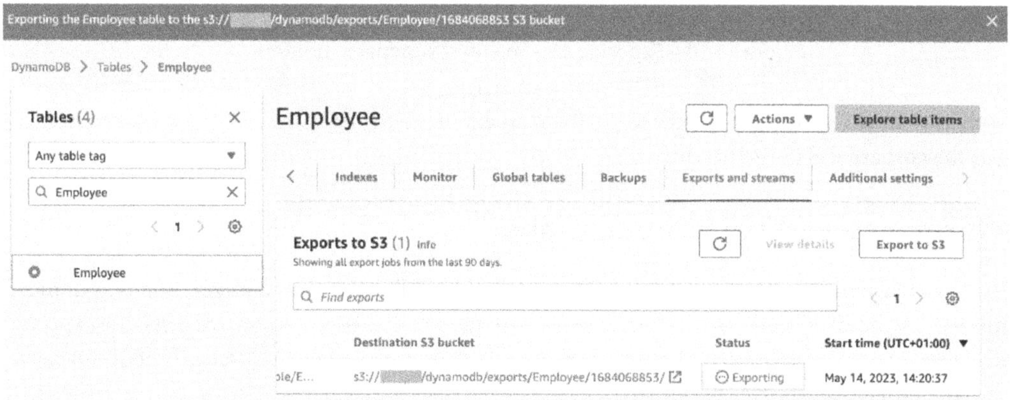

Figure 11.8 – AWS Management Console – Export to S3 – Status

5. On reviewing the output files, you will notice that all the data files are under the `/data/` prefix. In the case of the `Employee` table, there was only a single data file generated, due to the table data size being small. Along with the data file, there is also a bunch of metadata or manifest files generated as part of the export. The following figure shows all the files generated:

```
1 aws s3 ls s3://blooper/dynamodb/exports/Employee/1684068853/ --recursive
2 2023-05-14 14:20:48          0 dynamodb/exports/Employee/1684068853/AWSDynamoDB/01684070437624-dfbcc29c/_started
3 2023-05-14 14:22:12        362 dynamodb/exports/Employee/1684068853/AWSDynamoDB/01684070437624-dfbcc29c/data/6ubev7vkua6jzemaecezuzt5r4.json.gz
4 2023-05-14 14:22:49        232 dynamodb/exports/Employee/1684068853/AWSDynamoDB/01684070437624-dfbcc29c/manifest-files.json
5 2023-05-14 14:22:49         24 dynamodb/exports/Employee/1684068853/AWSDynamoDB/01684070437624-dfbcc29c/manifest-files.md5
6 2023-05-14 14:22:49        658 dynamodb/exports/Employee/1684068853/AWSDynamoDB/01684070437624-dfbcc29c/manifest-summary.json
7 2023-05-14 14:22:49         24 dynamodb/exports/Employee/1684068853/AWSDynamoDB/01684070437624-dfbcc29c/manifest-summary.md5
```

Figure 11.9 – Export to S3 – output files

6. On reviewing one of the manifest files in `manifest-summary.json`, you can see metadata about the export job:

```
 1 $ cat manifest-summary.json | jq
 2 {
 3     "version": "2020-06-30",
 4     "exportArn": "arn:aws:dynamodb:eu-west-2:012345678910:table/Employee/export/01684070437624-dfbcc29c",
 5     "startTime": "2023-06-14T13:20:37.624Z",
 6     "endTime": "2023-06-14T13:22:48.272Z",
 7     "tableArn": "arn:aws:dynamodb:eu-west-2:012345678910:table/Employee",
 8     "tableId": "40b3aadb-41ea-aaaa-bbbb-df5b58988ddf",
 9     "exportTime": "2023-06-14T13:20:37.624Z",
10     "s3Bucket": "mybucket",
11     "s3Prefix": "dynamodb/exports/Employee/1684068853",
12     "s3SseAlgorithm": "AES256",
13     "s3SseKmsKeyId": null,
14     "manifestFilesS3Key": "dynamodb/exports/Employee/1684068853/AWSDynamoDB/01684070437624-dfbcc29c/manifest-files.json",
15     "billedSizeBytes": 925,
16     "itemCount": 9,
17     "outputFormat": "DYNAMODB_JSON"
18 }
```

Figure 11.10 – Export to S3 – manifest summary contents

Now that we have learned about how to perform an export to S3, let us review the performance as well as the cost aspects of the feature.

Performance and cost implications

Every export job roughly results in 1 GB (when uncompressed) per output data file, and output files themselves are GZIP-compressed with an approximate ratio of 4:1. Based on several tests, the exports almost always complete in under 30 minutes, with table sizes under a few hundred GBs, but this may not be true for all exports ever performed.

In terms of costs, exports are charged per GB of table size by DynamoDB. You may also incur additional costs of S3 PUTs (1 PUT per 64 MB of data) made by DynamoDB to write the exported files on your behalf. Finally, if you choose to export the data into a cross-region bucket, additional data transfer costs may apply for data going out of the region where the table resides.

To help understand the overall costs associated with an export, consider the following example where an export is to be requested for a 20 GB DynamoDB table:

- **Number of S3 objects**: (data size / object size) = 20 GB / 1 GB = 20

- **S3 storage size**: (data size / compression ratio) = 20 GB / 4 = ~5 GB

- **S3 PUT calls**: (S3 storage size/ 64 MB) = 5 GB / 64 MB = ~80

- **DynamoDB costs (London/eu-west-2 region, $0.11886 per GB)**: 0.11886 x 20= ~$2.37

- **S3 costs (London/eu-west-2 region, $0.0053 per 1000 PUTs)**: 0.0053 x 80 = ~$0.42

- **Overall costs**: $2.79

More information about Export to S3 can be found in the AWS documentation (9), including an AWS blog (10) that covers more about the feature.

Now that we have learned about the performance and cost aspects associated with the native exports to S3, let us see how the use cases, when using the native export, may not be the optimal choice and what alternatives are commonly used for supporting complex analytics on DynamoDB data.

Limitations

While the native Export to S3 feature for DynamoDB is a powerful tool, it does come with some limitations. One major limitation is that the exported data is in a raw format, which may not be optimal for direct use in some analytical tools such as Amazon Redshift or Amazon Athena. This means that users may need to perform some additional processing on the data before it can be used in their analysis. Additionally, the native Export to S3 feature only supports full exports of a DynamoDB table, which may not be necessary or desirable in all cases.

Additionally, the native Export to S3 feature is not designed for real-time data synchronization or replication. It is a batch-oriented process that exports the data periodically based on a predefined schedule or a manual trigger. Therefore, it may not be suitable for use cases that require real-time or near-real-time access to the data.

Lastly, the native Export to S3 feature does not provide any advanced data transformation or cleansing capabilities. Users may need to implement their own processing pipeline or use third-party tools to prepare the data for analysis. In terms of alternatives, AWS Glue is one of the most used options that could overcome several of the limitations of the native export feature. AWS Glue could transform the data as desired and also support partial dumps into S3 instead of the full data dump each time. To process DynamoDB changes in real-time or near-real-time, a solution around DynamoDB Streams or Kinesis Data Streams for DynamoDB may be better suited. To learn more about streaming options for DynamoDB, see *Chapter 12, Streams and TTL*.

In this section, we explored the native Export to S3 feature of DynamoDB, which allows users to perform a full data dump of their DynamoDB tables to their specified S3 bucket and directory. We discussed how this feature does not incur any performance overhead on DynamoDB tables and can be used for analytical purposes. We also provided a step-by-step guide to using the AWS Management Console to perform a native Export to S3, including screenshots.

Moving forward, let's explore the process of natively importing data from S3 into DynamoDB. In the next section, we will discuss the benefits of importing data from S3, the supported data formats, and how to perform an import using the AWS Management Console. We will also discuss any limitations and best practices for importing data from S3.

Import from S3

Importing data into a DynamoDB table can be a challenging task, especially when dealing with large amounts of data. In some cases, these ingestions could be nightly bulk-loading jobs or part of migrating data into DynamoDB from different sources or cloud service providers. However, the native **Import from S3** feature in DynamoDB simplifies the process by enabling users to import data directly from an S3 bucket into a DynamoDB table. This feature is particularly helpful when there is a requirement to ingest significant amounts of data into DynamoDB tables without the need for setting up complex data pipelines or inefficient ingestion systems.

Sample use case

For example, consider a scenario where a team wants to create a new DynamoDB table to store product information for an e-commerce application. They have the product data available in CSV format, stored in an S3 bucket.

Instead of manually writing code to parse the CSV file and insert the data into the DynamoDB table, the team can leverage the native Import from S3 feature. They can configure the import settings to specify the S3 bucket, data format (CSV), and delimiter. By following the step-by-step process, the team can seamlessly import the CSV data into the DynamoDB table.

This use case demonstrates how the native Import from S3 feature allows users to easily seed a new DynamoDB table without the need for custom code or complex data transformations. It simplifies the process and saves time, especially when dealing with large datasets. The team can focus on defining the table schema and quickly import the data from S3, kickstarting their application's development without getting into the intricacies of data parsing and insertion.

Importing from S3 in action

Let us import the same data we exported using the native export feature earlier in this chapter into a new DynamoDB table using the native Import from S3 feature:

1. To start, we need to locate the output of the native export. You can either go to the DynamoDB console and find the exported path on the table you chose to export or browse to the exported path via the Amazon S3 console if you have an idea of the export location. The data files will be located under /data/prefix in the S3 bucket directory chosen while performing the native export. Use the **Copy S3 URI** option to copy the path to the data files so that we can use it when requesting a new DynamoDB import of this data.

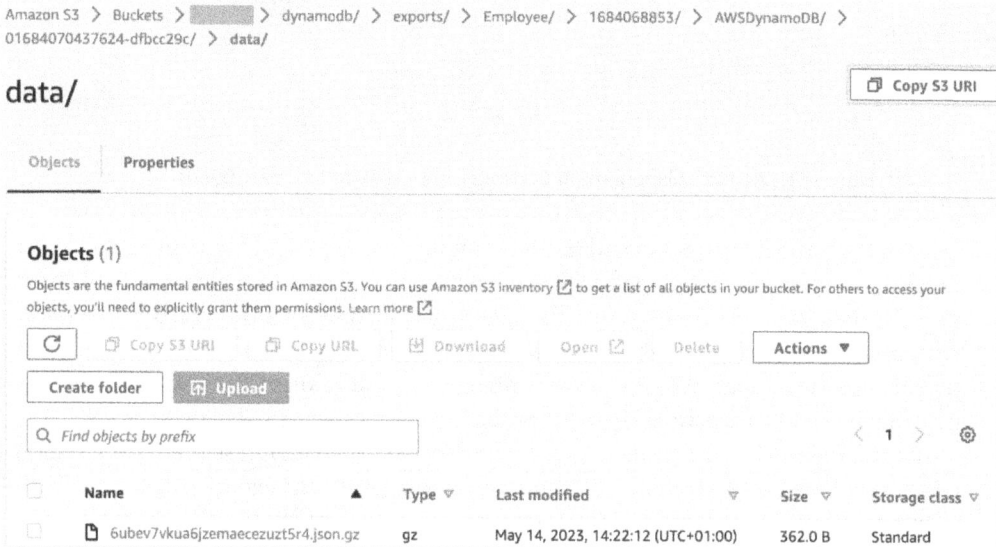

Figure 11.11 – AWS Management Console – exported S3 data location

2. Navigate to the DynamoDB console and select **Imports from S3** from the navigation pane on the left. This will take you to a page where you can view past import jobs that may have been requested. To initiate a new import, click on the **Import from S3** button to open the **Import from S3** wizard:

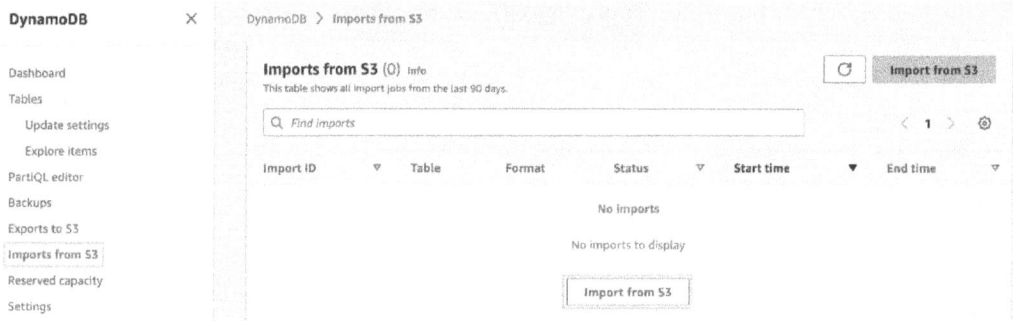

Figure 11.12 – AWS Management Console – Import from S3 – dashboard

3. *Step 1* of the **Import from S3** wizard involves entering the S3 path of the source data. To get the S3 path, you can use the **Copy S3 URI** option from the S3 console, which you should have copied earlier. Paste the contents into the appropriate field in the **Import from S3** wizard. Next, you will need to specify the owner of the S3 bucket where the source data resides. This can either be the current AWS account or a different one. Regardless, the user requesting the native import must have permission to access the S3 bucket being passed in the source S3 URL. Finally, you will need to specify the source data compression and import settings. For instance, in the case of the native DynamoDB export we created earlier, the exported data was GZIP-compressed DynamoDB JSON. Therefore, we will select **DynamoDB JSON** as the source data type and **GZIP** as the compression format. The following figure is a screenshot of the first step of the **Import from S3** wizard where you need to specify the S3 path and other relevant details.

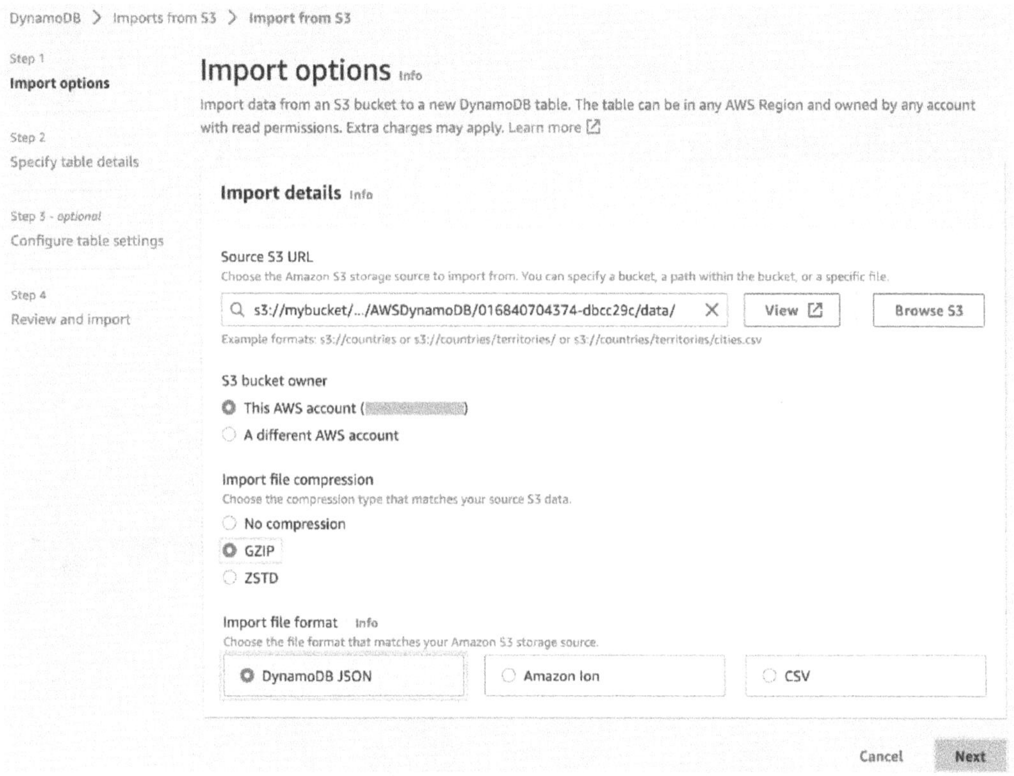

DynamoDB > Imports from S3 > Import from S3

Step 1
Import options

Step 2
Specify table details

Step 3 - *optional*
Configure table settings

Step 4
Review and import

Import options Info

Import data from an S3 bucket to a new DynamoDB table. The table can be in any AWS Region and owned by any account with read permissions. Extra charges may apply. Learn more [↗]

Import details Info

Source S3 URL

Choose the Amazon S3 storage source to import from. You can specify a bucket, a path within the bucket, or a specific file.

| Q s3://mybucket/.../AWSDynamoDB/016840704374-dbcc29c/data/   ✕ | View [↗] | Browse S3 |

Example formats: s3://countries or s3://countries/territories/ or s3://countries/territories/cities.csv

S3 bucket owner

◉ This AWS account (▨▨▨▨▨▨▨▨)

○ A different AWS account

Import file compression

Choose the compression type that matches your source S3 data.

○ No compression

◉ GZIP

○ ZSTD

Import file format   Info

Choose the file format that matches your Amazon S3 storage source.

| ◉ DynamoDB JSON | ○ Amazon Ion | ○ CSV |

Cancel   **Next**

Figure 11.13 – AWS Management Console – Import from S3 – Import options

4.   Next, we specify the target table schema including the table name. Since the exported data was from the Employee table, we must use the same schema as the Employee table to see a successful import. The table had only a partition key in LoginAlias of the String type and no sort key. Enter the same details and hit **Next**.

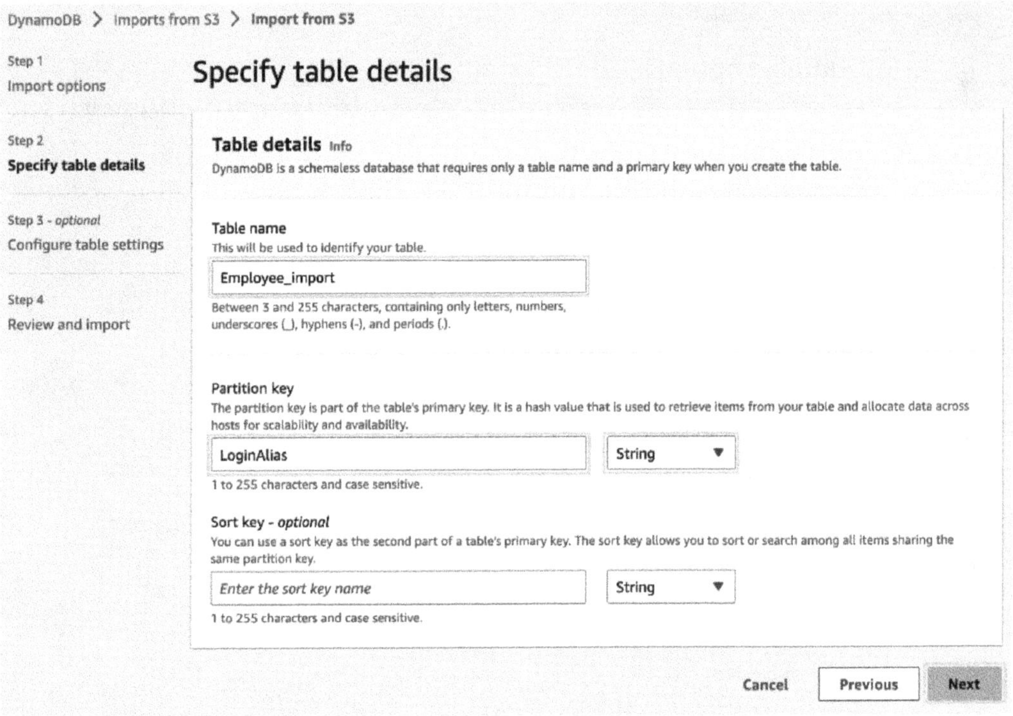

DynamoDB > Imports from S3 > **Import from S3**

Step 1 Import options	# Specify table details

Table details Info

DynamoDB is a schemaless database that requires only a table name and a primary key when you create the table.

Step 2
Specify table details

Step 3 - *optional*
Configure table settings

Table name
This will be used to identify your table.

```
Employee_import
```

Between 3 and 255 characters, containing only letters, numbers, underscores (_), hyphens (-), and periods (.).

Step 4
Review and import

Partition key
The partition key is part of the table's primary key. It is a hash value that is used to retrieve items from your table and allocate data across hosts for scalability and availability.

```
LoginAlias
```
String ▼

1 to 255 characters and case sensitive.

Sort key - optional
You can use a sort key as the second part of a table's primary key. The sort key allows you to sort or search among all items sharing the same partition key.

```
Enter the sort key name
```
String ▼

1 to 255 characters and case sensitive.

Cancel   Previous   **Next**

Figure 11.14 – AWS Management Console – Import from S3 – target table schema

5. After configuring the target table schema, we set the target table's capacity mode and other settings. By default, the target table gets created in the provisioned capacity mode with five `WriteCapacityUnits` and `ReadCapacityUnits`. Auto scaling is enabled on the target table by default and encryption-at-rest is done using DynamoDB-owned keys. The following figure shows the default table settings as is, with the option to customize them. Once these table settings are configured, hit **Next** to proceed.

DynamoDB > Imports from S3 > Import from S3

Step 1
Import options

Step 2
Specify table details

Step 3 - *optional*
Configure table settings

Step 4
Review and import

Configure table settings - *optional*

Table settings

○ **Default settings**
The fastest way to create your table. You can modify these settings now or after your table has been created.

○ **Customize settings**
Use these advanced features to make DynamoDB work better for your needs.

Default table settings
These are the default settings for your new table. You can change some of these settings after creating the table.

Setting	Value	Editable after creation
Capacity mode	Provisioned	Yes
Provisioned read capacity	5 RCU	Yes
Provisioned write capacity	5 WCU	Yes
Auto scaling	On	Yes
Local secondary indexes	-	No
Global secondary indexes	-	Yes
Encryption key management	Owned by Amazon DynamoDB	Yes
Table class	DynamoDB Standard	Yes
Deletion protection	Off	Yes

Cancel   Previous   Next

Figure 11.15 – AWS Management Console – Import from S3 – Table settings

The next screen will show you a summary of the configurations set, including the target table settings and source data S3 path. Hit **Import** to kick off the **Import from S3** process. This is where DynamoDB would try to access the source data, create the target DynamoDB table, and attempt importing data into the table from the source S3 files.

Performance and cost implications

Although the Import feature supports importing data into the DynamoDB table along with secondary indexes as part of the import, the overall import process may be fastest if secondary indexes were not being created and populated as part of the import job. You can always create global secondary indexes after the initial native import is completed. From a cost perspective, any secondary indexes that may be created as part of the import process are populated free of cost.

Tiny item sizes (typically below 200 bytes) in source data may see longer import durations than larger item sizes. Also, import speed may be most optimal when the data in source files is randomly shuffled across files as opposed to being in sorted order of partition and sort key combination.

More about how the Import from S3 feature works can be found in the AWS documentation (11), including the exhaustive list of best practices (12). The imports are charged at a flat per GB rate of source data in S3 when uncompressed (13).

Limitations

At the time of writing, the native Import from S3 feature in DynamoDB only allows for importing data into a new DynamoDB table. This means that the target table is created as part of the import from the S3 job, and it is not possible to natively import data into an existing DynamoDB table. However, this may change in the future as the feature evolves. If you need to import data into an existing table, you may need to use custom applications, such as AWS Glue-based ingestion workflows. The feature supports importing data in CSV, DynamoDB JSON, and Amazon Ion formats.

When using the native Import from S3 feature to import CSV data into a DynamoDB table, all attributes other than the primary key attributes of the table, including those of any secondary indexes being created as part of the import, are considered to be DynamoDB strings. This may not always be the best option for use cases where attributes are stored and accessed using non-scalar data types such as `Map` and `List` or scalar types such as `Number` and `Binary`.

An alternative for the native Import from S3 feature would be a fully managed, serverless solution involving **AWS Glue**, but with the added cost of implementation of the Glue script.

> **Important note**
> Consider this as a reminder to delete any resources you may have created while going through this chapter in your AWS account.

Summary

In this chapter, we've comprehensively addressed DynamoDB backup and restoration. We started by discussing the various backup and restore options in DynamoDB, highlighting their respective advantages and limitations. Subsequently, we explored the native backup and restore functionalities, including PITR, on-demand backup, and Export to S3/Import from S3 features.

We also delved into best practices for DynamoDB backup and restore, emphasizing the importance of well-designed partition keys and efficient data distribution across them to optimize restore times. Additionally, we stressed the significance of documenting restore times to establish clear recovery objectives for your applications.

Toward the chapter's end, we gained insights into native export and import features, accompanied by sample use cases, feature limitations, and step-by-step guides for executing export and imports.

In our upcoming chapter, *Streams and TTL*, we'll dive into two pivotal DynamoDB features: DynamoDB Streams and TTL. We'll discuss their benefits, practical use cases, configuration, and management, and share best practices for seamless implementation.

References

1. Write-ahead logging: https://en.wikipedia.org/wiki/Write-ahead_logging

2. AWS Backup pricing: https://aws.amazon.com/backup/pricing/

3. Copying a backup of a DynamoDB table with AWS Backup: https://docs.aws.amazon.com/amazondynamodb/latest/developerguide/CrossRegionAccountCopyAWS.html

4. Using IAM with DynamoDB backup and restore: https://docs.aws.amazon.com/amazondynamodb/latest/developerguide/backuprestore_IAM.html

5. DynamoDB managed restores:

    A. https://docs.aws.amazon.com/amazondynamodb/latest/developerguide/CreateBackup.html#CreateBackup_HowItWorks-restore

    B. https://docs.aws.amazon.com/amazondynamodb/latest/developerguide/CreateBackup.html

    C. https://docs.aws.amazon.com/amazondynamodb/latest/developerguide/CreateBackup.html

6. Using DynamoDB backup and restore: https://docs.aws.amazon.com/amazondynamodb/latest/developerguide/backuprestore_HowItWorks.html

7. Backup and restore with AWS Backup:

    A. https://docs.aws.amazon.com/amazondynamodb/latest/developerguide/backuprestore_HowItWorksAWS.html

    B. https://docs.aws.amazon.com/amazondynamodb/latest/developerguide/CrossRegionAccountCopyAWS.html

    C. https://docs.aws.amazon.com/amazondynamodb/latest/developerguide/CreateBackup.html

8. Amazon Ion: https://amazon-ion.github.io/ion-docs/

9. DynamoDB data export to Amazon S3: how it works: `https://docs.aws.amazon.com/amazondynamodb/latest/developerguide/S3DataExport.HowItWorks.html`

10. New – Export Amazon DynamoDB Table Data to Your Data Lake in Amazon S3, No Code Writing Required: `https://aws.amazon.com/blogs/aws/new-export-amazon-dynamodb-table-data-to-data-lake-amazon-s3/`

11. DynamoDB data import from Amazon S3: how it works: `https://docs.aws.amazon.com/amazondynamodb/latest/developerguide/S3DataImport.HowItWorks.html`

12. Best practices for importing from Amazon S3 into DynamoDB: `https://docs.aws.amazon.com/amazondynamodb/latest/developerguide/S3DataImport.BestPractices.html`

13. Pricing for Provisioned Capacity: `https://aws.amazon.com/dynamodb/pricing/provisioned/`

14. GitHub repository: `https://github.com/PacktPublishing/Amazon-DynamoDB---The-Definitive-Guide`

12
Streams and TTL

One of the popular architectural patterns that's used in modern applications is event-driven downstream processing. This involves performing work as a reaction to something that happened upstream. This also means working only when needed, as opposed to having "always on" systems that may spend a considerable amount of time in an idle state. Event-driven architectural patterns align well with the **pay-as-you-go** model of cloud computing.

Although the crux of DynamoDB usage is storing and retrieving data efficiently, real-life use cases often require more than just that. An application backed by DynamoDB may be highly performant in terms of OLTP operations, but often, there is a need to perform additional actions with the data. Some of these actions may involve tracking/auditing database activity, archiving data in a data lake, downstream processing the data into an analytics tool, or triggering actions downstream based on the changed state of data in the table. Having access to a change log for a database can unlock several possibilities, depending on the use case and nature of the application. In terms of DynamoDB, two options are available for accessing a change log – **DynamoDB Streams** and **Kinesis Data Streams (KDS) for DynamoDB**.

Another functionality that's supported by DynamoDB is called **Time to Live** (**TTL**), which allows us to delete data from a table that may have lost its relevance over time. TTL can help us manage storing (and thus storage costs) a DynamoDB table automatically by clearing out data that is no longer needed. Utilizing TTL is also beneficial as the deletes made by TTL do not consume any throughput and are free of cost.

In this chapter, we will review the stream options for DynamoDB tables, learn how they can be utilized, and the functional differences between the two. We will also learn about the TTL functionality and best practices for using TTL.

By the end of this chapter, you will understand how streams work, how to choose between the two streaming options for DynamoDB, know about common consumers for streams, and be able to utilize TTL for different use cases.

In this chapter, we're going to cover the following main topics:

- DynamoDB Streams
- KDS integration
- AWS Lambda as a consumer for streams
- **Kinesis Client Library** (**KCL**)-based consumer applications
- DynamoDB TTL

DynamoDB Streams

DynamoDB Streams provides a change log for all mutations that take place on a DynamoDB table. Such a change log in databases is also referred to as **change data capture** (**CDC**). A mutation could be a new item being inserted or an existing item being updated or deleted in the table. Whenever a put, update, or delete is successfully performed on a DynamoDB table, a stream event is generated in the DynamoDB stream associated with the table describing the change. If multiple mutations take place as part of a batch operation, each action will result in a stream event describing the individual item change.

You can configure the information that will be available in the stream events with one of the following options:

- KEYS_ONLY: Only key attributes will be available in the stream events
- NEW_IMAGE: The complete updated version of an item after it has been modified
- OLD_IMAGE: The complete original version of the item before the modification
- NEW_AND_OLD_IMAGES: Both the updated version and the previous version of the item

If the stream is configured to contain NEW_AND_OLD_IMAGES, a stream event that's generated as part of a new item being written would contain a new image, but not the old image. Similarly, a stream event that's generated as part of an item being deleted would contain its old image, but not the new image. If an item is being modified/updated, both the new and old images will be available in the corresponding stream event.

Feature characteristics

DynamoDB Streams comes with certain guarantees relating to the nature of stream event delivery:

- Exactly once delivery
- In-order delivery of item-level modifications

Whenever a mutation is successfully made to a DynamoDB item, exactly one event is generated and made available in the stream. This can be beneficial in building downstream processing systems that must react to every mutation without fail.

DynamoDB Streams retains change events for about 24 hours. This means that you could go back and forth within a stream to re-process events if needed. The maximum time you can go back to the past is 24 hours. The retention period for DynamoDB Streams is not configurable.

How it works

A stream record is a fundamental component of a DynamoDB stream. Multiple stream records may be grouped into **shards**. A shard could contain zero to an unlimited number of stream records. Every stream has a set of logical shard chains. Shard chains are more of a virtual group than something that can be explicitly called out anywhere, so consider these as a learning aid. Each shard chain has a 1:1 mapping with the DynamoDB table partitions themselves. The following figure demonstrates this mapping between table partitions and a logical group of shards – that is, via shard chains:

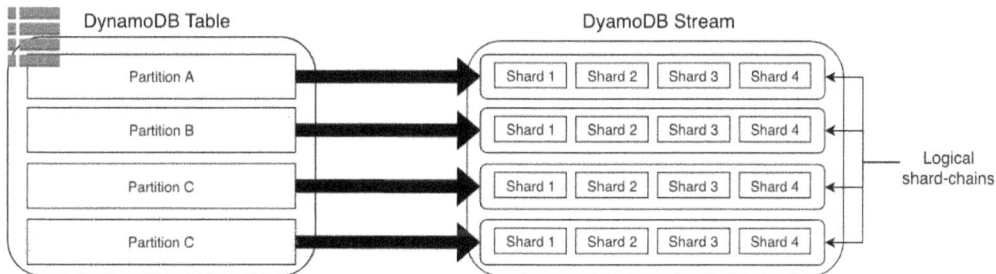

Figure 12.1 – DynamoDB stream mapping with a table

Each shard chain has a lineage (parent-child relationship) of shards, where each shard is created and stays open for approximately 4 hours. Only open shards actively receive new stream records when the corresponding table partition sees item mutations. After the shard has been open for 4 hours, it is marked as closed, and a new open shard is created in the same shard chain, which starts receiving new stream records. Irrespective of a shard being open or closed, it can still be read from, so long as the stream records in the shards are within the stream retention period of 24 hours. The following figure illustrates a single shard chain and its shard lineage:

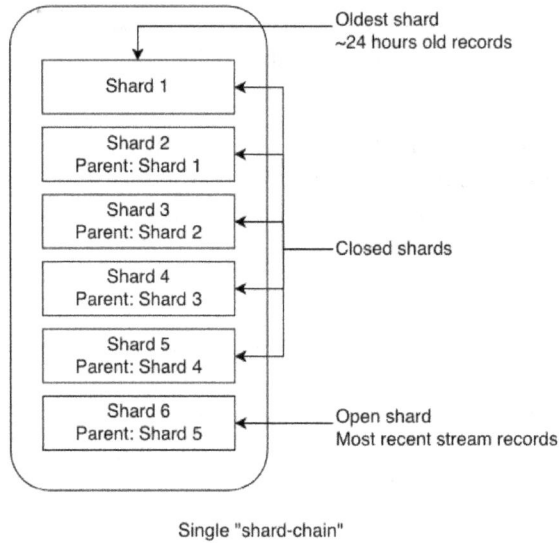

Figure 12.2 – Zooming into a single shard chain

It is also possible that a shard is closed in less than 4 hours when the shard chain's corresponding DynamoDB table partition splits. In the case of a table partition split, the corresponding DynamoDB stream shard chain also splits into two child chains. This also increases the potential parallelism that can be achieved by processing two child chains simultaneously instead of a single parent chain. The following figure illustrates the effect of a DynamoDB table partition split on its corresponding DynamoDB stream:

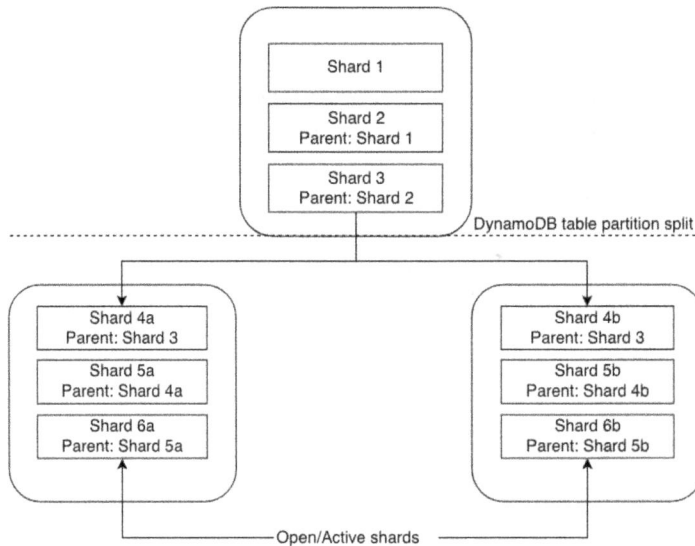

Figure 12.3 – The effect of a DynamoDB table partition split on a stream

Since the retention period of records in the stream is 24 hours from the time they were generated and considering the ~4-hour open life of a shard, at any given time, there may be 6–7 shards in each shard chain, all in a parent-child lineage. This 4-hour shard rotation behavior may change in the future as this is not documented much, but the parent-child lineage will always remain between old shards that split into new ones.

Consuming records from a DynamoDB stream

Reading or consuming events from a DynamoDB stream requires you to interact with multiple service APIs. We'll look at the available DynamoDB stream APIs individually first and then learn about the complete flow. The API specifications are available in the AWS docs (*1*).

DynamoDB stream APIs

Let's take a look at a few DynamoDB stream APIs that are available:

- **ListStreams**: This API lists all the DynamoDB streams associated with an AWS account in a particular AWS region.

- **DescribeStream**: This API returns metadata about the stream, the DynamoDB table it is associated with, and a paginated list of shards that can be consumed.

- **GetShardIterator**: This API returns a shard iterator that is valid for 15 minutes after it's returned to the requestor. A shard iterator can be thought of as a positional element within a shard that may or may not hold any stream records. The initial position within the shard that needs to be read from can be determined by the `ShardIteratorType` parameter, which can be one of the following:

  - `AT_SEQUENCE_NUMBER`: Start reading exactly from the position denoted by a specific sequence number. Each record arriving in the stream is assigned a unique sequence number.

  - `AFTER_SEQUENCE_NUMBER`: Start reading right after the position denoted by a specific sequence number.

  - `TRIM_HORIZON`: Start reading from the oldest available position within the shard. The oldest a record could be is around 24 hours, which is also the stream retention period.

  - `LATEST`: Start reading from the tip of the stream – that, is after the most recent stream record in the shard. For an open shard, new records can be expected when using `LATEST` as the shard iterator type. In the case of a closed shard, records may not be returned when using `LATEST` as the shard iterator type.

  - `GetRecords`: Returns an array of stream records from a given shard. It may also contain `NextShardIterator`, which is a shard iterator that points to the consequent page of records. It requires `ShardIterator`, which could either be obtained via a `GetShardIterator` call or from the response of a previous `GetRecords` call made against the same shard.

The flow of API calls to consume stream records

Reading from the stream would require calling all the APIs listed in the previous section. A typical consumer application would need to make the API calls in the following order:

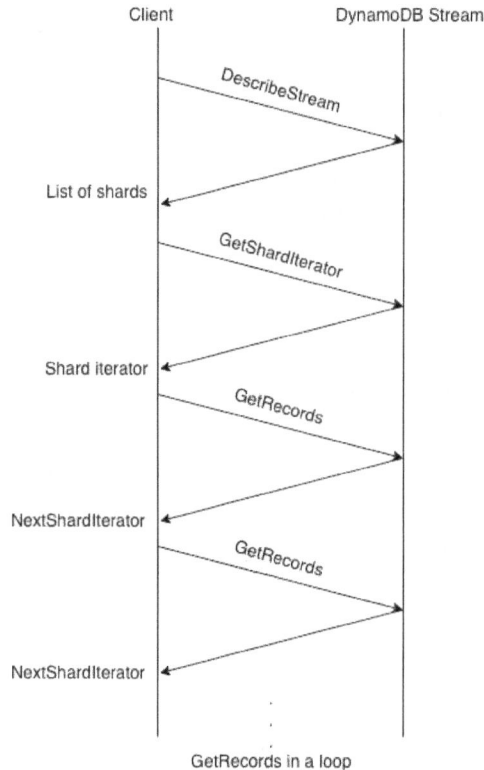

Figure 12.4 – API flow for consuming from a DynamoDB Stream

As you can see, the client would need to call DescribeStream to obtain the list of shards that exist in the stream. This may also require calling ListStreams first if the stream's **Amazon Resource Name (ARN)** isn't known. For a shard returned from the DescribeStream call, the client must then call GetShardIterator and specify the position within the shard that the client needs to start reading stream records from. The response of GetShardIterator would contain a shard iterator string that needs to be passed as a parameter in the GetRecords call. The GetRecords response would return zero or more stream records in an array, along with a NextShardIterator property that must be passed in the subsequent GetRecords call. The GetRecords calls from here on out can be made in a loop so that you can keep reading from the shard by utilizing the NextShardIterator values from the previous responses. When the client has reached the end of a shard and no more records are unread in the same shard, the NextShardIterator property will have a null value. This indicates that the client has reached the end of the shard.

Parent-child shard lineage is important to ensure that the stream records are consumed and processed in the same order as they were generated. A child shard must only be read from after its parent shard is completely processed.

Reading from a DynamoDB stream using the AWS CLI

Let's try to read stream events from a DynamoDB stream via the AWS CLI. This is for demonstration purposes only, so the parent-child lineage is not being respected.

First, I'll create a table in provisioned capacity mode and enable streaming at table create time with NEW_AND_OLD_IMAGES as the `StreamViewType`, as shown here:

```
{
    "AttributeDefinitions": [{
        "AttributeName": "PK",
        "AttributeType": "S"
    }],
    "TableName": "ddb_stream_table",
    "KeySchema": [{
        "AttributeName": "PK",
        "KeyType": "HASH"
    }],
    "BillingMode": "PROVISIONED",
    "ProvisionedThroughput": {
        "ReadCapacityUnits": 5,
        "WriteCapacityUnits": 5
    },
    "StreamSpecification": {
        "StreamEnabled": true,
        "StreamViewType": "NEW_AND_OLD_IMAGES"
    },
    "TableClass": "STANDARD"
}
```

I saved the preceding JSON in a file called `create_table_input.json` and referenced it in a `create-table` AWS CLI command, as shown here:

```
$ aws dynamodb create-table --cli-input-json file://create_table_
input.json --region eu-west-2
```

If the request is successful, you should see an output like the following. Make sure you note down the `LatestStreamArn` value. The same information can be obtained via the DynamoDB console:

```
 1 {
 2     "Table": {
 3         "AttributeDefinitions": [
 4             {
 5                 "AttributeName": "PK",
 6                 "AttributeType": "S"
 7             }
 8         ],
 9         "TableName": "ddb_stream_table",
10         "KeySchema": [
11             {
12                 "AttributeName": "PK",
13                 "KeyType": "HASH"
14             }
15         ],
16         "TableStatus": "CREATING",
17         "CreationDateTime": "2022-09-05T17:18:45.960000+01:00",
18         "ProvisionedThroughput": {
19             "NumberOfDecreasesToday": 0,
20             "ReadCapacityUnits": 5,
21             "WriteCapacityUnits": 5
22         },
23         "TableSizeBytes": 0,
24         "ItemCount": 0,
25         "TableArn": "arn:aws:dynamodb:eu-west-2:XXXX:table/ddb_stream_table",
26         "TableId": "ff042a46-4a8a-abab-a321-52d6350cb35a",
27         "StreamSpecification": {
28             "StreamEnabled": true,
29             "StreamViewType": "NEW_AND_OLD_IMAGES"
30         },
31         "LatestStreamLabel": "2022-09-05T16:18:45.960",
32         "LatestStreamArn": "arn:aws:dynamodb:eu-west-2:XXXX:table/ddb_stream_table
    /stream/2022-09-05T16:18:45.960",
33         "TableClassSummary": {
34             "TableClass": "STANDARD"
35         }
36     }
37 }
```

Figure 12.5 – CreateTable output JSON

Next, we will write some dummy data to the table so that it appears in the stream and can be consumed. Here's how I'm creating five items in the table while using the AWS CLI in a bash loop:

```
TABLE=ddb_stream_table;
for PK_VALUE in {1..5};
do
aws dynamodb put-item --table-name $TABLE \
--item '{"PK": {"S": " '$PK_VALUE' "}, "ATTR_1": {"S": "This is a
static attribute"}}' \
--region eu-west-2 ;
done
```

As described in the previous subsection, reading from the DynamoDB stream involves using `DescribeStream` to get the list of shards. Let us look at the AWS CLI command in the following snippet and the output of the command for the `DescribeStream` call in *Figure 12.6*:

```
STREAM_ARN="arn:aws:dynamodb:eu-west-2:XXXX:table/ddb_stream_table/
stream/2022-09-05T16:18:45.960";

aws dynamodbstreams describe-stream \
--stream-arn "$STREAM_ARN" \
--region eu-west-2
```

```
Output:
{
    "StreamDescription": {
        "StreamArn": "arn:aws:dynamodb:eu-west-2:XXXX:table/ddb_stream_table/stream/2022-09-05T16:18:45.960",
        "StreamLabel": "2022-09-05T16:18:45.960",
        "StreamStatus": "ENABLED",
        "StreamViewType": "NEW_AND_OLD_IMAGES",
        "CreationRequestDateTime": "2022-09-05T17:18:45.960000+01:00",
        "TableName": "ddb_stream_table",
        "KeySchema": [
            {
                "AttributeName": "PK",
                "KeyType": "HASH"
            }
        ],
        "Shards": [
            {
                "ShardId": "shardId-00000001662394728997-aabb",
                "SequenceNumberRange": {
                    "StartingSequenceNumber": "100000000003532443589"
                }
            }
        ]
    }
}
```

Figure 12.6 – The DescribeStream command and its output

As you can see, the stream only has a single shard in the `Shards` array, but there can be multiple shards in a DynamoDB stream. Notice that there is a `StartingSequenceNumber` property associated with the shard, but there isn't a corresponding `EndingSequenceNumber`. The presence of `StartingSequenceNumber` and the absence of `EndingSequenceNumber` indicate that the shard is an open shard and can actively receive events from corresponding mutations in the table. Also, there is no `ParentShardId`, which indicates the first shard in the stream. Run the same command after a few hours and the first shard will have an `EndingSequenceNumber` property, and the stream may have multiple shards with a `ParentShardId` property referring to this first shard. Next, we must get a shard iterator for this first shard. Note the value of `ShardId` from the output of your `DescribeTable` command and use it, as follows:

```
STREAM_ARN="arn:aws:dynamodb:eu-west-2:XXXX:table/ddb_stream_table/
stream/2022-09-05T16:18:45.960";
SHARD_ID="shardId-00000001662394728997-aabb"

aws dynamodbstreams get-shard-iterator \
--stream-arn "$STREAM_ARN" \
--shard-id "$SHARD_ID" \
--shard-iterator-type TRIM_HORIZON \
--region eu-west-2
```

```
Output:
{
    "ShardIterator": "arn:aws:dynamodb:eu-west-2:XXXX:table/ddb_stream_table/stream
/2022-09-05T16:18:45.960|1|AAAAAAAAAASomeRandomString=="
}
```

Figure 12.7 – The GetShardIterator command and its output

The output of this command will contain a long string of random-appearing characters, which is the shard iterator we need to consume from the shard at the position of `TRIM_HORIZON`. This is the start of the shard and may or may not contain the oldest available stream records. Finally, we can call `GetRecords` by using the shard iterator string, as follows:

```
SHARD_ITERATOR="arn:aws:dynamodb:eu-west-2:XXXX:table/ddb_stream_
table/stream/2022-09-05T16:18:45.960|1|AAAAAAAAAASomeRandomString==";

aws dynamodbstreams get-records \
--shard-iterator "$SHARD_ITERATOR" \
--region eu-west-2
```

```
Output:
{
    "Records": [],
    "NextShardIterator": "arn:aws:dynamodb:eu-
west-2:XXXX:table/ddb_stream_table/stream
/2022-09-05T16:18:45.960|1|AAAAAAAAAAAAAAnotherRandomString=="
}
```

Figure 12.8 – The GetRecords command and its output

As you can see, the first `GetRecords` call I made didn't return any records but returned a value for `NextShardIterator`. A shard iterator represents a position within the shard that may or may not store records. It is possible that if you are following along, you see records in the first call itself. Nonetheless, I used the shard iterator provided in the preceding output and called `GetRecords` again. The following is a partial output of the same:

```
 1 {
 2      "Records": [
 3          {
 4              "eventID": "7de3041dd709b024af6f29e4fa13d34c",
 5              "eventName": "INSERT",
 6              "eventVersion": "1.1",
 7              "eventSource": "aws:dynamodb",
 8              "awsRegion": "eu-west-2",
 9              "dynamodb": {
10                  "ApproximateCreationDateTime": "2022-09-05T17:33:30+01:00",
11                  "Keys": {
12                      "PK": {
13                          "S": "1"
14                      }
15                  },
16                  "NewImage": {
17                      "PK": {
18                          "S": "1"
19                      },
20                      "ATTR_1": {
21                          "S": "This is a static attribute"
22                      }
23                  },
24                  "SequenceNumber": "100000000003532767505",
25                  "SizeBytes": 38,
26                  "StreamViewType": "NEW_IMAGE"
27              }
28          }
29      ],
30      "NextShardIterator": "arn:aws:dynamodb:eu-west-2:XXXX:table/ddb_stream_table/stream
    /2022-09-05T16:18:45.960|1|AAAAAAAAAAAAAYetAnotherRandomString=="
31 }
```

Figure 12.9 – The GetRecords command's output with a stream record

As you can see, a stream record was returned on re-running the `GetRecords` command with the new shard iterator value. This stream record represents one of the `PutItem` calls I performed earlier. Notice that the information is available separately from the item itself. `eventName` is `INSERT`, indicating a new put, `ApproximateCreationDateTime`, pointing to the timestamp of when the stream record was created in the stream. The stream record only contains `NewImage`, despite the stream being configured to record both new and old images, which again alludes to the fact that this was a new item being put into the table.

For demonstration purposes, I used the `UpdateItem` operation on `PK=1`. The corresponding stream record is shown here:

```
1 {
2     "Records": [
3         {
4             "eventID": "e2de3841dd709b024af6f29e4fa13d375",
5             "eventName": "MODIFY",
6             "eventVersion": "1.1",
7             "eventSource": "aws:dynamodb",
8             "awsRegion": "eu-west-2",
9             "dynamodb": {
10                "ApproximateCreationDateTime": "2022-09-05T18:21:47+01:00",
11                "Keys": {
12                    "PK": {
13                        "S": "1"
14                    }
15                },
16                "NewImage": {
17                    "PK": {
18                        "S": "1"
19                    },
20                    "ATTR_1": {
21                        "S": "This static attribute was modified"
22                    }
23                },
24                "OldImage": {
25                    "PK": {
26                        "S": "1"
27                    },
28                    "ATTR_1": {
29                        "S": "This is a static attribute"
30                    }
31                },
32                "SequenceNumber": "600000000003533828985",
33                "SizeBytes": 80,
34                "StreamViewType": "NEW_IMAGE"
35            }
36        }
37    ],
38    "NextShardIterator": "arn:aws:dynamodb:eu-west-2:XXXX:table/ddb_stream_table/stream
/2022-09-05T18:18:45.960|1|AAAAAAAAAAAAGnuMoreRandomString=="
39 }
```

Figure 12.10 – The GetRecords record with eventName set to MODIFY

As you can see, the stream record that was generated on updating an existing item within the table generates a record with `eventName` set to `MODIFY` and both `NewImage` and `OldImage` representing the state of the item before and after the mutation, respectively. Note that a stream record that's generated by overwriting an item with `PutItem` would still have `eventName` set to `MODIFY`.

Common DynamoDB stream consumers

You may have found consuming the DynamoDB stream using the AWS CLI trivial, particularly because we ignored the shard lineage, and my table only had a single open shard. Production DynamoDB tables may have hundreds and thousands of shards at any given point, and it is often a requirement to read from the stream to respect the parent-child shard lineage. Failure handling and periodic checkpointing are other crucial factors to consider when consuming a stream.

Common consumers include **AWS Lambda** and applications that use **DynamoDB Streams Kinesis Adapter** (2). Both options take care of the undifferentiated heavy lifting and the complexities of stream processing so that you can focus on what matters to you the most – the business logic. Since both consumer options have an overlap with the other streaming option, **KDS**, we will dive deeper into these two consumer types later in this chapter.

Cost implications for utilizing DynamoDB streams

From a pricing perspective, you are charged for reading data from the DynamoDB streams in terms of read request units. Every single one of those `GetRecords` API calls, whether it's returning records or not, is billed as a stream read request unit and returns up to 1 MB of data from DynamoDB Streams. These read request units are different from the DynamoDB table read request units. None of the other DynamoDB Streams APIs are charged, nor is having a DynamoDB stream enabled without making any `GetRecords` API calls against it. Purely pay-as-you-go! If the DynamoDB table is part of global tables, any calls made by the DynamoDB service to consume the stream to perform global table replication aren't charged. Moreover, should you choose AWS Lambda as your consumer on the DynamoDB stream, any `GetRecords` calls made by it are completely free of charge. This can be a key factor in choosing Lambda as the consumer as opposed to other kinds of consumers, in addition to being completely serverless and fully managed.

In terms of non-Lambda consumers, DynamoDB Streams is only charged on the number of `GetRecords` API calls that are made, with a monthly free tier of the first 2.5 million calls across all DynamoDB Streams in an AWS region per AWS account. More about the costs associated with DynamoDB Streams are documented on the DynamoDB pricing page (3).

KDS integration

The other streaming option for DynamoDB tables is native integration with KDS. The same kind of stream records that are generated in DynamoDB Streams are encapsulated with Kinesis-specific metadata and made available in a Kinesis data stream when opting to use KDS for your DynamoDB table. The motivation behind supporting KDS integration for DynamoDB was to allow users to effortlessly connect their DynamoDB tables to the Kinesis suite of services. The Kinesis suite includes KDS for your stream processing workloads, Kinesis Firehose Delivery Stream for aggregating and batch loading of streaming data into various other downstream services such as an S3 bucket or an OpenSearch domain, among others, and finally Kinesis Data Analytics for enhanced analytical capabilities on the streaming data.

Like DynamoDB Streams, KDS can record any mutations that happen on its associated DynamoDB table. Unlike DynamoDB Streams, KDS needs to be created and managed by the user separately. Once a Kinesis data stream is associated with a DynamoDB table, DynamoDB will start writing stream records to the stream. These stream records contain `NEW_AND_OLD_IMAGES` equivalent information of a DynamoDB stream record. KDS supports two capacity modes – on-demand mode and provisioned mode. As their names suggest, you can choose to either let Kinesis scale the stream on your behalf with on-demand mode or choose to manage the scaling of the stream manually with provisioned mode.

In addition to its seamless integration with the Kinesis suite of services, KDS offers several advantages over DynamoDB Streams. These advantages include a configurable data retention period, enhanced fan-out consumption to enable increased parallelism in stream consumption, and support for a higher number of simultaneous consumers. For instance, KDS allows for a minimum data retention period of 24 hours, which can be extended up to 1 year. Unlike DynamoDB Streams, KDS supports enhanced fan-out consumption, making it suitable for use cases that require near real-time stream processing. With enhanced fan-out, each consumer application is allocated read throughput per shard, with a capacity of up to 2 MB per second.

Feature characteristics

In terms of the workings of a Kinesis data stream itself, there are few key differences compared to DynamoDB streams. KDS is also made up of shards, which may or may not be determined by the user depending on the capacity mode, but do not scale automatically with the DynamoDB table like DynamoDB streams do. Unlike DynamoDB Streams, KDS does not auto-rotate shards every few hours, but shards can split while respecting the parent-child lineage when stream shards scale in or out.

A significant difference between KDS and DynamoDB Streams is that KDS does not guarantee exactly once delivery or in-order delivery of item-level modifications. What this means is that with KDS, it is possible to observe duplicate events within the stream. It is also possible to see out-of-order delivery for item-level modifications with KDS, although rarely. If you're using KDS to process stream events from DynamoDB, you must ensure that the downstream systems are idempotent in processing duplicates or out-of-order events for the same item. In most cases, this may be possible when designing the application. If not, you may need to use DynamoDB Streams due to the guarantees it comes with.

A comprehensive list of distinctions between KDS for DynamoDB and DynamoDB Streams can be found in the developer guide (4). Now that we've delved into the integration of KDS with DynamoDB tables as an alternative streaming option, let's delve into the various stream consumers that can be configured so that we can poll from these streaming options, beginning with AWS Lambda as a stream consumer.

AWS Lambda as a stream consumer

AWS Lambda is an event-driven, serverless computing platform that eliminates the need for you to manage servers. It enables you to run custom code for a wide range of applications or backend services, and you're only charged for the execution time of your code. Lambda also allows you to parallelize the execution of the same function code across multiple simultaneous invocations, each operating independently.

Lambda serves as a common stream consumer for both KDS and DynamoDB Streams. When Lambda functions are configured as stream triggers, the Lambda service regularly polls the respective stream with a predefined, non-configurable frequency. It invokes your function code only when stream events are detected during polling.

Lambda stream triggers provide various configurations and features. For instance, you can set a batch size while specifying how many stream events should accumulate before invoking the Lambda function. This allows you to control the granularity of function invocations, from processing individual stream records to handling hundreds or thousands of events at once. You can also establish time-based thresholds to accumulate events over a certain period before invoking the function code.

Regarding scalability, Lambda efficiently polls multiple shards simultaneously, buffers stream records, and invokes functions based on user-defined thresholds such as batch sizes and batch windows. This parallelism ensures that Lambda can adapt to varying levels of load on the DynamoDB table. Moreover, you have control over concurrency to manage Lambda costs if necessary.

Lambda as a stream trigger supports setting up a **dead-letter queue (DLQ)** to handle processing failures gracefully. In cases of critical processing failures due to application bugs or resource constraints, a DLQ can be configured to capture information about failed stream records and pass them to an SQS queue. This allows processing to continue without data loss, all while considering the stream's data retention period. Failed records can be reviewed later, or alerts can be set up for human intervention.

When to use Lambda as a stream consumer

Consider employing AWS Lambda as a stream consumer in scenarios where processing requirements are relatively lightweight or when the Lambda function doesn't need to communicate with other organization-level subsystems that have diverse network or authentication requirements. It's important to note that configuring Lambda to interact with other AWS services is straightforward.

Additionally, Lambda can be an appealing choice for stream consumption if you aim to avoid the burden of managing infrastructure and the ongoing maintenance associated with it.

> **Important note**
> AWS Lambda polls multiple DynamoDB stream shards at a base rate of four times per second. In the case of KDS, Lambda polls its shards at a base rate of once per second. In both cases, Lambda may poll more times than the base rate if it continues to receive records from the respective shard.

Architectural patterns that are unlocked with streams

Now that we've discussed AWS Lambda as a prevalent stream consumer, let's explore the various patterns that can be implemented through this integration. Event-driven architecture stands out as a popular design pattern or model for application design as it offers efficient resource utilization. In event-driven solutions, work is only executed in response to actual pending tasks.

Think about the contrast between having continuously running application servers, irrespective of pending work, versus an event-driven approach. Running servers continuously, even during periods of no traffic, not only leads to unnecessary costs but also necessitates periodic maintenance. While there are cases where continuously running servers make sense, most workloads can benefit significantly from adopting event-driven patterns.

With event-driven architecture patterns, you only incur costs for the time spent on valuable work. When there's no application traffic, there's no work to be done, translating to no costs incurred. For instance, consider an application that manages user data using DynamoDB. If there's a business requirement to send a welcome email when a user registers, this can be implemented efficiently in an event-driven manner using AWS Lambda and DynamoDB streams. Work is only pending when a new user item is created in the DynamoDB table, and Lambda only performs this work when it detects a relevant stream record indicating the creation of a new item.

The following diagram illustrates the diverse event-driven patterns that can be implemented using either of the streaming options for DynamoDB:

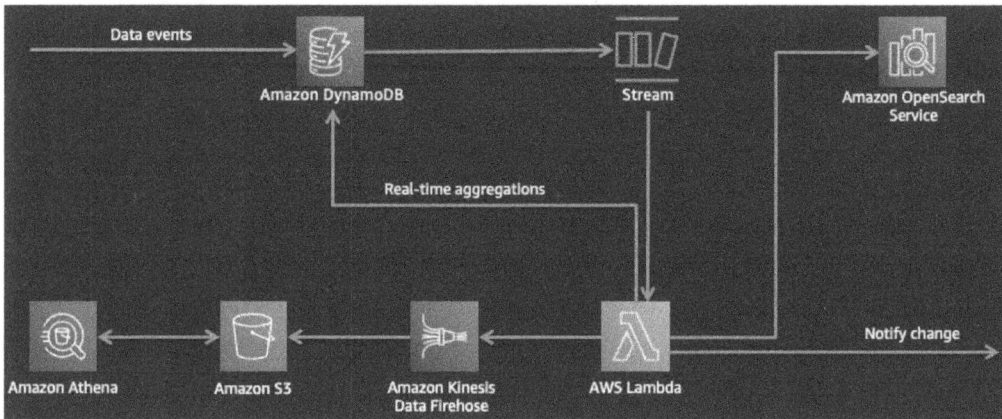

Figure 12.11 – Event-driven architecture patterns with streams

In the preceding diagram, **Stream** can be either a DynamoDB stream or a Kinesis data stream. As you can see, a wide range of use cases can be addressed effectively by employing an event-driven pattern with Lambda and streaming options to capture mutations occurring in a DynamoDB table. We'll dive into some of those next.

Real-time aggregations

Considering that a Lambda consumer application gets triggered when records arrive in a stream based on mutations from a corresponding DynamoDB, you can choose to perform real-time aggregations and support application access patterns where counts may be required for certain entities.

Let's consider a popular music streaming application as an example. Imagine that you have a DynamoDB table that stores information about every song you've liked, favorited, or added to your playlist. Each of these actions may require aggregated counts from the application's perspective. When a user interacts with the app and such interactions result in DynamoDB items, stream records are generated. These stream records trigger Lambda functions. Moreover, each Lambda invocation could process numerous stream records simultaneously.

In this scenario, the Lambda logic can iterate over batches of stream records, calculate aggregations within each batch, and update the sums into a single item within the same or another DynamoDB table. These count updates can be achieved without needing to know the existing values, thanks to the use of atomic counters provided by the UpdateItem API.

When the application needs to access these aggregated results later, it can do so by reading a single DynamoDB item, regardless of the number of likes or favorites generated by users of the music app. To prevent potential infinite loops when updating aggregated counts back to the same table, it's a good practice to include an entity-type attribute on every item, representing the set of information stored for each item. The Lambda logic can then either filter stream records based on this entity-

`type` attribute or utilize Lambda event filtering to efficiently exclude unwanted entity stream records, all without incurring performance or cost penalties. You can find more details about Lambda event filtering in the developer guide (5).

Without employing an event-driven pattern such as this, the traditional database approach to supporting aggregations would involve running a query for every application request. This query would go through all the rows related to likes or favorites and perform real-time counts. However, this approach may not scale well, particularly when the application is popular and generates such events at high rates.

Another alternative to achieve aggregations is to run nightly jobs that perform aggregations. While this approach can handle high event rates, it comes with the trade-off of accepting that the results will always be stale until the nightly job updates them.

Downstream message processing

Like the real-time aggregation use case, event-driven patterns can be used for use cases where downstream systems must react to mutations on the DynamoDB table data. These use cases could range from sending stream records to other data stores for supporting additional functionality on the application and passing messages based on the stream records to other subsystems or microservices to simply archiving data for compliance reasons.

Consider an eCommerce application as an example. A DynamoDB table can be used as a product catalog for storing information about all products, including links to images, customer reviews, product prices, descriptions, and so on. If a user visits the web store to order a product with `SKU: ABC123ZZ`, it may be trivial for the application to make a single `GET` on the table to get all information about the product, but that is not how it works. One of the web store's most important functionalities is to allow users to run full-text searches about product names or descriptions and land them a list of results that are as close to the desired products as possible. Several times throughout this book, we've stated that full-text searches are not particularly OLTP or cost-effective on a database built for OLTP access. Hence, an event-driven pattern can be implemented where any mutations that occur on the product catalog's DynamoDB table could be indexed downstream into an OpenSearch cluster that's been purpose-built for highly efficient full-text searches. The Lambda consumer for a stream associated with the DynamoDB table can perform this indexing into the OpenSearch cluster asynchronously, and in batches, with proper failure handling setup. Now, the full-text search functionality on the web store can be supported by the OpenSearch cluster and detailed information about any products can still be supported by the DynamoDB table.

Another common use case that can benefit from an event-driven pattern with streams is archiving table data or capturing state change information for compliance or complex analytics. Similar to the eCommerce example, a Lambda consumer can be used with a DynamoDB table's stream to relay information from the stream records.

This concludes our current exploration of AWS Lambda as a stream consumer. Next, we'll shift our focus to the other prevalent type of consumer application based on the open source KCL provided by AWS. This library can be employed in situations where AWS Lambda may not be the most suitable choice for certain use cases.

KCL-based consumer applications

Another recommended approach for processing change events from either a DynamoDB or Kinesis data stream is to utilize the **Kinesis Client Library** (**KCL**). It's worth noting that both DynamoDB Streams and KDS are intentionally designed with similar APIs to simplify the learning curve when dealing with these services separately.

KCL is an open source library that's provided by AWS and is designed to abstract much of the coordination and orchestration required for reliable stream processing in a distributed environment. When working with DynamoDB Streams, you can employ the KCL-based DynamoDB Streams adapter (6). Conversely, for building consumer applications with KDS, KCL itself can be used (7).

Let's delve into how KCL-based consumer applications operate, their compatibility with various streaming options, and the scenarios where choosing them over other stream consumer options is advantageous.

How it works

When a consumer application built using KCL starts, it will initialize a worker thread. This worker thread will then connect to the stream that's been specified, enumerate shards within the stream, and coordinate shard associations with other workers in case there are multiple instances of the consumer application running in a distributed fashion. For scalability, resiliency, and high availability, it is recommended to have more than a single application instance running across hosts. Such configuration can also allow high performance for near-real-time processing requirements.

The worker thread would also spawn multiple other threads for each shard it's responsible for processing. These threads would be responsible for polling records from the shards, passing those records to other threads running your record-processing logic, and checkpointing successful processing progress such that if an application or host failure occurs, another worker could pick up the processing from the last checkpoint to avoid reprocessing stream records. Since KCL intends to be resilient to host failures too, checkpointing cannot be done on the host locally but is available to every other worker. Hence, KCL uses a separate DynamoDB table for checkpointing and managing shard-worker associations. Each worker spawns local threads when a record is processed and put in this DynamoDB table.

In addition to checkpointing record-processing progress, the workers also record heartbeats on the KCL lease DynamoDB table to let every other worker know that it is actively processing a shard. In other words, every worker also periodically monitors this DynamoDB table to keep track of other workers and their respective shard associations to find shard leases that may not be processed by another worker anymore. In such cases, workers also re-balance shard-worker associations between them to pick the shard lease that hasn't seen a heartbeat update in a while. Worker-spun threads also periodically poll the stream and compare the shard information between the stream and the KCL lease's DynamoDB table to remove records of shards that may have finished processing or add new shard records in case of shard splits. This periodic operation is referred to as shard sync.

Each time a KCL-based consumer application is initialized, the worker checks for an existing DynamoDB lease table for the application and creates one if there isn't one. If the table already exists, the worker begins to scan the table to get information about available leases and leases that other workers are heartbeating and checkpointing for to get an idea of what leases it can pick up. It must ensure they're either not being processed by other workers or steal leases from other workers to balance the consumption load across the different workers that may be processing the stream.

KCL versions and compatibility

The KCL library is available in two major versions, v1.x and v2.x. The DynamoDB Streams Kinesis Adapter (6) uses v1.x only, whereas KCL-based consumer applications for KDS could use either version. The major difference between the versions is that v2.x utilizes AWS Java SDK v2 (8), which allows for non-blocking I/O, including support for plugging in a different HTTP implementation at runtime. This benefits consumer applications with extremely low end-to-end latency consumption, along with enhanced fan-out capabilities. Since enhanced fan-out support is also optionally provided by KDS, using KCL v2.x in consumer applications can be performant for near-real-time processing scenarios. Irrespective of the KCL versions, the library also supports additional configurations that can be fine-tuned as per the requirements.

KCL is primarily implemented in Java, but to make it easier for developers to write record processor business logic in other programming languages, KCL also provides a Java-based daemon called **MultiLangDaemon** that does all the heavy lifting. You can have your record-processing logic in your choice of programming language with this MultiLangDaemon running in the background, which calls the record-processing logic sub-process. To run MultiLangDaemon, you still need to have Java set up on the host running the consumer application. AWS also has open source KCL interfaces with MultiLangDaemon for Python (9), Ruby (10), .NET (11), and NodeJS (12).

When to use KCL-based consumer applications

When comparing KCL-based applications to the more common AWS Lambda-based stream consumers, it's important to note that KCL applications should run on hosts that support specific dependencies. These hosts could be either self-managed EC2 servers or containers, such as those running on AWS-managed hosts. In both scenarios, users are responsible for managing these dependencies. Additionally, using KCL necessitates an extra DynamoDB lease table, which can add to both management and usage expenses.

You should consider using KCL-based consumer applications in scenarios where record processing is just one part of the application's tasks, and you're also handling additional complex processing or serving end user requests. Furthermore, KCL-based applications are a suitable choice when your organization faces networking constraints that limit the use of Lambda functions. These constraints might involve seamless communication with internal subsystems or specific authentication requirements that could be challenging to implement with AWS Lambda functions. Lastly, opting for KCL-based consumer applications is ideal when you require greater control over tuning parameters for stream processing speed or wish to extend the KCL implementation for custom tasks. While Lambda functions for stream consumption offer some tuning capabilities, KCL-based applications provide more flexibility in this regard.

Now that we've explored KCL-based consumer applications and their suitability for various scenarios, let's shift our focus to another DynamoDB feature: TTL.

DynamoDB (TTL)

One of the features of DynamoDB is its ability to automatically delete or discard data from a table that you may no longer need. This feature is called TTL. In DynamoDB, you can set a per-item timestamp, which serves as a criterion for determining when an item becomes obsolete. Following the specified timestamp's date and time, DynamoDB automatically removes the item from your table, all without utilizing any write throughput. DynamoDB stores table items on hot SSDs and charges a per-GB monthly price on storage based on the amount of data stored in a table. Since TTL-induced deletes are not charged, this makes TTL a free-of-cost automatic function to delete items from your tables that may have lost relevance over time and allow you to manage the storage size – and, thus, costs – for your DynamoDB tables.

Common use cases for using TTL

Before getting into how the feature works and the key considerations for using TTL, let's look at some of the potential use cases where TTL could be super beneficial.

Deleting financial instruments data after a certain period

TTL becomes especially valuable when you store items that become irrelevant once a certain amount of time has passed. Imagine a use case that deals with financial instrument data where you may store enriched granular stock information in a DynamoDB table so that your end users can consume it for actionable insights. While this data may be written to the table at high throughputs and accessed by end users for a few hours or even days, after a certain point of time, this data would no longer be accessed by any users. This period could be from a few minutes to months and years. It would make sense to not be paying for storing this data after this period when it becomes irrelevant. Using TTL for such a use case would be useful in managing the storage size and costs associated with the table as DynamoDB would keep deleting the older items as their TTL is reached. First, you don't get charged for these deletes happening in the background; second, you stop being charged for storing items that are no longer relevant to you or your end users.

Archiving data to a data lake or colder storage to optimize storage costs

Imagine a use case where the data in your tables never completely loses its relevance or becomes infrequently accessed after a certain period. This period may be predictable or not. In those cases, you can choose to expire items from the current table and move them into colder storage to optimize storage costs so that when you need to access that data on an off request, this is possible. There are two common destinations where this almost irrelevant data can land:

- Amazon S3, which also serves as a data lake for all archived data

- Another table that contains DynamoDB's **Standard-Infrequent Access (SIA)** table class

TTL item deletion changes are also available in the table's streams, just like an application-induced delete. However, here, distinguishable properties indicate that the deletion was made by TTL and not by the application. The old image of the record (the new image being empty for a delete action) can then be either moved to your data lake or a different table with an SIA table class.

How can we decide between a data lake and another table containing the SIA table class? Well, if you'd still like older data to be served with the same single-digit millisecond latencies that the new data is being served with, the SIA table class would be suitable. In most cases, you may be okay with serving the old data with higher latencies, in which case you could choose a data lake to be the destination. Although both the table with the SIA table class and an Amazon S3-based data lake would have different cost profiles, they would still be much more optimized than the DynamoDB Standard table class, which will always have the more recent or hot data stored.

Complying with data regulations

Like the financial instruments data use case, if you have a use case involving data that needs to be retained only for a fixed period so that it remains compliant with local data regulations, TTL can help delete that data automatically from your table before the maximum retention period that's allowed by those regulations. Regardless of the industry, certain geographical regions may have regulations to only retain users' PII data for a few months or years, and DynamoDB TTL can be useful to remain compliant with the respective regulations.

How it works

When you activate TTL on a DynamoDB table, it's essential to specify an attribute name that the service will use to identify whether an item qualifies for automatic expiration. It isn't required to have all or any items with the attribute in them when enabling or disabling TTL.

After enabling TTL on the table and specifying the attribute's name, DynamoDB initiates background processes on each partition containing table data. These processes continually scan the data and remove items with a TTL attribute value greater than the system time. It's important to note that these background processes have no adverse impact on the table's performance and only operate when the partitions have available system-reserved resources.

Only when TTL deletes the expired items from the table is a change event generated in the table's streams, if any. Also, when TTL deletes an expired item, the item is deleted from any local or global secondary indexes the table may have. In the case of global tables, TTL deletes in one region may be deleted in the other regions eventually based on the global table's standard eventual replication across regions. More about how TTL works is provided in the AWS docs (*13*). You will learn more about global tables in *Chapter 13*.

Key considerations

The following are some of the key considerations when it comes to using TTL in DynamoDB. Believe me – many customers wished they knew these before using TTL in production in ways that the feature isn't designed to be used:

- TTL attributes must have the same name across any items that need to be deleted by TTL on expiry. Items without the set TTL attribute name will be ignored.

- The TTL attribute value must be of the DynamoDB Number data type. Items with TTL attribute values of any other data type will be ignored.

- The TTL attribute value must be in Unix Epoch time format (*14*) in seconds (for example, 1645119622). Any values not in this format will be ignored.

- TTL deletion can take from a few minutes to up to a few days (usually, it doesn't take this long) after expiry (when the attribute's value is less than the current time) before DynamoDB deletes the item.

- Enabling, disabling, or modifying TTL settings for a table typically requires around 1 hour for the changes to propagate and permit any subsequent TTL-related actions to take effect.

- When enabling TTL, the deletions may not be immediately visible, but if you've configured everything correctly, you should start seeing deletions occurring shortly. Ensure that the TTL attribute name you specify when enabling TTL matches exactly with how it appears in the table items and has no unintended leading or trailing whitespaces.

Important note

DynamoDB TTL deletion can take up to a few days after expiry before DynamoDB deletes the item.

Expired items, such as those in which the TTL attribute's value has passed the current time, may still appear in reads and the client must ignore/discard these items from read results in the application itself.

Since there are no strict guarantees on DynamoDB deleting an expired item immediately after its expiry, applications must avoid depending on DynamoDB deleting expired items immediately.

Streams and TTL

You can archive items that have been deleted by TTL into a destination of your choice by utilizing either of the streaming options available for DynamoDB tables. Since TTL delete actions are DynamoDB item deletes, these generate stream events that are distinguishable from normal application deletes. A stream consumer can forward the TTL deleted items to a separate destination that's been specifically set up for archiving data that was expired.

Here's an example of the distinguishing portion of a TTL delete stream event that's absent from a stream event generated by an application delete. This sample is for a DynamoDB Streams event; however, it's like an event that's been generated when a Kinesis data stream is associated with a DynamoDB table:

```
1 "Records": [
2     {
3         ...
4
5         "userIdentity": {
6             "type": "Service",
7             "principalId": "dynamodb.amazonaws.com"
8         }
9
10        ...
11
12    }
13 ]
```

Figure 12.12 – Part of a single stream record for a TTL delete

As you can see, the userIdentity part of a stream record has a type that corresponds to Service, indicating a DynamoDB TTL-induced delete. For an application deleted item, the userIdentity part of the stream record is null. This can be used to filter out TTL deletes from application deletes and archive them to an appropriate data store. If you're using AWS Lambda as the stream consumer, you can use Lambda event filtering to only trigger Lambda invocations with records that were TTL deleted based on the presence of the userIdentiy.type value as Service.

A common event-driven pattern to archive TTL deleted items into an Amazon S3 data lake involves using DynamoDB streams that have been consumed by a Lambda function (with event filtering). This Lambda can then sanitize records before sending them in batches to a Kinesis Firehose delivery stream that can buffer, batch, aggregate, and encrypt data before writing them to an S3 prefix with default timestamp-based or custom partitioning logic. Detailed steps for implementing this pattern can be found in an AWS blog (*15*).

During **re:Invent** 2022, which is AWS's annual conference that takes place in Las Vegas, USA, **Amazon EventBridge Pipes** was announced as a service that can also help achieve this archival pattern with little to no code. This is because it supports DynamoDB Streams as well as KDS as a source, and a Kinesis Firehose delivery stream as a target. More about using EventBridge pipes can be found in the AWS docs (*16*).

Backfilling a TTL attribute on existing table data

As an advanced or experienced DynamoDB user, you may already have DynamoDB tables in production with gigabytes to potentially petabytes of data. If you have come across a requirement to backfill the existing data with a TTL attribute such that data gets deleted by TTL based on a period greater than an existing attribute in the items, say an item creation timestamp attribute that your application adds to the items, you have multiple options to achieve this.

Using a sweeper process

A simple way to perform backfilling would involve scanning through all the table items and updating every item without a TTL attribute, with a TTL attribute and a value based on any other timestamp-related attribute within the same item. Such a script would make `Scan` API calls and iterate over each page of DynamoDB table items to perform updates to each item via the `UpdateItem` API or `ExecuteStatement` PartiQL API. The `UpdateItem` and `ExecuteStatement` APIs can also perform conditional updates where you can set a condition on the item's partition key being non-existent, indicating that the item may have been deleted between the scan calls and the update calls.

As an advanced user, you can implement a multi-threaded sweeper script that performs parallel scans, followed by rate-limited update calls. Rate limiting is helpful in cases where the table is also being actively accessed in production. Rate limiting can be done based on consumed write throughput instead of the number of requests made per thread. For limiting throughput, the update calls can have an additional parameter passed, `ReturnConsumedCapacity=INDEXES`, which would return consumed write throughput per update call in its response separately for the table and any secondary indexes the table may have. The sums of write throughput that are consumed by the table, as well as all local secondary indexes, can be used to rate limit the update calls. This would provide room for write throughput consumption by the production application and prevent any throttling errors, while safely proceeding with the backfilling process. The same `ReturnConsumedCapacity` parameter is available for the `Scan` API as well, but in the case of backfilling, the update calls will likely be more time-consuming operations due to the network round trips required for each scanned page of items. To learn more about the `ReturnConsumedCapacity` parameter for the `UpdateItem` API, take a look at *17* in the *References* section at the end of this chapter.

Compute for such a sweeper could be anything from a periodic scheduled Lambda to an EC2 server or a container-based setup.

Using ETL

Backfilling can also be performed efficiently using AWS Glue ETL or Hive/Spark on Amazon EMR. Backfilling jobs using any of these would essentially work in the same manner as using a sweeper process but they can support high parallelism and rate limiting in a distributed environment natively.

The advantage of using Glue ETL over Amazon EMR would be Glue's serverless design and no infrastructure management. With Amazon EMR, additional flexibility and tuning capabilities are supported at the cost of having to manage infrastructure.

To use Glue ETL to interact with DynamoDB tables along with rate-limiting configurations, see the AWS docs (*18*). To perform backfilling using Hive on EMR, along with best practices, see *19* and *20* in the *References* section.

> **Important note**
> Consider this as a reminder to delete any resources you may have created while going through this chapter in your AWS account.

Summary

In this chapter, we introduced various CDC methods and why they can be important for a database. We also learned about the different streaming options that are available for DynamoDB tables to interact with CDC events. Diving deeper, we learned about how DynamoDB streams work concerning shards and what information may be available in the stream records that are generated in a DynamoDB stream. Next, we learned about the process of reading stream records from a DynamoDB stream. After, we looked at the common consumers of a DynamoDB stream and the cost implications of using one.

We also examined the alternative streaming option, which is KDS for DynamoDB. During our discussion, we explored the functional distinctions between KDS and DynamoDB Streams, as well as differences in their features. For a comprehensive guide on cost optimization strategies related to each streaming option, please refer to *20* in the *References* section.

We also gained insights into the two prevalent building blocks for consumer applications with these streaming options: AWS Lambda and KCL. We explored scenarios where you might opt for either option and the various configurations that each consumer service can accommodate. I recommend that new AWS or DynamoDB users pick Lambda as their consumer of choice and only build KCL-related consumer applications if they know exactly what they're doing.

Finally, we dove into DynamoDB TTL and its significance in automatically removing or archiving data that has become obsolete. This discussion encompassed architectural strategies for archiving and various methods for populating the TTL attribute in pre-existing tables with substantial data volumes.

In the next chapter, we will explore a crucial DynamoDB feature known as global tables. Global tables offer multi-region, multi-write capabilities to applications, enhancing their resilience to complete regional outages and enabling seamless failovers with a minimal **recovery point objective** (RPO). Additionally, global tables can optimize low-latency local read and write operations for geographically dispersed user bases.

References

1. *DynamoDB Streams API actions:* https://docs.aws.amazon.com/amazondynamodb/latest/APIReference/API_Operations_Amazon_DynamoDB_Streams.html

2. *Using KCL Adapter to consume DynamoDB Streams:* https://docs.aws.amazon.com/amazondynamodb/latest/developerguide/Streams.KCLAdapter.html

3. *DynamoDB Pricing:* https://aws.amazon.com/dynamodb/pricing/provisioned/

4. *Streaming options for DynamoDB:* https://docs.aws.amazon.com/amazondynamodb/latest/developerguide/streamsmain.html#streamsmain.choose

5. *Lambda event filtering:* https://docs.aws.amazon.com/lambda/latest/dg/invocation-eventfiltering.html

6. *DynamoDB Streams Kinesis Adapter:* https://github.com/awslabs/dynamodb-streams-kinesis-adapter

7. *Kinesis Client Library:* https://github.com/awslabs/amazon-kinesis-client

8. *AWS Java SDK v2:* https://docs.aws.amazon.com/sdk-for-java/latest/developer-guide/home.html

9. *KCL Python:* https://github.com/awslabs/amazon-kinesis-client-python

10. *KCL Ruby:* https://github.com/awslabs/amazon-kinesis-client-ruby

11. *KCL .NET:* https://github.com/awslabs/amazon-kinesis-client-net

12. *KCL NodeJS:* https://github.com/awslabs/amazon-kinesis-client-nodejs

13. *TTL: How it works:* https://docs.aws.amazon.com/amazondynamodb/latest/developerguide/howitworks-ttl.html

14. *Unix Epoch time format:* https://en.wikipedia.org/wiki/Unix_time

15. *Archiving data to S3 using TTL:* https://aws.amazon.com/blogs/database/automatically-archive-items-to-s3-using-dynamodb-time-to-live-with-aws-lambda-and-amazon-kinesis-firehose/

16. *Amazon EventBridge Pipes:* https://docs.aws.amazon.com/eventbridge/latest/userguide/eb-pipes.html

17. *The UpdateItem API:* https://docs.aws.amazon.com/amazondynamodb/latest/APIReference/API_UpdateItem.html

18. *Glue ETL DynamoDB Options:* `https://docs.aws.amazon.com/glue/latest/dg/aws-glue-programming-etl-connect.html#aws-glue-programming-etl-connect-dynamodb`

19. *Backfilling TTL using Hive on EMR Blog I:* `https://aws.amazon.com/blogs/database/backfilling-an-amazon-dynamodb-time-to-live-ttl-attribute-with-amazon-emr/`

20. *Backfilling TTL using Hive on EMR Blog II:* `https://aws.amazon.com/blogs/database/part-2-backfilling-an-amazon-dynamodb-time-to-live-attribute-using-amazon-emr/`

13
Global Tables

Geographically distributed applications are gaining popularity as companies expand globally, requiring data storage and access from multiple regions. However, managing cross-region self-managed databases can be a complex and challenging task, particularly for teams lacking expertise in this area. Configuring and maintaining conflict resolution, replication consistency, and patching pose significant challenges for developers building and maintaining these systems.

Amazon DynamoDB global tables offer a fully managed solution, simplifying the creation and management of globally distributed applications. With **global tables**, developers can effortlessly replicate data across multiple regions, enabling the construction of highly available, fault-tolerant applications resilient to failures in any region. This solution provides a scalable and durable framework for distributed databases, facilitating easy data replication across multiple regions and allowing applications to be set up to read and write data from any region.

Global tables primarily serve two use cases: **serving a geographically distributed user base** and business continuity or **disaster recovery**. Both use cases will be explored in detail later in this chapter.

One of the key features of global tables is automatic multi-master replication. When data is written to a table in one region, the changes are automatically propagated to all other regions. This ensures that all regions have consistent data, even in the event of a regional failure. Global tables also provide automatic conflict resolution with a *last writer* wins strategy, ensuring that data conflicts are resolved consistently across all replicas. With global tables, developers no longer need to worry about configuring and managing cross-region replication, conflict resolution, or patching. This allows them to focus on building applications and delivering value to their customers.

In this chapter, we will explore the various features and benefits of DynamoDB global tables. We will cover how global tables work, including how data is replicated and how conflicts are resolved. We will also discuss the various use cases for global tables, including how they can be used to build highly available and fault-tolerant applications that can serve users across the world, and how they can help businesses plan for disaster recovery. Additionally, we will provide a detailed guide on how to operate and manage global tables, including best practices for monitoring and troubleshooting issues.

By the end of this chapter, you will have a thorough understanding of how DynamoDB global tables work and how they can be used to build highly available and scalable distributed databases. You will also know how to operate and manage global tables, including how to monitor and troubleshoot issues. With this knowledge, you will be able to confidently use global tables to build distributed applications that can serve users across the world and ensure business continuity in the event of a regional outage.

In this chapter, we are going to cover the following main topics:

- Going over global table basics

- Operating and managing global tables

- Understanding consistency and conflict resolution

- Best practices for multi-region architectures

Going over global table basics

In this section, you will learn about the motivations behind organizations requiring a globally replicated database, the two versions of global tables and the differences between them, and how global tables replicate data. Finally, you will be provided with a step-by-step guide on how to convert a single-region table into a global one.

Why global tables?

Global tables have become increasingly crucial as businesses expand globally, and users access applications from various regions. The growing demand for globally available applications necessitates a solution that offers data access from anywhere with low latency and high availability. Global tables provide a fully managed solution for distributed databases, streamlining the development and management of globally distributed applications. This approach delivers a highly scalable and durable solution, simplifying data replication across multiple regions and enabling applications to read and write data from any region through a single endpoint. As mentioned in this chapter's introduction, the use cases for global tables can be broadly categorized into two areas: serving a geographically distributed user base and providing disaster recovery.

In scenarios where applications cater to users across multiple regions, it is crucial to ensure that data is stored and accessed from the nearest region. Global tables facilitate data replication across multiple regions, allowing applications to read and write data from any region using a single endpoint. This ensures optimal performance and user experience, regardless of their geographical location, addressing the first use case of serving a geographically distributed user base.

The second use case for global tables is business continuity or disaster recovery. Disaster recovery is paramount for businesses that require continuous application availability, even in the face of regional failures. Global tables offer a highly durable and scalable solution for disaster recovery, allowing businesses to implement a mechanism with virtually the lowest possible **recovery point objective**

(RPO) and **recovery time objective (RTO)** against region-wide outages. In the event of a regional failure, applications can seamlessly read and write data from alternative regions, enabling uninterrupted business operations.

Global tables for a geographically distributed user base

Global tables can be used to serve a geographically distributed user base with low latency and local reads and writes. This is particularly important for applications that serve users in multiple regions.

For example, a retail company may have an e-commerce application that serves users in North America, Europe, and Asia. With global tables, the company can replicate data across multiple regions, allowing the application to read and write data from any region with a single endpoint. This ensures that users receive the best possible performance and experience, regardless of where they are located. The following figure illustrates a global table with three replicas that serve local reads and writes while automatically replicating the data across all replicas:

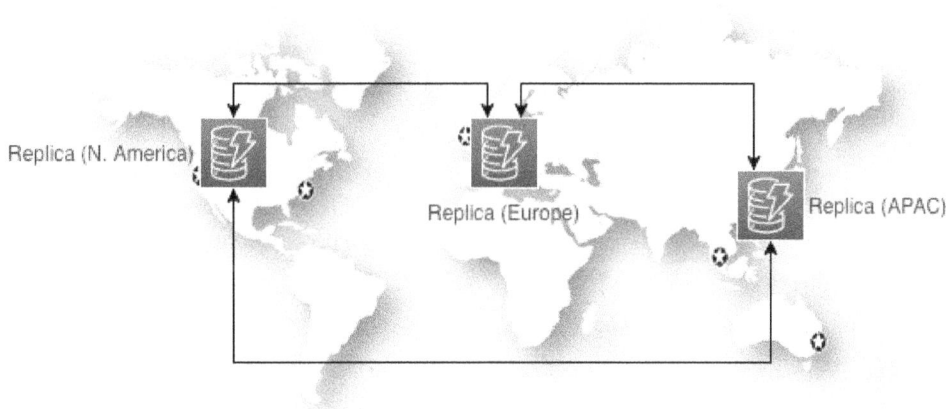

Figure 13.1 – Geographically distributed user base use case

The following are additional scenarios where utilizing DynamoDB global tables is advantageous for a distributed user base:

- **Social media**: A social media platform with users in multiple countries can use global tables to ensure low-latency access to user data globally. By replicating data across multiple regions, global tables provide users of the platform with the best possible performance and experience.

- **Online gaming**: An online gaming company offering multiplayer games can use global tables to ensure low-latency access to game data for all players worldwide. By replicating game data across multiple regions, global tables allow players to enjoy consistent, low-latency gameplay, regardless of their location.

- **Video streaming**: A video streaming service can use global tables to ensure that videos are available to users worldwide with low latency. By replicating video data across multiple regions, global tables help the service deliver optimal performance and a seamless viewing experience to all users.

- **Travel booking**: A travel booking platform offering global services can use global tables to provide low-latency access to booking data for all users. By replicating this data across multiple regions, global tables enable users to book travel quickly and consistently, regardless of their location.

- **Financial services**: A financial services company operating globally can use global tables to provide low-latency access to financial data for all users. By replicating data across multiple regions, global tables ensure users can access their financial information quickly and consistently, regardless of where they are located.

Global tables offer automatic active-active replication, ensuring seamless propagation of data changes to all regions. Next, we'll explore how to utilize global tables in the context of disaster recovery.

Global tables for disaster recovery

Global tables can also be used for disaster recovery. Disaster recovery is critical for businesses that need to ensure their applications are always available, even in the event of a regional failure.

In the financial industry, global tables can be used to ensure that data related to financial transactions, such as account balances and payment history, is always available, even in the event of a regional outage. This is critical to ensure that the business can continue to process transactions and maintain customer trust.

For example, suppose a financial institution has its application stacks deployed in North America, as well as in Europe. If one region experiences an outage, a global tables-based application can automatically redirect traffic to the other region, ensuring that users can still access and interact with their financial assets and accounts. This ensures that the business can continue to operate without disruption and maintain customer satisfaction.

The following figure illustrates a global tables-based multi-region deployment that enables users in San Francisco, California, to interact with their financial assets and accounts, even when an entire region in North America is experiencing an outage that disrupts access to the microservices hosted there. This setup ensures that users can continue to access and manage their financial assets, providing a seamless experience and uninterrupted access to their data. However, it is important to note that this approach assumes compliance with data privacy laws, as certain regulations may restrict the transfer of financial information across country borders:

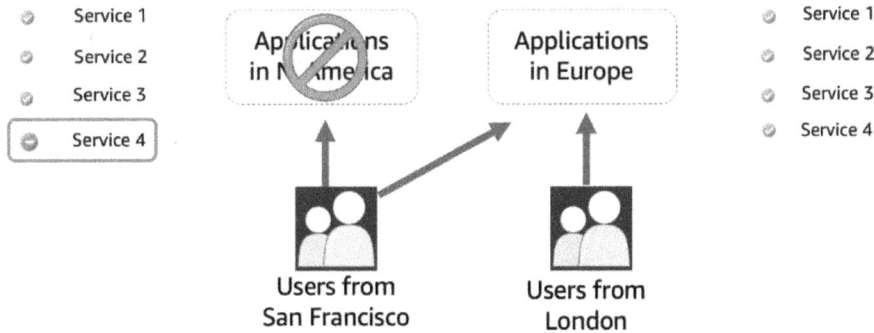

Figure 13.2 – Business continuity/disaster recovery use case

Similar to the financial industry example illustrated previously, the following are examples of other industries where global tables are – and can be – leveraged for business continuity:

- **Healthcare**: Global tables can be an asset in the healthcare industry to ensure that critical patient data, such as medical records and treatment plans, is always available. This is essential to ensure that patients receive proper care and treatment, especially in emergencies.

- **Transportation**: In the transportation industry, global tables can be used to guarantee that real-time data on vehicle location and status is accessible. This feature enables companies to optimize routes and schedules for the timely delivery of goods and services, which is critical for fleet management and logistics.

- **Manufacturing**: In the manufacturing industry, global tables can be used to ensure that production data, such as inventory levels and production schedules, is always available, even in the event of a regional outage. This is important to ensure that manufacturing operations continue to run smoothly and efficiently, and that supply chain disruptions are minimized.

These are a few examples of how global tables offer a highly durable and scalable solution for disaster recovery, allowing businesses to prepare for region-wide outages with minimal RPO and RTO. In the following section, we will explore how global tables perform seamless cross-region replication.

How it works

At the time of writing, DynamoDB global tables replicate data asynchronously across regions. This approach allows local updates to persist with low latencies, typically in the single-digit millisecond range. The replicas eventually converge to the same state, ensuring data consistency across regions. Asynchronous replication means that data changes are propagated across regions without the need for immediate consistency, which is useful for applications that require low-latency writes and local reads. With global tables, updates are automatically propagated to all replicas, ensuring that data is eventually consistent across regions. This allows applications to deliver fast, responsive experiences to users in multiple regions while maintaining data integrity.

Before exploring how global tables enable cross-region replication, it is essential to note that global tables can only create replicas within the same AWS partition. AWS partitions (1) define boundaries between AWS Regions. At the time of writing, there are three partitions:

- **aws**: Commercial AWS Regions
- **aws-cn**: AWS China Regions
- **aws-us-gov**: AWS GovCloud (US) Regions

For instance, you can only create cross-region global table replicas within the aws-cn partition, but not between a region in the aws partition and one in the aws-cn partition. With this clarified, let's learn about DynamoDB Streams' crucial role in driving cross-region replication.

DynamoDB Streams plays a crucial role in cross-region replication within global tables. To create a global table or convert a non-global table into a global one, DynamoDB Streams should be enabled on the table(s), and the stream view type should be NEW_AND_OLD_IMAGES. DynamoDB Streams provides a change log of mutations that occur on a DynamoDB table and guarantees **exactly once delivery** and **in-order delivery of item-level changes**. By utilizing DynamoDB Streams, global tables can perform cross-region replication and propagate data changes across all replicas automatically. This ensures that the data remains consistent across regions and can be accessed with low latencies, providing fast and responsive experiences to users. To learn more about DynamoDB Streams and its available stream view types, see *Chapter 12, Streams and TTL*. The following figure illustrates how global tables utilize DynamoDB Streams to perform cross-region replication:

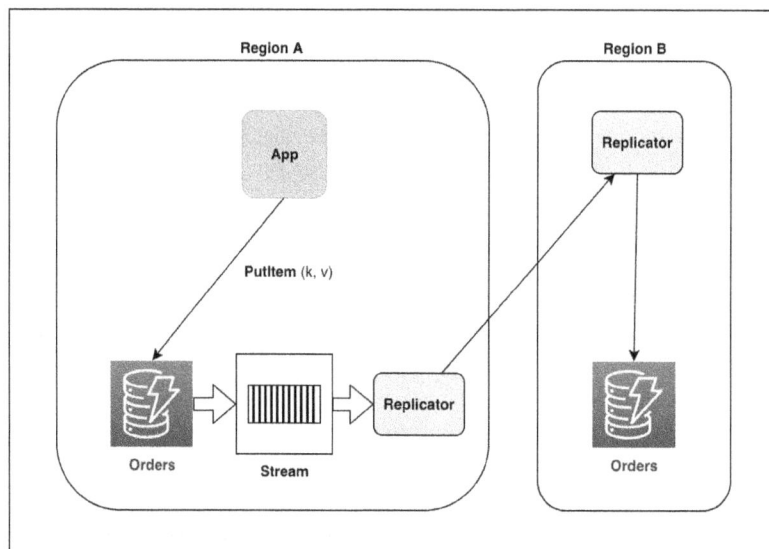

Figure 13.3 – Global table replication overview

According to the preceding figure, when a customer application writes to one of the global table replicas in *Region A*, a change record is generated in the table's DynamoDB Stream, typically within a few milliseconds. An invisible service component called a replicator reads this change record from the stream and determines that the write was originally made to the *Region A* replica. With knowledge of all cross-region replicas that are part of the global table, the replicator writes the change record to the other replica tables that are part of the global table. The *Region B* replica would also have a DynamoDB Stream where a change record would be generated due to the replicator's write on the *Region B* table. To avoid an infinite loop of updates going across regional replica tables, the replicators perform conditional updates to replicated regions based on the source region name and the source region commit timestamp. If the destination region replica already had a write for that key with a newer commit timestamp, then the replicator's write attempt would encounter a condition check failure, and the loop would be closed.

Both the source region name and source region commit timestamp are system attributes (at the time of writing) and are not available to customers to consume if you are using the 2019.11.21 version of global tables. However, they are available in the earlier 2017.11.29 version. Talking about versions, let's learn about the two available global table versions and their differentiating factors.

Global table versions

Two versions of DynamoDB global tables exist – version 2019.11.21 (current) and version 2017.11.29 (legacy). It is recommended that customers prioritize using the 2019.11.21 (current) version due to its enhanced flexibility, improved efficiency, and reduced write capacity consumption compared to the 2017.11.29 (legacy) version. While the legacy version is still available and being used by many customers, there are differences between the two versions that should be noted. The 2019 version is referred to as **GTv2**, while the 2017 version is referred to as **GTv1**; we'll use this reference throughout the rest of this chapter.

While it should be clear that both perform automatic cross-region replication of data in a fully managed way, the two main differences between the two versions are in terms of replication attributes and costs.

Replication attributes are those that help global tables perform cross-region replication reliably. GTv1 (the 2017 version global table) uses at least three attributes that help DynamoDB replicate the item reliably across the replicas:

- `aws:rep:updateregion`
- `aws:rep:updatetime`
- `aws:rep:deleting`

Every user write to a GTv1 global table is followed by an additional DynamoDB update that adds or updates the preceding three attributes to the item. These attributes are visible to users but must not be modified as doing so could disrupt cross-region replication for the item. Since every user/customer write is accompanied by this additional update, write costs are effectively doubled compared to GTv2 tables, where DynamoDB does not issue these additional updates.

In GTv2 (2019 version), replication attributes visible in GTv1 are hidden as system attributes. These attributes still exist in a similar form, but they are not available to users when interacting with items. This change was likely made to prevent issues caused by customers relying on or modifying these attributes from their applications unknowingly, which caused many issues with replication for the earlier global table versioned tables. Unlike GTv1, every user write is not followed by a DynamoDB update, which means that the write costs for simply making a new write are half of what GTv1 would charge. You will learn more about the cost implications and considerations in the next section of this chapter, *Operating and managing global tables*.

Apart from costs and replication attributes, there are minor differences between GTv1 and GTv2. One of those differences is that with GTv2, all replicas must either be in on-demand capacity mode or have auto-scaling enabled for writes in provisioned capacity mode. This requirement is based on lessons learned from GTv1, where the lack of such enforcement led to issues. Configuring highly varying write capacity units on global table replicas in GTv1 could cause synchronization problems, resulting in out-of-sync replicas across regions.

Another difference between the two versions is that with GTv2, you can convert a non-global, single-region table into a global table, regardless of whether the table already contains data. This flexibility allows you to add new replica regions on the fly for a GTv2 global table. In contrast, to convert a non-global, single-region table into a GTv1 global table, the table must be brand new and empty. You cannot convert a table into GTv1 if it has ever contained data. However, such a table can be converted into GTv2 seamlessly. We will learn how to convert a non-global table into a GTv2 table in the next section of this chapter, *Operating and managing global tables*.

Both versions of global tables differ in how they handle replica deletion. For GTv1 tables, a table can be made non-global without deleting its data. This stops automatic replication and converts the removed replica into a standalone, single-region DynamoDB table. In contrast, with GTv2, deleting a replica results in the deletion of the table and its data. The remaining replicas continue to function with automatic replication as usual, but the data in the deleted replica will no longer be accessible.

It is recommended to use the latest version – that is, GTv2 – if you plan to convert your single-region tables into global tables. If you already have a GTv1 deployment in production, you can migrate it to GTv2 natively via the AWS Management Console. However, be aware that GTv2 does not use the same replication attributes as GTv1. Therefore, before migrating, ensure that your application does not rely on the replication attributes specific to GTv1. For more information, please refer to the AWS docs (2).

Now that we understand why global tables are useful in our applications and have gone through the basics of global tables, it's time to learn about operating global tables, including some of their functional aspects.

Operating and managing global tables

In this section, we will provide a step-by-step guide on how to convert a single-region DynamoDB table into a global table with at least one replica. We will also cover some of the functional and cost aspects of using global tables, as costs should always be a factor in designing solutions. Additionally, we will review the monitoring support that's available with global tables, in addition to the standard monitoring provided for non-global, single-region DynamoDB tables.

Converting a single-region table into a global table

If you have been through earlier chapters of this book, you should know that we occasionally performed several tests using an `Employee` table. In case you haven't, it is a single-region, non-global DynamoDB table with a handful of items and can be thought of as a table that contains information about employee aliases, their full names, their designation, skills, and line manager aliases. The table's key schema only has a partition key, with its attribute name set to `LoginAlias`. One of the items from that table looks like this:

```
 1 {
 2      "ManagerLoginAlias": {
 3          "S": "marthar"
 4      },
 5      "LoginAlias": {
 6          "S": "amdhing"
 7      },
 8      "FirstName": {
 9          "S": "Aman"
10      },
11      "LastName": {
12          "S": "Dhingra"
13      },
14      "Designation": {
15          "S": "Architect"
16      },
17      "Skills": {
18          "SS": [
19              "software"
20          ]
21      }
22 }
```

Figure 13.4 – Employee table sample item

Now, let's convert this single-region table into a GTv2 global table. This single-region table currently resides in the eu-west-2/London region. We shall add a new cross-region replica to this table and, in turn, make the table part of a global table. When we add a replica to a non-global table, DynamoDB will provision the infrastructure required to set up the new table in the other region, perform an initial copy of any existing table data from the source region to the replica region, and set up online replication via DynamoDB Streams before allowing us to access or interact with this new replica. Follow these steps to learn how to create a new replica for our non-global Employee table:

1. Log in to the AWS Management Console and select **DynamoDB**.

2. In the DynamoDB dashboard, select the table you want to convert into a global table. In our case, this would be the Employee table.

3. Select the **Global tables** tab on the **Tables** page, as shown in the following figure:

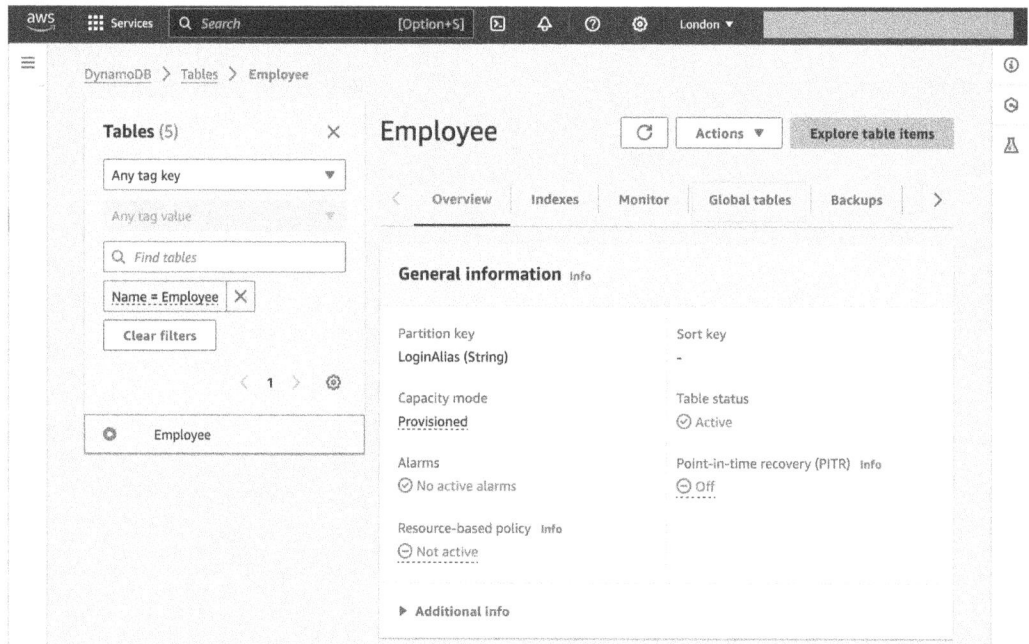

Figure 13.5 – DynamoDB | The Employee table page

4. As expected, the list of replicas will show nothing since the table is currently a standalone single-region table. Next, select **Create replica**, as shown in the following figure:

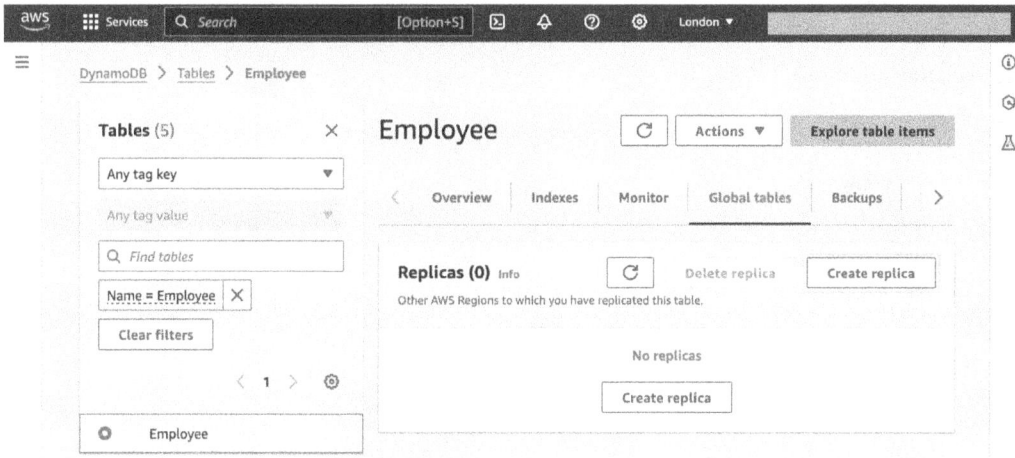

Figure 13.6 – DynamoDB | The Global tables tab

5. This is where you will get the option to choose the region where you would like this table data to be replicated and the global table set up. Let's say my organization has grown globally and I need to support employees in Asia while using the same application stack and database. I am going to add a new replica in the `ap-southeast-1/Singapore` region. The following figure shows the **Create replica** wizard showing the same. DynamoDB also displays the **AWS Identity and Access Management (IAM)** service role it will use to perform replication. Note that you may see calls being made to either replica using this IAM role. You cannot modify or manage this IAM role since it is used by the service for carrying out tasks in a fully managed manner, on your behalf. Once you've finished selecting the replication region, hit **Create replica**, as shown in the following figure:

DynamoDB > Tables > Employee > **Create replica**

Create replica

Replication settings

Current Region
Europe (London)

Available replication Regions
You can replicate your table to one of these Regions.

| Asia Pacific (Singapore)   ▼ |

IAM role
This service-linked role is used for replication.

AWSServiceRoleForDynamoDBReplication

ⓘ For replication to work, DynamoDB Streams will be activated automatically for new and old images.

Cancel   **Create replica**

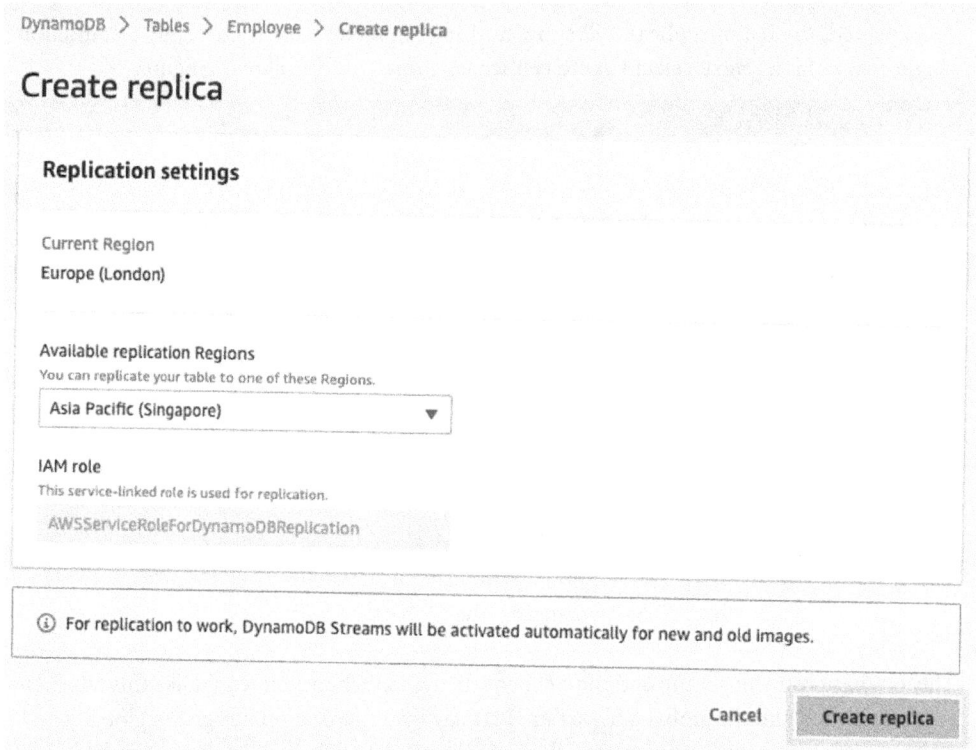

Figure 13.7 – DynamoDB | The Create replica wizard

6. As you click on **Create replica**, you will see a colored balloon at the top of the AWS console that acknowledges that your request to create a replica has been received and work has begun by the DynamoDB service to copy the table data, followed by setting up online replication both ways. Notice that the console mentions that I am using global tables version 2019.11.21. This is GTv2, which is the default now for new tables. I would only get the option to create a 2017.11.29 or GTv1 replica if my table was empty and was never written to. The following figure shows my console while I wait for my `Singapore` or `ap-southeast-1` replica to show up in the **Replicas** list:

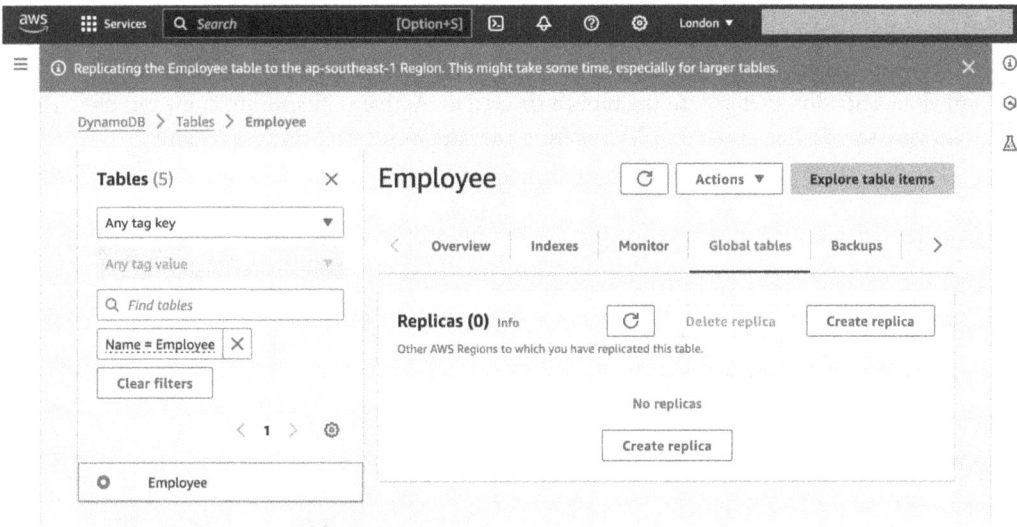

Figure 13.8 – DynamoDB | The Global tables tab

7.   After a few seconds, on refreshing the console page, I was able to see the Singapore replica in the **Creating** state, listed in the **Replicas** section of the AWS console. I am still working out of the London region in my AWS console. Here's a screenshot showing this:

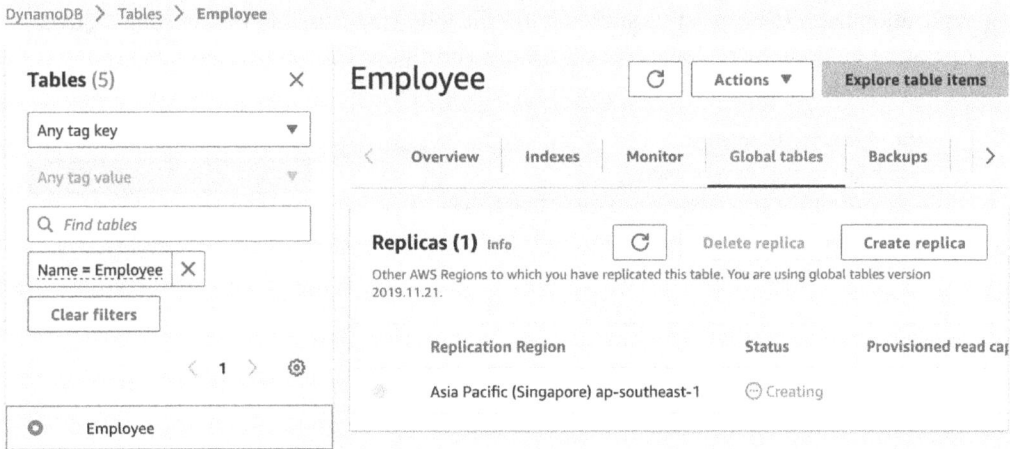

Figure 13.9 – DynamoDB | The Replicas list

8. Depending on the table size, it may take from a few minutes to several hours for DynamoDB to copy the data from the initial table to the new replica. In my case, the Employee table only had a handful of items, so the replica showed an **Active** status within a few minutes. The following screenshot shows this. This is from the eu-west-2/London region:

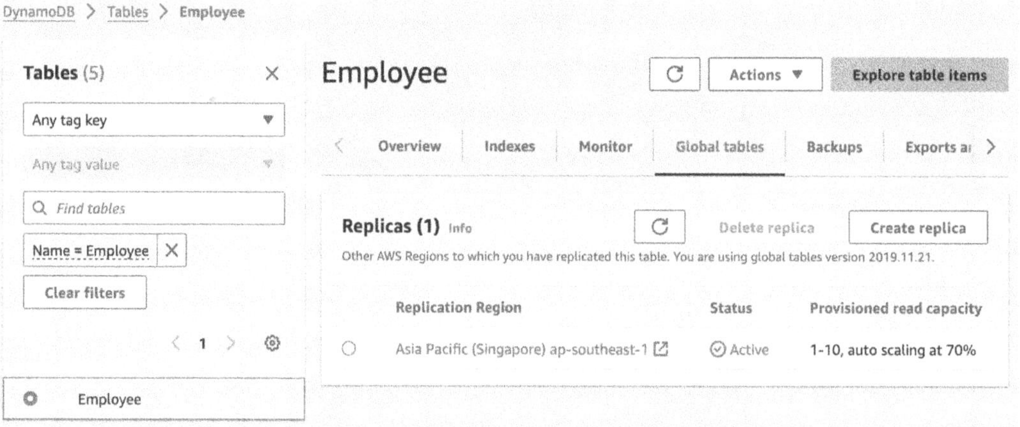

Figure 13.10 – DynamoDB | Active global table replica

9. Once the replica is in the **Active** state, it indicates the table can now be accessed. This also means that all the data that existed in the initial table has been copied to this new replica and online replication has also been set up successfully. The following figure shows running a Query operation on both the regional replicas for the same item. The following snippet shows the Query operation that was issued against the initial replica in eu-west-2/London:

```
aws dynamodb query \
--table-name Employee \
--key-condition-expression "LoginAlias = :pk_val" \
--expression-attribute-values '{":pk_val": {"S": "amdhing"}}' \
--region eu-west-2 | jq -rc
```

Here's the output:

```
1 {"Items":[{"ManagerLoginAlias":
  {"S":"marthar"},"LoginAlias":{"S":"amdhing"},"FirstName":
  {"S":"Aman"},"LastName":{"S":"Dhingra"},"Designation":
  {"S":"Architect"},"Skills":{"SS":
  ["software"]}}],"Count":1,"ScannedCount":1,"ConsumedCapaci
  ty":null}
```

Figure 13.11 – Output of the AWS CLI Query operation on the London replica data

Now, perform the same operation in the Singapore/ap-southeast-1 replica:

```
aws dynamodb query \
--table-name Employee \
--key-condition-expression "LoginAlias = :pk_val" \
--expression-attribute-values '{":pk_val": {"S": "amdhing"}}' \
--region ap-southeast-1 | jq -rc
```

Here's the output:

```
1 {"Items":[{"ManagerLoginAlias":
  {"S":"marthar"},"LoginAlias":{"S":"amdhing"},"FirstName":
  {"S":"Aman"},"LastName":{"S":"Dhingra"},"Designation":
  {"S":"Architect"},"Skills":{"SS":
  ["software"]}}],"Count":1,"ScannedCount":1,"ConsumedCapaci
  ty":null}
```

Figure 13.12 – Output of the AWS CLI Query operation on the Singapore replica data

You can also seamlessly delete a regional replica if you no longer need it. Regardless of the version of the global table, deleting or removing a replica of a replica from a global table group will not impact any of the other replicas or their ability to replicate data to other replicas that are still active.

Now that we've learned how to convert a single-region table into a global one, let's explore how to manage global tables in terms of costs.

Cost considerations and implications

To understand the costs associated with standard read/write activity on a global table replica, it's important to note that the **time to live** (**TTL**) feature behaves differently for GTv2 global tables. Specifically, TTL deletes on the first replica table are free of cost. However, when these TTL deletes are replicated to other replica tables, they are charged like user-issued global table write requests. The first replica table could be any of the replicas where the TTL system identified and issued a delete of an expired item. Subsequent replicas that replicate this delete event will incur charges for the same item being deleted in their local DynamoDB tables.

Global table writes are charged as **replicated write capacity units** (**rWCUs**) as opposed to the WCU charged by non-global table writes. On the other hand, reads are charged as standard **read capacity units** (**RCUs**), regardless of a global or non-global table. To learn about the costs associated with each version of the global tables, imagine that you had two replicas as part of the global table and each write was charged as 1 rWCU. The following table compares rWCUs charged by each version, depending on the operation type and the savings of using GTv2 over GTv1:

	GTv1 rWCU	GTv2 rWCU	Savings (GTv2 over GTv1)
New Put	2+2	1+1	50%
Delete	1+2	1+1	33%
Update	2+1	1+1	33%
Put, then TTL Delete	2+2, 0+0	1+1, 0+1	25%

Table 13.1 – Global table versions cost comparison

As shown in the preceding table, a new Put operation on a GTv1 table could cost 2 rWCUs on the source replica and two additional rWCUs on the other replica, totaling 4 rWCUs for a single Put operation. In contrast, a GTv2 table would incur a total of 2 rWCU: one for the source replica and another for replicating the item to the other replica table.

Similarly, if a new Put operation was made to a GTv1 replica and relied on TTL to automatically delete the item, the TTL delete itself would incur no rWCU cost on either replica. Thus, the total cost for the Put and TTL delete would be 4 rWCUs. In contrast, for a GTv2 replica, the initial write would cost 2 rWCU for both replicas. The TTL delete on the initial replica incurs no cost, but when replicated to the other replica, it costs 1 rWCU to delete the item. This results in an overall cost of 3 rWCUs for a new Put and TTL delete operation on a GTv2 global table with two replicas, making it 25% more cost-efficient than GTv1.

Another aspect of using global tables compared to non-global DynamoDB tables is that global table writes are charged as rWCU, and therefore, upfront capacity reservations do not apply to global table writes. This means that writes to any global table replica are not eligible for upfront capacity reservations, which are available only for non-global DynamoDB table writes. Capacity reservations are upfront commitment-based pricing models that can reduce DynamoDB throughput costs by **54%**

to 77% compared to standard pricing for provisioned capacity mode reads and writes. Although global table writes do not benefit from these reservations, writes to global secondary indexes on global table replicas are covered by these capacity reservations.

Now that we've learned about the cost implications of using global tables, let's learn how DynamoDB global tables ensure consistency across its replica tables and handle conflicting writes to the same item across global table replicas simultaneously.

Understanding consistency and conflict resolution

In this section, we'll understand the consistency model of DynamoDB global tables, followed by the conflict resolution protocols.

Cross-region consistency

DynamoDB global tables use asynchronous replication to propagate updates across regions. When a write operation is performed on a global table, it is first written to the local table of the region where the write operation was initiated. Then, the change is asynchronously propagated to the replica tables in other regions. This asynchronous replication allows for lower latencies and higher availability as updates can be processed locally without the need to wait for confirmation from other regions.

A common analogy for asynchronous replication is sending a letter through the mail. When you send a letter, you drop it off at your local post office, and it is then delivered to the recipient's local post office, where it is finally delivered to the recipient. The process is asynchronous as you do not wait for confirmation that the letter has been delivered before continuing with your day. Similarly, with asynchronous replication in DynamoDB global tables, updates are sent to local tables without the need to wait for confirmation from other regions, allowing for faster processing and lower latencies.

Asynchronous replication in DynamoDB global tables implies that data consistency across replicas is eventual. When a write is made to a local replica, there may be a delay before it is replicated to cross-region replicas. Consequently, if a read is performed on a cross-region replica before the replication is complete, the item may not be visible or may display an older state if a previous version exists in the global table replicas.

Eventual consistency matters in globally deployed applications because users may temporarily see different states of the same item while replicas converge to a consistent state. Global table replication typically ranges from 0.5 to 1.5 seconds, with the convergence time depending on factors such as network latency and the distance between regions. For example, replication between Ohio and Sydney may take longer than between Ireland and London. If your use case cannot tolerate such eventual consistency and requires strong read-after-write consistency across regions, you may need to use technologies that provide this guarantee. These technologies might impact write performance as writes may not be acknowledged until all cross-region replicas have persisted the data. With eventually consistent replication, you are trading off consistency for improved performance and availability.

When using DynamoDB global tables in a disaster recovery scenario, be aware that the most recent writes to a region may not be immediately available to all other replicas if the region becomes unavailable just after persisting the writes locally. In other words, the RPO might be minimal but not zero. This consideration is crucial when developing a disaster recovery strategy to address full region outages.

Achieving zero RPO in cross-region replication may require synchronous replication, which could increase write latencies beyond the single-digit milliseconds typically supported by DynamoDB. I am sure there may be use cases for zero RPO, such as financial services or healthcare, but at the time of writing, such a feature is not available in DynamoDB, or any AWS databases for that matter.

Conflict resolution

When writes across global table replicas are asynchronous, conflicts can arise if multiple replicas receive conflicting writes to the same item simultaneously. To resolve these conflicts, DynamoDB global tables employ a **last writer wins** strategy. In this approach, the most recent write to an item is used to resolve any discrepancies between replicas.

For instance, if a global table with three replicas in different regions has a user in Region A and another in Region B updating the same item at the same time, both updates are propagated asynchronously to the other replicas. When these updates arrive at the other replicas, a conflict occurs because each replica holds a different version of the item.

DynamoDB resolves this conflict using the timestamp associated with each write. The replica with the most recent timestamp updates the item's value, and other replicas discard their version of the update. This ensures that the latest value is propagated to all replicas. The timestamp that's used for resolution is highly granular, internal to the service, and not visible to users.

Now that we understand how DynamoDB global tables handle conflicts and the nature of eventual consistency in asynchronous replication, let's explore best practices for managing multi-region architectures with DynamoDB global tables.

Best practices for multi-region architectures

Now that we understand most aspects of the DynamoDB global table technology itself, let's review some best practices for using global tables in multi-region, globally available applications.

Avoid introducing cross-region dependencies

One essential best practice for maintaining a robust multi-region architecture in DynamoDB global table-based applications is to proactively prevent the introduction of cross-region dependencies. This principle underscores the importance of achieving regional isolation by routing requests to the appropriate regional endpoint of the application as early in the technology stack as possible. This strategic routing approach mitigates the need for cross-region database calls, which can inadvertently introduce latency and elevate the risk of errors stemming from network disruptions. Instead, by isolating requests to the specific regional endpoint of the application, you ensure the application's self-sufficiency and minimize dependencies on other regions.

To implement this functionality, one effective technology to leverage is **content delivery network (CDN)** edge servers. These servers are strategically distributed across the globe and facilitate routing to the nearest edge server, helping identify the optimal region where your applications are deployed. Alternatively, you can explore other approaches, such as utilizing **latency-based routing** coupled with comprehensive health checks on your individual regional application endpoints. In this context, **deep health checks** involve conducting successful application operations rather than merely pinging a server or endpoint to receive a basic response. Another viable option is to utilize global load balancers that support latency-based routing algorithms in addition to standard load balancing.

Choose the right regions

While adding or removing DynamoDB global table replicas is flexible, selecting regions should be done with care. Choose regions that provide geographical diversity while ensuring low-latency access for your users. The goal is to balance meeting your users' application-level latency **service-level agreements (SLAs)** with managing global table costs effectively. Regions closer to your users can reduce latency and improve their experience, but this may increase costs due to higher replication traffic and greater storage needs. Therefore, being able to carefully evaluate these trade-offs is crucial if you wish to align with your business requirements and budget.

Monitor replication lag

Monitoring and maintaining replication consistency is crucial for a DynamoDB global table setup. One of the key metrics to track is the replication lag between the replicas to ensure that data is consistent across all the replicas within acceptable timeframes. DynamoDB provides CloudWatch metrics to monitor replication lag and identify any potential issues.

These metrics can be used to set up alarms to be investigated or escalated to AWS in case the replication lag exceeds the typical range of 0.5s – 1.5s on average for a prolonged period. For GTv1, `ReplicationLatency` and `PendingReplicationCount` are two Amazon CloudWatch metrics that provide insight into the replication performance between any two replicas of a global table. GTv2, on the other hand, only provides the `ReplicationLatency` metric at the time of writing and is used for monitoring cross-region replication. It is important to track these metrics and take corrective actions promptly to ensure data consistency and avoid data loss. You can read more about these CloudWatch metrics for GTv1 in (3) and GTv2 in (4).

Test your disaster recovery strategy and document it regularly

If you're using DynamoDB global tables for disaster recovery purposes, you should typically prepare against a complete regional outage. One way to test your disaster recovery strategy is to use technologies such as **AWS Fault Injection Simulator** (5) to introduce artificial interruptions in your application on different components, such as servers or private networks. Regardless of the technology that's used to perform chaos engineering (6), it is important to test the strategy for failing over regularly to ensure that it can be relied upon when real interruptions occur.

Additionally, the recovery time should be documented and reviewed to set expectations with stakeholders and identify opportunities for improvement. By regularly testing and improving your disaster recovery strategy, you can better ensure the availability of your application in the face of unexpected events.

Those were some of the important best practices associated with multi-region applications based on DynamoDB global tables. The AWS docs also contain some other best practices for GTv1 (7) and GTv2 (8). This is recommended reading.

> **Important note**
>
> Consider this as a reminder to delete any resources you may have created in your AWS account while going through this chapter.

With that, we can conclude this chapter. Let's recap what we've learned.

Summary

In this chapter, we thoroughly examined DynamoDB global tables and their benefits for supporting geographically dispersed user bases and ensuring disaster recovery readiness. We explored the inner workings of global tables, including the different versions available. We also discussed the complexities of operating and managing global tables, such as converting a single-region table into a global table, and the associated cost considerations. Additionally, we addressed data consistency and conflict resolution, focusing on cross-region consistency and DynamoDB's conflict resolution mechanisms. Finally, we provided a detailed overview of best practices for designing multi-region architectures, stressing the importance of minimizing cross-region dependencies and actively monitoring replication lag.

In summary, DynamoDB global tables provide an effective way to scale applications globally, ensuring high availability and minimal latency for users. By following best practices and properly configuring your tables, you can achieve optimal performance and cost-efficiency for your global applications. For further guidance, refer to the AWS docs (9), which offers valuable insights into how various AWS customers have successfully implemented global tables in multi-region setups and strategies for managing user requests during region failures.

In the next chapter, we will delve into the necessity of in-memory caching for modern applications and explore **DynamoDB Accelerator** (**DAX**). DAX is a fully managed, in-memory caching service that is API-compatible with DynamoDB, enabling both read-through and write-through caching for DynamoDB tables. We will also examine the myriad benefits of caching and take a closer look at other popular caching technologies, complementing our exploration of DynamoDB DAX.

References

1. AWS Docs: AWS Partitions: `https://docs.aws.amazon.com/whitepapers/latest/aws-fault-isolation-boundaries/partitions.html`

2. AWS Docs: Global Tables: `https://docs.aws.amazon.com/amazondynamodb/latest/developerguide/GlobalTables.html`

3. AWS Docs: Global Tables GTv1 Monitoring: `https://docs.aws.amazon.com/amazondynamodb/latest/developerguide/globaltables_monitoring.html`

4. AWS Docs: Global Tables GTv2 Monitoring: `https://docs.aws.amazon.com/amazondynamodb/latest/developerguide/V2globaltables_monitoring.html`

5. AWS Docs: AWS Fault Injection Simulator: `https://docs.aws.amazon.com/fis/latest/userguide/what-is.html`

6. Chaos Engineering: `https://en.wikipedia.org/wiki/Chaos_engineering`

7. AWS Docs: Global Tables GTv1 Best Practices: `https://docs.aws.amazon.com/amazondynamodb/latest/developerguide/globaltables_reqs_bestpractices.html`

8. AWS Docs: Global Tables GTv2 Best Practices: `https://docs.aws.amazon.com/amazondynamodb/latest/developerguide/V2globaltables_reqs_bestpractices.html`

9. AWS Docs: Global Tables – Prescriptive Guidance: `https://docs.aws.amazon.com/prescriptive-guidance/latest/dynamodb-global-tables/introduction.html`

10. GitHub repository: `https://github.com/PacktPublishing/Amazon-DynamoDB---The-Definitive-Guide`

14
DynamoDB Accelerator (DAX) and Caching with DynamoDB

As modern applications continue to evolve and become more complex, the need for efficient data storage and retrieval becomes increasingly important. These modern applications could be backing end user websites, mobile apps, or even high-throughput internal systems serving several end user applications.

The importance of fast page load times for websites cannot be understated. Studies have shown that slow page load times can have a significant negative impact on businesses. For example, a study by Akamai Technologies (1) found that a delay of just one second in page load time can result in a 7% reduction in conversions. The same study also highlights that a 2-second delay in web page load time increases bounce rates by 103%. Bounce rate represents the percentage of all user sessions on a website where the user exited the website after only viewing a single page and is one of the primary metrics that websites must track – the lower, the better. Slow mobile apps can lead to frustration and abandonment, resulting in lost revenue and damage to the business's reputation.

This highlights the need for businesses to prioritize the speed and performance of their applications. Caching can play a critical role in this regard, by allowing businesses to improve the speed and performance of their applications and deliver a better user experience.

Caching is a technique used to store frequently accessed data in a fast-access memory location, such as RAM, in order to reduce the need for slower disk access. This can significantly improve the performance of an application by reducing the number of expensive disk reads and writes that are required to access the data. In fact, with the use of caching, it is possible to achieve sub-millisecond latencies, as opposed to the millisecond latencies typically achievable with modern databases such as DynamoDB.

There are several potential benefits to implementing caching in applications, including increased speed and scalability, reduced load on the database, and improved user experience. Caching can help to improve the performance of mobile apps by reducing the amount of data that needs to be transmitted over the network and retrieved from the database, leading to a faster and more responsive user experience.

If not for write requests, you may need faster reads to be served at the lowest possible latencies without compromising on durability, scalability, and high availability of the data. For writes, you may find that single-digit millisecond response times may be acceptable if that means the data written is durable. Apart from latency improvements, in some cases, caching can also help prevent hot spots on the database, by serving requests off the cache instead of going back to the database for every end user request – especially when the data does not change frequently.

Throughout the earlier chapters in this book, we discovered that using a well-designed DynamoDB schema can allow for serving requests at any amount of throughput with predictable performance. This predictable performance would likely be in the order of single-digit milliseconds on the database level. However, in order to have lower bounce rates for your websites or applications in general, you may need sub-millisecond response times for serving data, and adding a caching layer in your application could certainly be a good solution.

This does not mean every application under the sun must include a caching layer from the get-go. Since caches store data in memory, they are not meant to be used as a durable persistence later, but rather as a complement to a primary datastore that is highly durable, scalable, and available. An important trade-off of using caches is the strong read-after-write consistency of data. If data updates frequently in your application and you cannot accept serving stale data that may have been accurate in the past, then caches may not be suitable. Caches also tend to be quite costly since memory isn't always as cost-effective as storing and retrieving data off disks.

In this chapter, we will dive into caching, **DynamoDB Accelerator (DAX)** – a fully managed, highly available, in-memory cache particularly built to supplement DynamoDB and optimize latencies by up to 10X from milliseconds to microseconds. We will learn how to use DAX with DynamoDB, and we will look at its benefits in more detail and discuss some of the challenges and considerations involved in implementing caching in your applications. We will also review other popular caching alternatives to DAX for DynamoDB-based applications and compare their features with DAX.

By the end of this chapter, you will understand the whys and hows of caching, how to use DAX to interact with DynamoDB, scaling DAX and DynamoDB-based applications, and when to consider some of the popular alternatives to DAX for caching DynamoDB data.

In this chapter, we are going to cover the following main topics:

- Basics and setting up DAX
- DAX and DynamoDB working together
- Using DAX for high-volume delivery
- Other in-memory caches

Basics and setting up DAX

In this section, you will learn about the different components of a DAX cluster and what's required to use DAX and get familiar with DAX cluster creation and interaction using DAX's Python SDK.

Reviewing DAX cluster components

Amazon DAX is a fully managed, in-memory cache for DynamoDB that delivers fast read and write performance. With DAX, you can use DynamoDB as the primary data store for your applications and benefit from faster read performance for frequently accessed data.

DAX is a cluster-based, **Virtual Private Cloud** (**VPC**)-only service that you can use to improve the performance of your DynamoDB tables. Unlike DynamoDB, a DAX cluster would be a server-based resource that would reside in your own AWS account's chosen VPC and across your choice of subnets in the VPC. A DAX cluster could have a minimum of a single node and, at the time of writing, a maximum of up to 11 nodes for high scalability and availability of the cached data. DAX clusters can typically only be accessed privately, that is, within the scope of the same VPC, or different peered VPCs.

The following figure illustrates the topology of a DAX cluster and its interaction with the DynamoDB web service at a basic level:

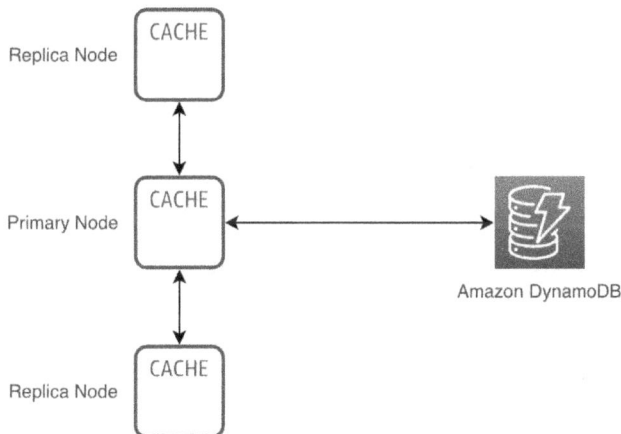

Figure 14.1 – DAX cluster topology

The cluster topology itself consists of a single primary node and zero or more replica nodes. The primary node is responsible for handling all the write requests that are made via the DAX cluster to DynamoDB. Whenever a write request is made through DynamoDB or a read request is made for which the DAX cluster cache does not already have data stored, the primary node reaches out to the DynamoDB web service to make the respective operations and fetch results. These results may then be cached within the cluster to serve subsequent requests for the same piece of data depending on how the cluster is configured. Both the primary and replica nodes can serve read requests for data cached in the cluster and perform cache eviction among other cluster management tasks.

The DAX cluster can interact with the DynamoDB web service either publicly, which is the default, or privately when a VPC endpoint for DynamoDB is part of the VPC networking setup. The following figure shows how applications can interact with a DAX cluster from within a VPC, and the DAX cluster further interacts with the DynamoDB web service publicly, which is the default means of connection:

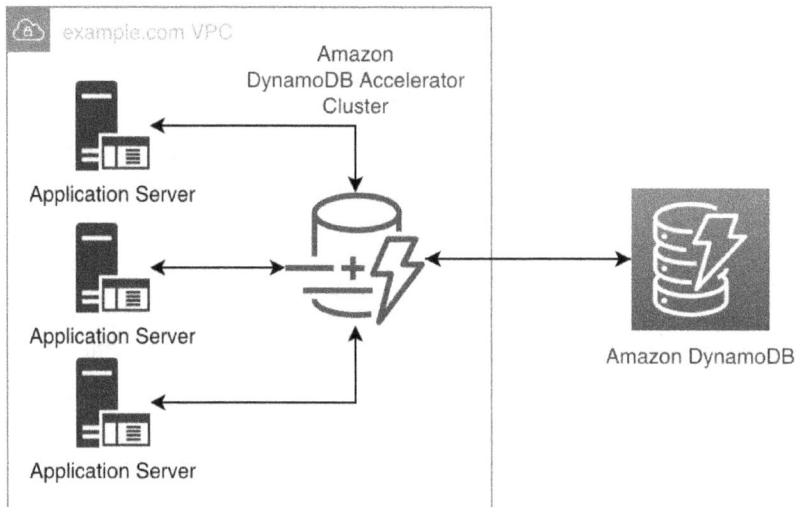

Figure 14.2 – Application interaction with DAX and DynamoDB

As per the preceding figure, the application simply interacts with the DAX cluster via a cluster endpoint, regardless of the number of nodes in the cluster. The DAX client SDK has built-in functionality to discover cluster topology, establish long-lived connections, and load balance requests across the cluster nodes transparently. The DAX cluster endpoint itself is similar to `my-cluster.foo.dax-clusters.eu-west-2.amazonaws.com` and it would resolve to IP addresses from anywhere in the world; however, the IP addresses themselves are private addresses, as per the VPCs CIDR ranges.

> **Important note**
> It is recommended to have at least three nodes in a production environment DAX cluster spanning three AWS Availability Zones for high availability and they should be fault tolerant.

DAX provides several additional features to enhance the management and maintenance of DAX clusters. One of these features is cluster parameter groups, which allow you to customize the configuration settings for your DAX cluster. These settings include the cache size, the number of threads to use for processing requests, and the maximum item size that can be stored in the cache. By modifying these settings, you can optimize the performance of your DAX cluster to meet the specific needs of your applications.

Additionally, DAX provides event notifications to keep you informed about important events related to your DAX cluster, such as when a new node is added or when the cluster is modified. You can receive these notifications via email or through an Amazon **Simple Notification Service** (**SNS**) topic. Finally, DAX also allows you to schedule maintenance windows during which the cluster will be automatically upgraded to the latest software of DAX along with any system or security patching that may be required. This ensures that your cluster is always running the latest software and provides the best performance and reliability. Maintenance windows are typically 60 minutes long and, during this period, cluster nodes typically get patched one by one to ensure cluster availability. Like any system, it is recommended to set the maintenance window in a period of least expected traffic on the application accessing the cluster.

Information about cluster components is also available in the AWS docs (2).

Creating a DAX cluster

To create a DAX cluster, you first need to navigate to the **DAX** dashboard in the AWS Management Console. Once there, click on the **Create cluster** button to get started. The following figure shows the DynamoDB DAX Management Console dashboard from where you would start to create your DAX cluster:

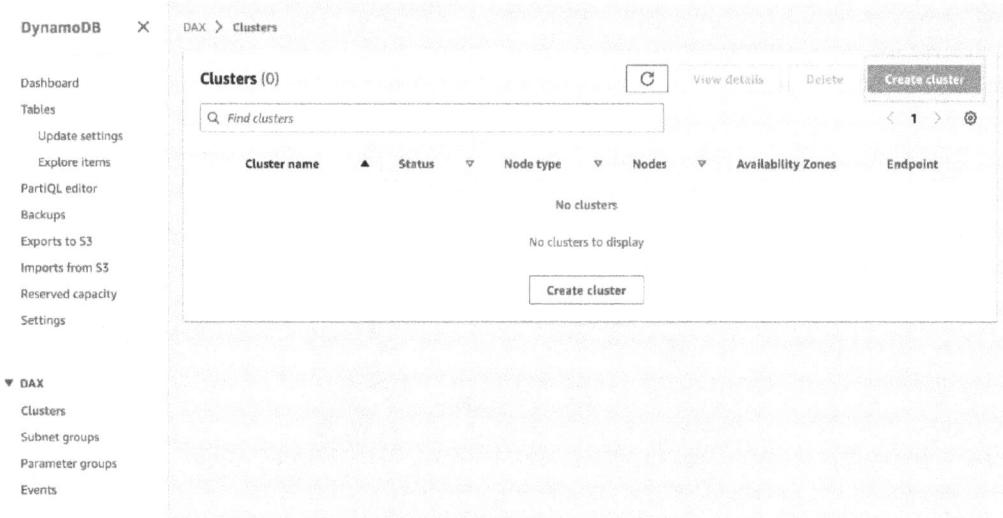

Figure 14.3 – DynamoDB DAX Management Console dashboard

Hitting the **Create cluster** button shown in the preceding figure would lead you to the cluster creation wizard. First up, you will provide the cluster with a name and description. Next, you select the cluster node type from several available node types. Cluster node types are like the standard EC2 instance types, but with the DAX system set up and managed for you. It is recommended to not use a T-type family of nodes for your DAX cluster in production as they are not best suited to be performant always and allow bursting of vCPU. The R-type family, however, is purpose-built for memory-intensive use cases and thus a good fit for production workloads needing a caching layer for DynamoDB data. T-type family nodes are an optimal choice for development or non-production environments in general.

The following figure shows the **Create Cluster** wizard where you provide the said details:

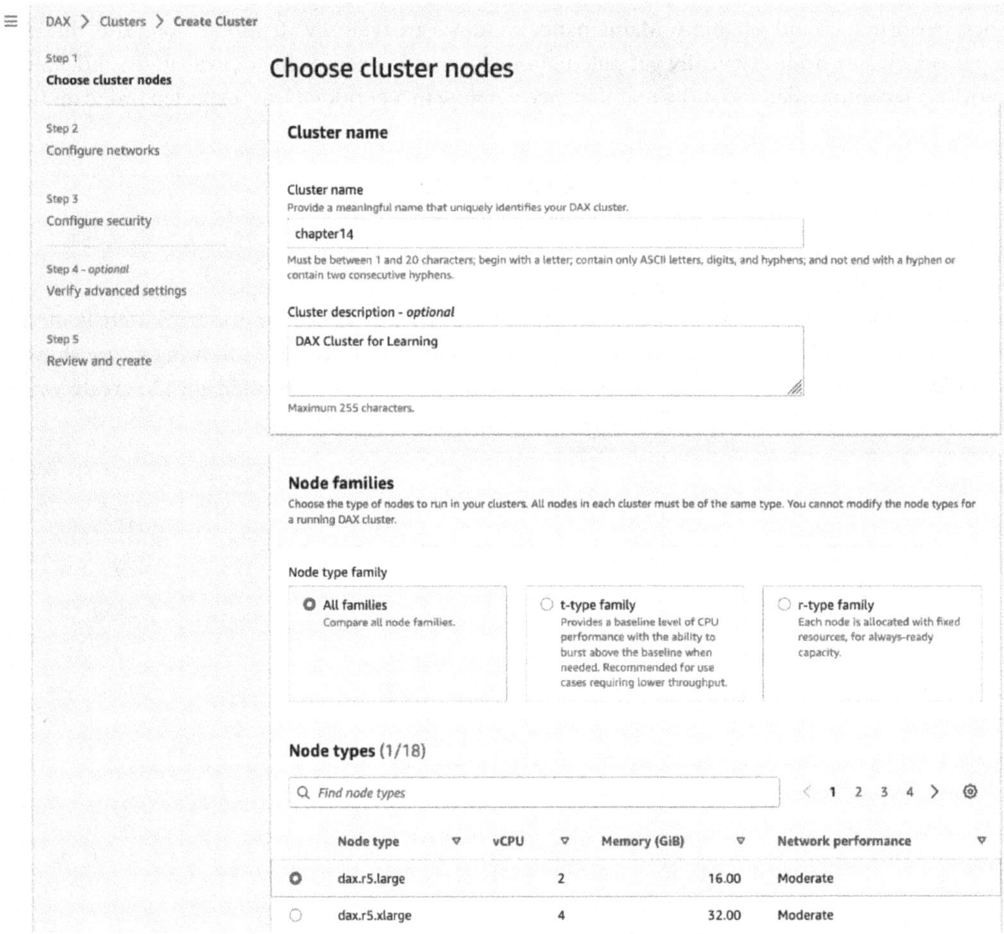

Figure 14.4 – The Choose cluster nodes step of DAX cluster creation: part I

On the same page as the preceding figure, you can also specify the number of nodes you want in the cluster. The following figure shows the part of the cluster creation wizard where you can configure the cluster size on creation:

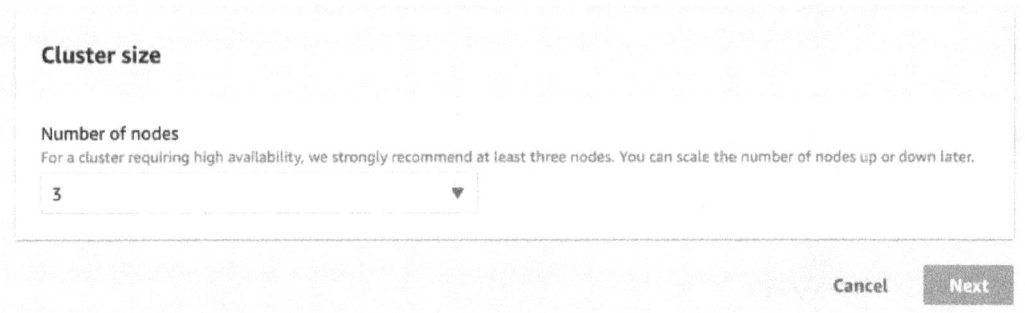

Cluster size

Number of nodes

For a cluster requiring high availability, we strongly recommend at least three nodes. You can scale the number of nodes up or down later.

| 3 | ▼ |

Cancel   **Next**

Figure 14.5 – The Choose cluster nodes step of DAX cluster creation: part II

After configuring the number and type of nodes for your DAX cluster, the next step of the **Create Cluster** wizard is to configure the network-related properties for the DAX cluster. This will include the VPC and subnets you want the DAX cluster nodes to be launched in. This step also includes providing the security group you want the cluster to be associated with for managing inbound and outbound network reachability. The following screenshot shows the next step of creating a DAX cluster via the AWS Management Console, where you can configure the network including the subnet and security groups:

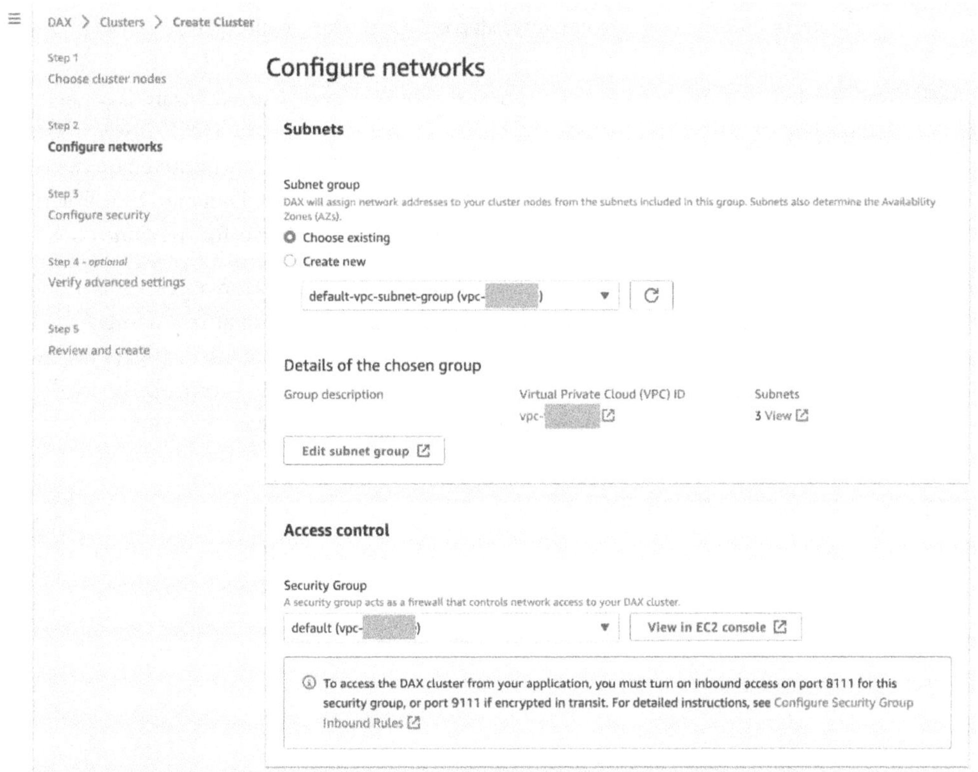

≡   DAX ⟩ Clusters ⟩ Create Cluster

Step 1
Choose cluster nodes

Step 2
Configure networks

Step 3
Configure security

Step 4 - *optional*
Verify advanced settings

Step 5
Review and create

Configure networks

Subnets

Subnet group

DAX will assign network addresses to your cluster nodes from the subnets included in this group. Subnets also determine the Availability Zones (AZs).

◉ Choose existing
◯ Create new

| default-vpc-subnet-group (vpc-▮▮▮▮▮) | ▼ | ↻ |

Details of the chosen group

| Group description | Virtual Private Cloud (VPC) ID | Subnets |
| | vpc-▮▮▮▮ ↗ | 3 View ↗ |

Edit subnet group ↗

Access control

Security Group

A security group acts as a firewall that controls network access to your DAX cluster.

| default (vpc-▮▮▮▮) | ▼ | View in EC2 console ↗ |

ⓘ To access the DAX cluster from your application, you must turn on inbound access on port 8111 for this security group, or port 9111 if encrypted in transit. For detailed instructions, see Configure Security Group Inbound Rules ↗

Figure 14.6 – The Configure networks step of DAX cluster creation: part I

Within the same **Configure networks** step of the **Create Cluster** wizard, you can also be very specific and configure the Availability Zone for each node within the cluster. By default, the DAX service will spread your cluster nodes evenly across Availability Zones to be highly available. Generally, you do not want to have to specify the Availability Zone for each node. The following figure shows the second half of the same **Configure networks** page (*Figure 14.6*) where you can see the option to allocate Availability Zones per cluster node:

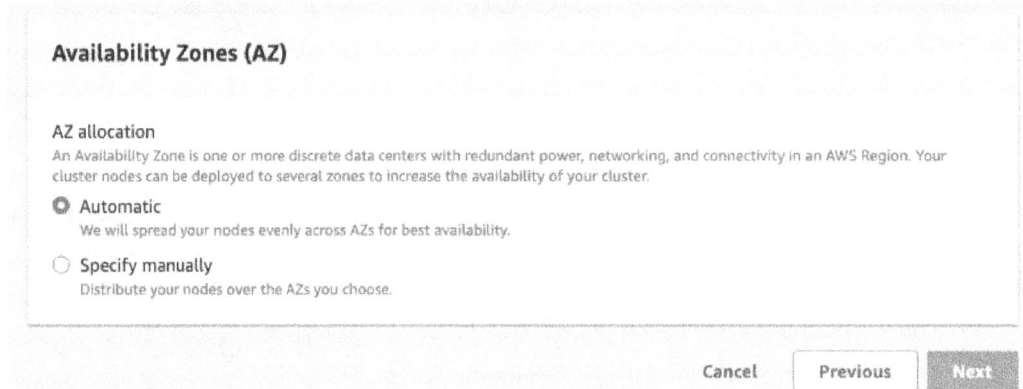

Figure 14.7 – The Configure networks step of DAX cluster creation: part II

Once the network-related configurations are set, the next step of the **Create Cluster** wizard is configuring security-related properties. These will include the IAM role that you want the DAX cluster to use to communicate with the DynamoDB web service. When any cache misses are encountered on the DAX cluster, the cluster's primary node will use this IAM role to interact with the DynamoDB web service on your behalf to serve the data while also caching it for subsequent requests for the same data.

On the same page, you can also configure your DAX cluster to use encryption at rest or, optionally, encryption in transit. Enabling encryption at rest would encrypt all data at rest using AES-256 encryption (3). Enabling encryption in transit on the cluster would encrypt all interaction between the application and the DAX cluster using the **Transport Layer Security** (**TLS**) protocol. The following is a screenshot of the **Configure security** page of the cluster creation wizard:

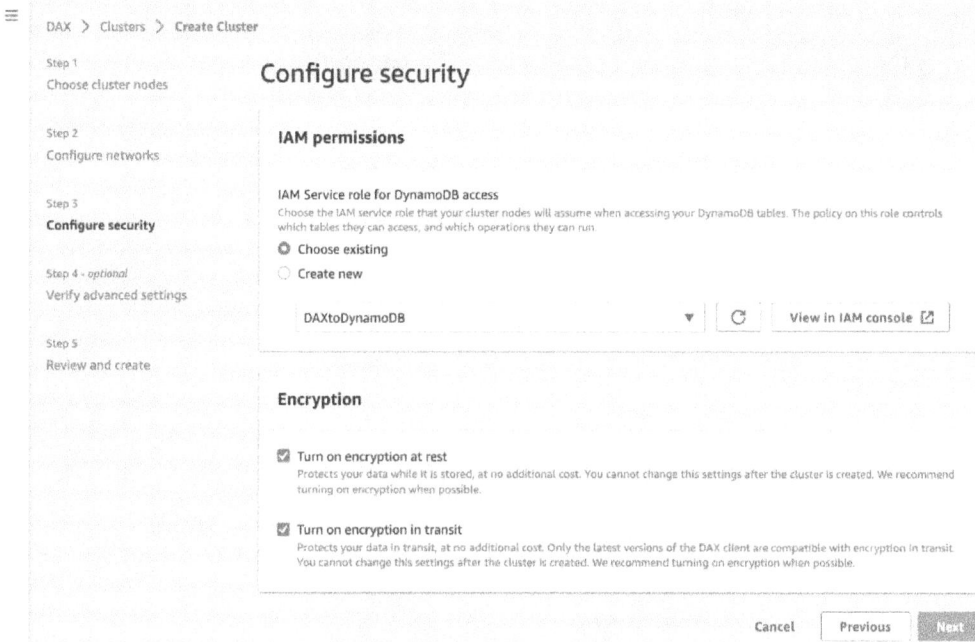

Figure 14.8 – The Configure security step of DAX cluster creation

After configuring the security-related properties of the soon-to-be DAX cluster comes yet another significant configuration – the **time-to-live** (TTL) values. Detailed information about these is later in this chapter in the *DAX and DynamoDB working together* section; however, the gist of it is that these values determine the amount of time data is cached in memory after being stored. Only when this time has elapsed will the data be invalidated from the cache and any subsequent request for the same piece of data will trigger DAX to reach out to the DynamoDB web service to get the data to serve to the application and then cache it again for the configured TTL duration. These TTL values can be set via a logical construct called a **parameter group**, which can be swapped even after a cluster has been created.

On the same page of the cluster creation wizard, DAX allows you to configure a maintenance window during which maintenance operations, such as upgrades and patches, can be performed on the DAX cluster without impacting application availability. It is essential to configure the maintenance window of a DAX cluster appropriately to minimize downtime and ensure that maintenance operations do not impact application performance.

To recall, best practices for configuring the maintenance window of a DAX cluster include selecting a time when application traffic is low, avoiding overlapping maintenance windows across multiple clusters, and notifying users of scheduled maintenance in advance. Additionally, it is recommended to monitor the DAX cluster's performance and availability during and after the maintenance window to ensure that the application continues to function correctly.

The following is a screenshot of the **Verify advanced settings** step of DAX cluster creation where you can specify the parameter group as well as configure the cluster maintenance window:

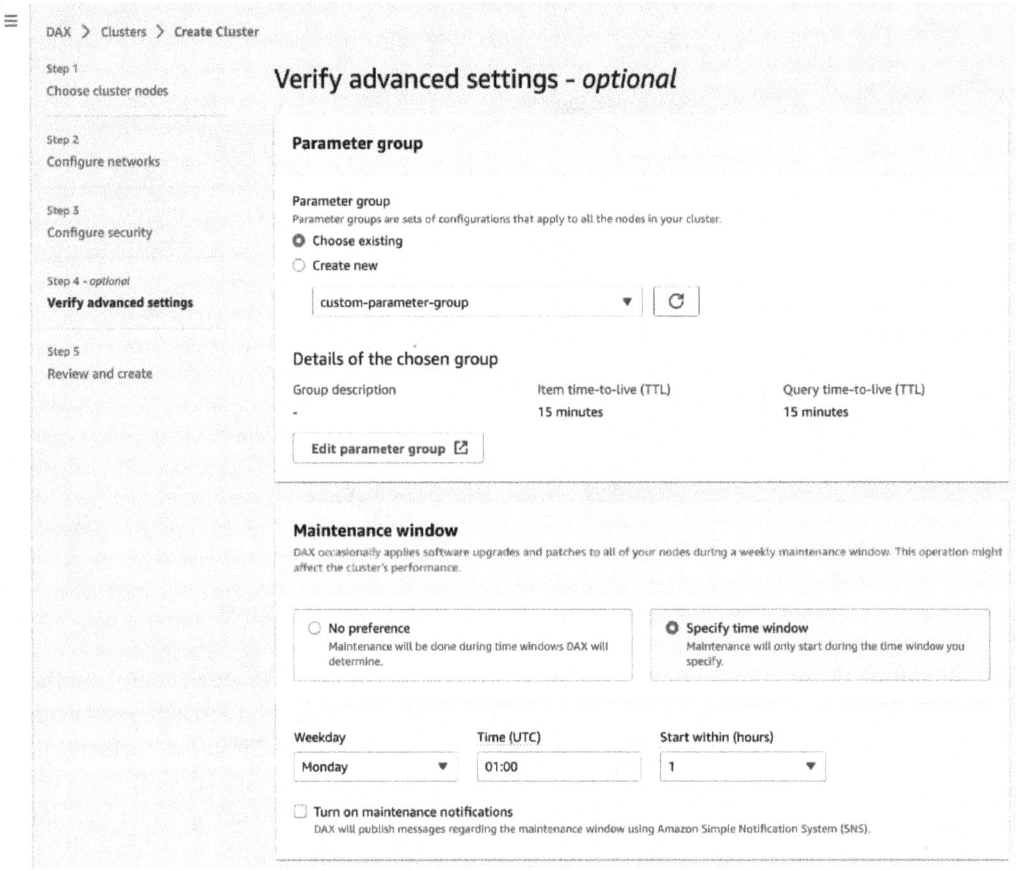

Figure 14.9 – The Verify advanced settings step of DAX cluster creation

Once all the configurations shown in the preceding figure are specified, you get an opportunity on the DAX Management Console to review all selected options in all the previous four steps, as partially shown in the following figure:

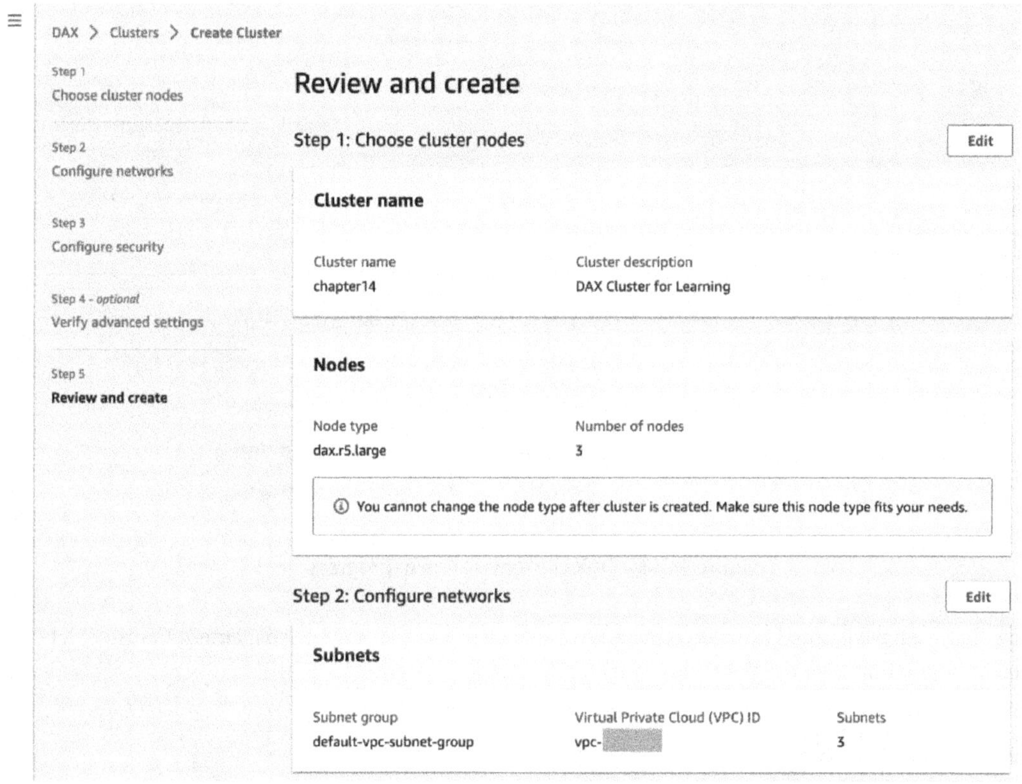

Figure 14.10 – The Review and create step of DAX cluster creation

Finally, click on the **Create DAX cluster** button to launch your cluster. It may take a few minutes for the cluster to become available. You can monitor the status of the cluster creation in the **DAX** dashboard by clicking on the cluster name and then selecting the **Events** tab:

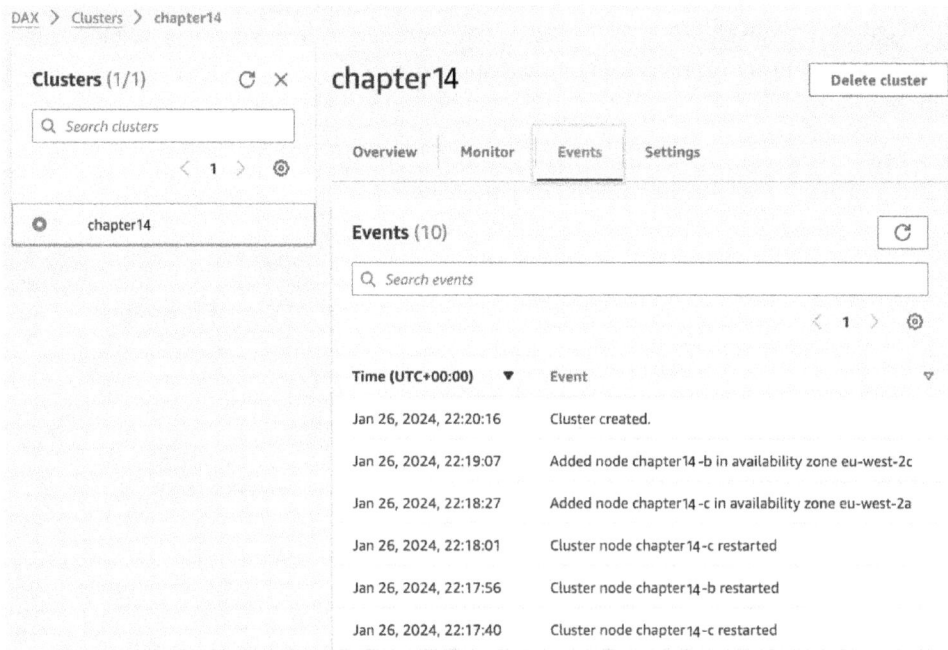

DAX > Clusters > chapter14

Clusters (1/1)   C ×   **chapter14**   Delete cluster

Q *Search clusters*

< 1 > ⚙   Overview   Monitor   Events   Settings

○   chapter14   **Events** (10)   C

                            Q *Search events*

                                                                < 1 > ⚙

                            Time (UTC+00:00) ▼   Event   ▽

                            Jan 26, 2024, 22:20:16   Cluster created.

                            Jan 26, 2024, 22:19:07   Added node chapter14-b in availability zone eu-west-2c

                            Jan 26, 2024, 22:18:27   Added node chapter14-c in availability zone eu-west-2a

                            Jan 26, 2024, 22:18:01   Cluster node chapter14-c restarted

                            Jan 26, 2024, 22:17:56   Cluster node chapter14-b restarted

                            Jan 26, 2024, 22:17:40   Cluster node chapter14-c restarted

Figure 14.11 – DAX cluster creation in progress

Now that we have learned how to create a DAX cluster using the AWS Management Console, let us look at interacting with the DAX cluster using the Python-based DAX SDK.

Querying a DAX cluster

To demonstrate how to use a DAX cluster to query a DynamoDB table, we will use the Amazon DAX client for Python (4). Remember that such code would need to be run with network access to the DAX cluster as the DAX cluster endpoint would always resolve to private IP addresses and, by default, the cluster nodes would only be reachable from within the VPC chosen for the DAX cluster. The following sample code shows how to perform a Query operation on a DAX cluster using the Amazon DAX client for a sample DynamoDB table:

```
from amazondax import AmazonDaxClient

dax_client = AmazonDaxClient(
    endpoint_url='daxs://xxxx.yyyy.dax-clusters.eu-west-2.amazonaws.
com',
    region_name='eu-west-2')

response = dax_client.query(
    TableName="Employee",
```

```
        KeyConditionExpression="LoginAlias = :pk_val",
        ExpressionAttributeValues={
            ":pk_val": {"S": "ripani"}
        }
    )
    print(response)
```

The preceding code first creates a DAX client using the `AmazonDaxClient` class from the `amazondax` module. It then uses this client to perform a `Query` operation on a DynamoDB table named `Employee` with a partition key value of `LoginAlias`. The response from the `Query` operation is then printed to the console. The following is the output of the same:

```
 1 {
 2     'Items': [{
 3         'LoginAlias': {
 4             'S': 'ripani'
 5         },
 6         'Designation': {
 7             'S': 'Developer'
 8         },
 9         'FirstName': {
10             'S': 'Lorenzo'
11         },
12         'LastName': {
13             'S': 'Ripani'
14         },
15         'ManagerLoginAlias': {
16             'S': 'marthar'
17         },
18         'Skills': {
19             'SS': ['software']
20         }
21     }],
22     'Count': 1,
23     'ScannedCount': 1
24 }
```

Figure 14.12 – Output of Python sample for querying via the DAX cluster

In this section, we have covered the basics of DAX, including DAX cluster components, how to create a DAX cluster using the AWS Management Console, and how to query a DynamoDB table using the Amazon DAX client for Python. With DAX, you can significantly improve the performance of your DynamoDB applications and easily scale to handle millions of requests per second.

Next up, we will learn how DAX and DynamoDB interact with each other to serve the data access requests.

DAX and DynamoDB working together

In this section, you will learn about how DAX and DynamoDB work together when your application needs to perform reads and writes on DynamoDB table data. There are a couple of caching strategies that may be used, depending on the use case and, mostly, the write-heavy or read-heavy nature of the application. Before learning about these caching strategies, let's learn more about a crucial feature of DAX called TTL.

TTL

DAX provides a TTL feature for both the item cache and query cache. The TTL allows you to set an expiration time for cached items and queries, ensuring that stale data is automatically removed from the cache. The item cache stores individual items retrieved from DynamoDB, while the query cache stores the results of queries executed against DynamoDB. The difference between the two caches matters because the TTL values may need to be different for each cache depending on the application's requirements. For example, the TTL for the item cache may be shorter than the TTL for the query cache if the application frequently updates data.

The TTL values for both item and query cache can be changed throughout the lifetime of a DAX cluster. This means that if the application's data access patterns change, the TTL values can be adjusted to optimize performance. However, changing the TTL values may impact cache performance and increase the risk of stale data being served from the cache. Therefore, it is important to carefully evaluate the impact of changing the TTL values before making any changes.

When deciding on the values of TTL for both the item and query cache, it is essential to consider the nature of the application and data access patterns. If the application frequently updates data, a shorter TTL for the item cache may be more appropriate to ensure that stale data is quickly removed from the cache. On the other hand, if the application performs a lot of queries that return the same results, a longer TTL for the query cache may be beneficial to improve performance. Additionally, it is important to consider the impact of cache invalidation and the cost of fetching data from DynamoDB when setting TTL values. A thorough analysis of the application's data access patterns and performance requirements can help determine appropriate TTL values for both the item and query cache in a DAX cluster.

Caching strategies with DAX

To achieve optimal performance with DAX, it is essential to select an appropriate caching strategy. DAX supports two main caching strategies: write-through and write-around.

Write-through caching

Write-through caching involves issuing writes to DAX, which, in turn, issues the write to DynamoDB before concluding the write as successful. When a read request is made for an item that is not in the cache, DAX fetches it from DynamoDB and writes it to the cache. The subsequent read requests for the same item are served from the cache.

All write requests first go to DAX, which then writes the data to DynamoDB before returning a response. This strategy ensures that the cache and DynamoDB are always in sync and reduces the likelihood of stale data in the cache. The following figure illustrates an application architecture where DAX is used with a write-through caching strategy:

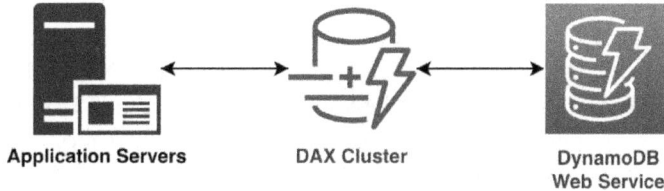

Figure 14.13 – Write-through caching strategy with DAX

The primary advantage of write-through caching is that it provides a high cache hit rate and consistency, making it an excellent choice for applications that require strong consistency and high availability. However, the downside of this strategy is that it increases write latency as it involves writing data to both the cache and DynamoDB. Additionally, write-through caching can increase the amount of data that needs to be cached on DAX, leading to higher costs.

Write-around caching

Write-around caching, on the other hand, involves writing data directly to DynamoDB and bypassing the cache. This strategy is useful for write-heavy applications, where write operations are more frequent than read operations. In this strategy, only read requests are served from the cache, and write requests are written directly to DynamoDB, avoiding the cache altogether. This approach ensures that the cache is not overloaded with data that may not be needed in the future, thus freeing up memory for more frequently accessed data. The following figure shows an application architecture where DAX is used with a write-around strategy:

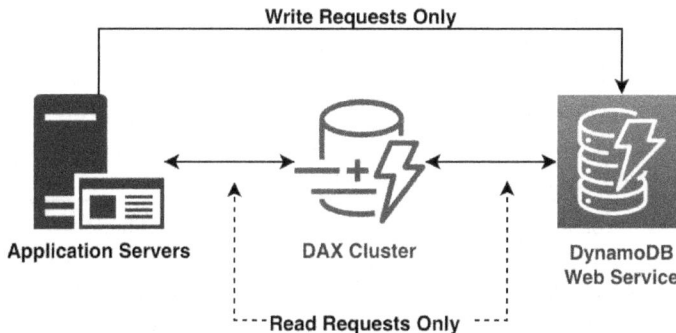

Figure 14.14 – Write-around caching strategy with DAX

The primary advantage of write-around caching is that it reduces write latency and the load on DAX, which can result in lower costs. However, the downside of this strategy is that it can result in a low cache hit rate, leading to higher latency for read requests. Write-around caching is suitable for applications that require low write latency and do not require strong consistency.

One approach to achieving a higher cache hit rate with write-around caching is to use a DynamoDB Stream and a Lambda trigger. This trigger forces the caching of recently inserted data by pre-emptively firing read requests onto the DAX cluster for every mutation seen in the DynamoDB stream. One consideration of this approach is that the data will be cached in memory for the TTL duration specified, regardless of whether it is required by the application to serve customer requests. This may prove to be costly if read activity on the data just written is not sufficient. The following figure illustrates this pattern of hydrating the cache via DynamoDB Streams and Lambda for an application that has implemented write-around caching:

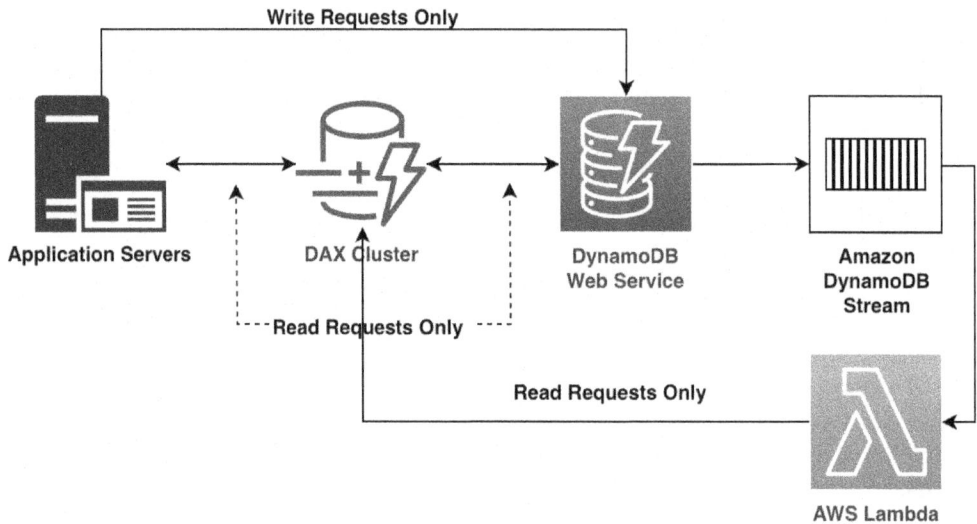

Figure 14.15 – Write-around caching with DynamoDB Streams and Lambda hydration

Choosing the appropriate caching strategy depends on the following specific needs of the application.

Write-through caching	Write-around caching
Strong read-after-write consistency	Cache staleness is acceptable
High availability of data in the cache	Lazy loading
Low read latency requirements (read-heavy OLTP)	Low write latency requirements (write-heavy OLTP)

Table 14.1 – Feasibility factors for write-through and write-around caching strategies

In conclusion, DAX provides an excellent way to improve the performance of read-heavy applications using DynamoDB. Choosing the right caching strategy is critical to achieving optimal performance and reducing costs. Write-through caching provides high consistency and availability but increases write latency and the load on DynamoDB. Write-around caching reduces write latency and the load on DAX but may result in a lower cache hit rate. Understanding the requirements of the application is essential to choose the right caching strategy.

Now that we have learned more about DAX in terms of TTL and commonly followed caching strategies for applications implemented with an in-memory caching layer in DAX, let's take a look at some performance numbers and how DAX can be used for high-volume, low-latency delivery of repeated read requests.

Using DAX for high-volume delivery

After learning about various aspects and concepts of DAX, let's compare its performance with reading from the DynamoDB web service directly. As a reminder, DynamoDB stores data on warm **Solid-State Drives (SSDs)** under the hood, and all reads require seeking these SSDs for data before decrypting them using your specified encryption context type (3). In contrast, with DAX, the cached data is stored in memory, which results in reading speeds that are tens of times faster.

For the purposes of this demonstration, I will be creating a new DynamoDB table called `chapter14` with a partition key attribute named `PK` and no sort key attribute. This means that each unique value of the `PK` attribute will identify a single item in the table. Using the same DAX cluster that I created in the previous section, I will measure the end-to-end latency of performing reads from an EC2 instance in the same VPC as the DAX cluster. While I am not a professional programmer, the following script snippets are sufficient for demonstrating the performance differences between repeatedly reading from DynamoDB versus using DAX.

The following Python code measures the performance of reading data from DynamoDB, a managed NoSQL database service provided by AWS, and from DAX, a managed in-memory cache for DynamoDB. This is also available in the book's GitHub repo (5). I will break down the script into several parts; you can put them one after the other to get the whole thing or, alternatively, get it from the GitHub repo. The code performs the following tasks:

1. Set up logging and AWS clients for DynamoDB and DAX as in the following code:

    ```
    import sys
    import time
    import logging
    import boto3
    import threading
    import random
    from amazondax import AmazonDaxClient
    ```

```python
from botocore.config import Config

NUM_WRITES = 100000
NUM_READS = 200000
READ_CYCLE_MINUTES = 10
NUM_THREADS = 2

# Set up logger
logger = logging.getLogger()
logging.basicConfig(
    stream=sys.stdout,
    level=logging.INFO,
    format='%(asctime)s %(levelname)s %(message)s'
)
logger.setLevel(logging.INFO)
TS_FORMAT = "%Y-%m-%dT%H:%M:%SZ"

# Custom client configuration
my_config = Config(
    region_name='eu-west-2',
    max_pool_connections=100,
    retries={'max_attempts': 10, 'mode': 'adaptive'})

# Set up the DynamoDB clients
# DynamoDB.Table resource for making batch writes
dynamodb = boto3.resource('dynamodb',
    region_name='eu-west-2',
    config=my_config)
table = dynamodb.Table('chapter14')

# DynamoDB.Client for making reads from DynamoDB
ddb_client = boto3.client('dynamodb',
    region_name='eu-west-2',
    config=my_config
)

# Set up the DAX client
logger.info("Setting up DAX Client")
dynamodb_dax = AmazonDaxClient(
    endpoint_url='daxs://xxxx.yyyy.dax-clusters.eu-west-2.
amazonaws.com',
    region_name='eu-west-2',
    config=my_config)
```

2. Generate dummy data and insert it into the DynamoDB table. Measure the time it takes to insert this dummy data. The following code illustrates this:

```python
# Generate dummy data
data = [
    {
        'PK': 'employee{}'.format(i),
        'Name': 'Employee {}'.format(i),
        'Age': random.randint(20, 50)
    } for i in range(NUM_WRITES)]

# Insert the dummy data into the DynamoDB table
def put_items():
    with table.batch_writer() as batch:
        for item in data:
            batch.put_item(Item=item)

# Measure the time it takes to insert the dummy data
logger.info(
    "Putting {} items into the DynamoDB table"
    .format(NUM_WRITES))
start_time = time.time()
put_items()
end_time = time.time()
logger.info(
    "Time to insert data: {} seconds at {} Writes per second"
    .format(end_time - start_time,
            NUM_WRITES / (end_time - start_time)))
```

3. Define a get_items function to perform GetItem requests on the DynamoDB table or the DAX cluster. This function in the following code takes a client as a parameter, so it can be reused to read from both sources:

```python
# Define a function to perform GetItem requests
def get_items(client, thread_index,
              latencies, op_count_list):
    op_count = 0
    start_t = time.time()
    target_time = start_t + READ_CYCLE_MINUTES * 60
    while time.time() < target_time:
        for i in range(100):
            try:
                key = {'PK': {'S': 'employee{}'.format(
                    random.randint(0, 9999))}}
```

```
                            response = client.get_item(
                                TableName='chapter14', Key=key)
                            op_count += 1
                        except Exception as e:
                            print(e, file=sys.stderr)
                    # time.sleep(0.01)
                end_t = time.time()
                total_t = end_t - start_t
                # adjusted_duration = (
                    # total_t - 0.01 * int(op_count / 100))
                adjusted_duration = total_t
                logger.info("Individual Thread Ended Execution. \
                    Number of Operations: {} at {} Reads per second"
                    .format(op_count, (op_count / adjusted_duration)))
                latencies[thread_index] = adjusted_duration / op_count
                op_count_list[thread_index] = op_count
```

4. Define a measure_latency function as in the following code to measure the time it takes to perform the GetItem requests for a duration of 10 minutes using multiple threads:

```
# Define a function to measure
# the time it takes to perform GetItem requests
def measure_latency(client, name):
    latencies = [None] * NUM_THREADS
    op_count_list = [None] * NUM_THREADS

    start_time = time.time()
    threads = [threading.Thread(
        target=get_items,
        args=(client, i, latencies, op_count_list)
    ) for i in range(NUM_THREADS)]
    for thread in threads:
        thread.start()
    for thread in threads:
        thread.join()
    end_time = time.time()
    duration = end_time - start_time

    logger.info("{} Latency Results: Total Time={} seconds, \
        Average Latency={} ms" .format(name, duration,
            1000 * (sum(latencies) / len(latencies))))
    return 1000 * (sum(latencies) / len(latencies))
```

5. Finally, call the `measure_latency` function, once for DynamoDB directly, and once for DAX. Calculate and print the percentage improvement of DAX over DynamoDB in terms of repeated reading. The following code does all of this:

```python
# Measure the latency of GetItem requests on the DynamoDB table
logger.info("Performing GetItem on same {} \
    items directly to DynamoDB for {} minutes"
    .format(NUM_WRITES, READ_CYCLE_MINUTES))
ddb_avg = measure_latency(ddb_client, "DynamoDB")

# Measure the latency of GetItem requests on the DAX cluster
logger.info("Performing GetItem on same {} \
    items via DAX for {} minutes"
    .format(NUM_WRITES, READ_CYCLE_MINUTES))
dax_avg = measure_latency(dynamodb_dax, "DAX")
logger.info("Repeated reading from DAX is {} \
    % better than from DynamoDB"
    .format(100 * (1 - (dax_avg / ddb_avg))))
```

By running the whole script with Python 3.7 and on a `c5.xlarge` EC2 instance type from the same VPC as the DAX cluster, the following results could be seen:

```
1 [ec2-user@ip-xx-xx-xx-xx dax-project]$ python3 measure_latency.py
2 2023-03-25 17:01:20,593 INFO Found credentials from IAM Role: XYZ
3 2023-03-25 17:01:20,635 INFO Setting up DAX Client
4 2023-03-25 17:01:20,648 INFO Found credentials from IAM Role: XYZ
5 2023-03-25 17:01:20,969 INFO Putting 100000 items into the DynamoDB table
6 2023-03-25 17:02:01,342 INFO Time to insert data: 40.37338376045227 seconds at 2476.87933697435 Writes per second
7 2023-03-25 17:02:01,343 INFO Performing GetItem on same 100000 items directly to DynamoDB for 10 minutes
8 2023-03-25 17:12:01,514 INFO Individual Thread Ended Execution. Number of Operations: 166400 at 277.2552535830826 Reads per second
9 2023-03-25 17:12:01,572 INFO Individual Thread Ended Execution. Number of Operations: 166000 at 276.56099371486766 Reads per second
10 2023-03-25 17:12:01,572 INFO DynamoDB Latency Results: Total Time=600.2295784950256 seconds, Average Latency=3.6113117893361957 ms
11 2023-03-25 17:12:01,572 INFO Performing GetItem on same 100000 items via DAX for 10 minutes
12 2023-03-25 17:22:01,599 INFO Individual Thread Ended Execution. Number of Operations: 690100 at 1150.115534930913 Reads per second
13 2023-03-25 17:22:01,619 INFO Individual Thread Ended Execution. Number of Operations: 689600 at 1149.2460692990262 Reads per second
14 2023-03-25 17:22:01,619 INFO DAX Latency Results: Total Time=600.0464720726013 seconds, Average Latency=0.8698067682310977 ms
15 2023-03-25 17:22:01,619 INFO Repeated reading from DAX is 75.91438183765963 % better than from DynamoDB
```

Figure 14.16 – Output of DAX versus DynamoDB end-to-end performance harness

As shown in the preceding figure, when running two threads performing repeated reads among 100,000 items stored in DynamoDB using a write-around caching strategy, reading data via DAX was approximately 75% faster than reading directly from the DynamoDB web service. In other words, on average, reading data via the DAX cluster was about four times faster than reading directly from DynamoDB. However, it's important to note that the results obtained from this test are not definitive and your own performance results may vary based on several factors such as the EC2 instance type used, network throughput of the EC2 instance, size of DynamoDB items, and DAX TTL values, among others.

Next, let's review some of the limitations of DAX and some popular alternatives.

DAX limitations

Having seen how powerful in-memory caching with DAX can be for your read-heavy or repeated read workloads, what if DAX is not suitable for your application due to one of the shortcomings of the service?

While DAX is a highly performant and reliable in-memory caching data store for DynamoDB, it has certain limitations when compared to other in-memory caching solutions. DAX could appear to be complex to set up and maintain, requiring significant expertise and resources to optimize its performance when compared with other popular non-DAX alternatives such as Amazon ElastiCache. Additionally, for applications that may need to cache heterogeneous forms of data, the fact that DAX can only cache DynamoDB data can be a potential limitation.

Furthermore, DAX has a limited set of features and functionality, which may not meet the needs of all applications. For instance, DAX does not support data partitioning or sharding, which can limit its scalability for large or complex datasets. Therefore, it is important to consider other in-memory caching solutions, such as Redis and Memcached, which offer a wider range of features and support for multiple data stores. These solutions are widely used in industry and have been proven to be highly scalable, reliable, and easy to use.

In the next section, we will explore other in-memory caching alternatives, including Redis and Memcached, in detail, highlighting their key features and benefits and comparing them to DAX to help you make an informed decision about which solution is right for your application.

Other in-memory caches

While DAX is an excellent choice for in-memory caching with DynamoDB, it is not the only option available. Depending on the requirements of the application, other caching data stores such as Redis or Memcached may be more suitable.

One of the primary considerations when choosing a non-DAX caching solution is the need to manage and implement the caching logic on your own. While DAX is fully managed (meaning that the caching logic is handled by AWS), Redis and Memcached require more involvement from the user. This can include setting up and configuring the cache cluster, implementing cache eviction policies, and monitoring cache performance.

Redis is a popular in-memory cache that provides a range of data structures, such as strings, hashes, lists, sets, and sorted sets, allowing for more complex data models than DAX. Redis also supports features such as pub/sub messaging, Lua scripting, and transactions. Redis is often used for real-time applications, such as chat apps or analytics, where low latency is crucial.

Memcached, on the other hand, is a simple, high-performance key-value store that provides fast access to cached data. Unlike Redis, Memcached does not support data structures or advanced features such as pub/sub messaging or transactions. Memcached is often used for web applications that require fast response times, such as e-commerce or social media platforms.

While Redis and Memcached offer more flexibility than DAX, they also require more effort to set up and manage. Additionally, they do not provide integration with DynamoDB out of the box, meaning that you must implement the caching logic by yourself. This can be both time-consuming and complex, particularly for those without significant experience in caching or distributed systems.

One advantage of using DAX over Redis or Memcached is its integration with Amazon ElastiCache, a fully managed in-memory caching service that supports both Redis and Memcached. ElastiCache provides a range of features, such as automatic failover, backup and restore, and scaling, making it easier to manage and monitor the cache cluster. Additionally, ElastiCache integrates seamlessly with AWS services, such as EC2, RDS, and DynamoDB, allowing for efficient and optimized performance.

Another recent development in the in-memory caching space is the introduction of Amazon MemoryDB, a fully managed, Redis-compatible, in-memory database service. MemoryDB offers the benefits of Redis, such as data structures and advanced features, but with the simplicity and scalability of a fully managed service. MemoryDB is designed to provide high performance, low latency, and low variability, making it an ideal choice for real-time applications.

Another alternative is Momento (6), a serverless caching alternative to DAX that enables in-memory caching for DynamoDB without the need for managing infrastructure or scaling. Momento provides a cost-effective solution for workloads with unpredictable traffic patterns, where the use of a dedicated caching layer such as DAX may not be feasible. Momento works by using a Lambda function to handle the DynamoDB requests and caching frequently accessed data in an in-memory cache that is shared across multiple invocations of the Lambda function. The cache is automatically managed by Momento, ensuring high availability and low latency for read-intensive DynamoDB workloads. In addition, Momento also supports automatic cache invalidation using DynamoDB Streams, ensuring that data is always up to date in the cache. With Memento, DynamoDB users can benefit from in-memory caching without the need for additional infrastructure, enabling them to achieve high performance at a lower cost. It is worth mentioning that Momento is a third-party offering and is available to use via the AWS Marketplace (7).

In conclusion, while DAX is an excellent choice for in-memory caching with DynamoDB, it is not the only option available. Depending on the requirements of the application, other caching data stores such as Redis, Memcached, or Momento may be more suitable. However, these alternatives require more effort to set up and manage and do not provide integration with DynamoDB out of the box. ElastiCache and MemoryDB offer a fully managed and scalable in-memory caching service for Redis and Memcached that provides integration with AWS services and advanced features, making them an excellent choice for applications that require high performance and low latency.

> **Important note**
>
> Consider this as a reminder to delete any resources you may have created while going through this chapter in your AWS account.

Summary

In this chapter, we delved into the fundamentals of DAX and its setup process. We provided a comprehensive overview of the key components that make up a DAX cluster and walked you through the step-by-step process of creating your own DAX cluster. Additionally, we demonstrated how to effectively query a DAX cluster and retrieve cached data.

Moving forward, we explored the synergy between DAX and DynamoDB in enhancing application performance. We examined the TTL feature of DAX, which automatically removes expired items from the cache, optimizing cache size and overall performance. Furthermore, we delved into various caching strategies compatible with DAX, including write-through and write-around, offering insights into their respective advantages and disadvantages. We also shed light on how DAX can facilitate high-volume data delivery by harnessing its automatic sharding and failover capabilities.

Toward the end of this chapter, we took a closer look at alternative in-memory caching solutions such as Redis and Memcached. While each of these solutions presents its own unique benefits and drawbacks, we introduced Momento as a serverless caching alternative to DAX specifically designed for use with DynamoDB. Momento offers a flexible and scalable caching solution that autonomously manages data expiration while seamlessly integrating with existing DynamoDB applications. Unlike DAX, which demands upfront capacity planning and management, Momento adapts dynamically based on application traffic and usage patterns.

In the upcoming chapter, we will explore a variety of analytical patterns commonly implemented in DynamoDB-based applications. DynamoDB excels in addressing mission-critical **Online Transaction Processing (OLTP)** workloads. However, modern applications often demand both OLTP and **Online Analytical Processing (OLAP)** capabilities. To fulfill these diverse requirements, we'll delve into additional DynamoDB features and analytical solutions that may incorporate other AWS services within the analytical space.

References

1. Akamai report on the impact of slow website load times on businesses – `https://www.akamai.com/newsroom/press-release/akamai-releases-spring-2017-state-of-online-retail-performance-report`

2. AWS docs: DAX cluster components – `https://docs.aws.amazon.com/amazondynamodb/latest/developerguide/DAX.concepts.cluster`

3. AWS docs: Encryption at rest – `https://docs.aws.amazon.com/amazondynamodb/latest/developerguide/EncryptionAtRest.html`

4. AWS docs: Installing the DAX client – `http://dax-sdk.s3-website-us-west-2.amazonaws.com/`

5. GitHub: Repo for online chapter assets – `https://github.com/PacktPublishing/Amazon-DynamoDB---The-Definitive-Guide`

6. Momento serverless cache – `https://www.gomomento.com/`

7. Momento: AWS Marketplace – `https://aws.amazon.com/marketplace/pp/prodview-tntv64zqgaqm2`

Part 5:
Analytical Use Cases and Migrations

The final part addresses analytical use cases and data migration strategies. Enhanced analytical patterns are discussed to leverage DynamoDB data for complex querying and data analysis tasks. The chapter on migrations provides comprehensive guidelines for moving data from other databases to DynamoDB, ensuring a smooth and efficient transition. These chapters equip you with the knowledge to handle sophisticated data processing tasks and integrate DynamoDB into your existing systems and workflows.

Part 5 has the following chapters:

- *Chapter 15, Enhanced Analytical Patterns*
- *Chapter 16, Migrations*

15
Enhanced Analytical Patterns

For any internet-scale application, serving end users with the most reliable, highly available, and rich experience is crucial. This means delivering high-quality services consistently, especially when an end user is waiting for a real-time response. We have already learned about these **Online Transactional Processing (OLTP)** needs and how DynamoDB can be a powerful part of the solution for achieving low-latency, consistent performance at scale. While OLTP is a major part of the requirements, it is often not the only requirement.

As the business owner of an e-commerce web store, you need a view of the business and its key performance indicators. You want to know which products are in high demand, which products have seen the lowest customer feedback, what the overall sales and profits look like, and much more. All these insights are available in the data, often across different data stores owned by different microservices and teams.

The same data helps in identifying trends, making re-stocking forecasts, and projecting end-of-year sales and revenues to determine acceleration opportunities and staff incentives. The same data also helps in generating recommendations for end users to improve the overall user experience.

All these use cases require performing data analytics. Order data may be needed along with the product catalog to generate user product recommendations. Current and historical order data for a particular region in a particular period of the year may be needed to identify trends and forecast demand when the same period of the year comes around again. It is not just that multiple sources of data are needed together; the data from one or more sources also needs to run through algorithms beyond what an OLTP database can and must be used to do. You may need to count the number of orders made in a particular time period in a region, the number of orders that were delayed in being fulfilled, the average number of products per order, and so on.

While OLTP is clearly important, **Online Analytical Processing (OLAP)** is also often an important requirement of modern internet-scale applications. It is necessary to share insights with business teams or simply to implement first-class features such as a product recommendations system.

While DynamoDB is designed to support OLTP with predictable performance at scale, it supports first-class functionalities that allow full-fledged analytics. Since DynamoDB is not your run-of-the-mill RDBMS, where a one-size-fits-all approach was made to work, the patterns of implementing analytics are different. The patterns are different because they consider that the table data could range from kilobytes to petabytes, and there must be no impact on live applications that are interacting with DynamoDB while the typically offline analytics jobs are crunching data.

In this chapter, you will learn about common analytical patterns you can use for your DynamoDB data. These are patterns that are already used and battle-tested across industries and geographies, by customers of different shapes and sizes. We will dig into several patterns and the best practices associated with them.

We will start by understanding the need for analytics and the level of complexity that certain analytical tasks can require. We will then dig into the patterns themselves. Of course, there may be multiple ways to reach the destination, so we will also learn about the factors you must consider when deciding on a particular analytical requirement. We will also learn about bulk processing patterns for DynamoDB. Finally, we will cover some key optimizations that can be made on DynamoDB or other services that will be part of the analytical pipeline.

By the end of this chapter, you will be fully equipped with the knowledge required to implement analytics on DynamoDB data. You will have a thorough understanding of the key functionalities within DynamoDB, as well as other integrating services, which will allow you to set up a robust data strategy for your DynamoDB data.

In this chapter, we are going to cover the following main topics:

- Need and complexity of analytics
- Diving into analytical patterns
- Bulk data processing
- Optimizing usage of DynamoDB and other AWS services for analytics

Need and complexity of analytics

Modern applications rely heavily on low-latency, high-throughput access characteristic of OLTP systems. Users, whether human or other applications, expect immediate responses from client applications or web interfaces. No customer ever says, *please slow down the performance; it is too fast for our end users, and we want you to take more money for database services while you are at it.*

DynamoDB excels at OLTP, making it an ideal choice for such workloads. However, data needs to be accessible not only to end users but also to internal business and data science teams. These teams require the same data for various analyses, demanding different formats and access methods to run algorithms that generate actionable insights. This necessity drives the need for analytics.

Technically, one could perform a full table scan on a DynamoDB table to retrieve data for reporting or machine learning pipelines. This naive approach, however, is impractical for large-scale workloads as it can degrade the performance of OLTP applications sharing the same table, leading to poor user experience. Implementing dynamic rate-limiting, exponential back-offs, and retries may mitigate some issues, but these solutions are suboptimal. Analytical queries on GBs or TBs of data can be costly and disruptive.

The following figure illustrates a high-level idea of OLAP queries issued directly on DynamoDB causing interruptions to the OLTP access and end user experience:

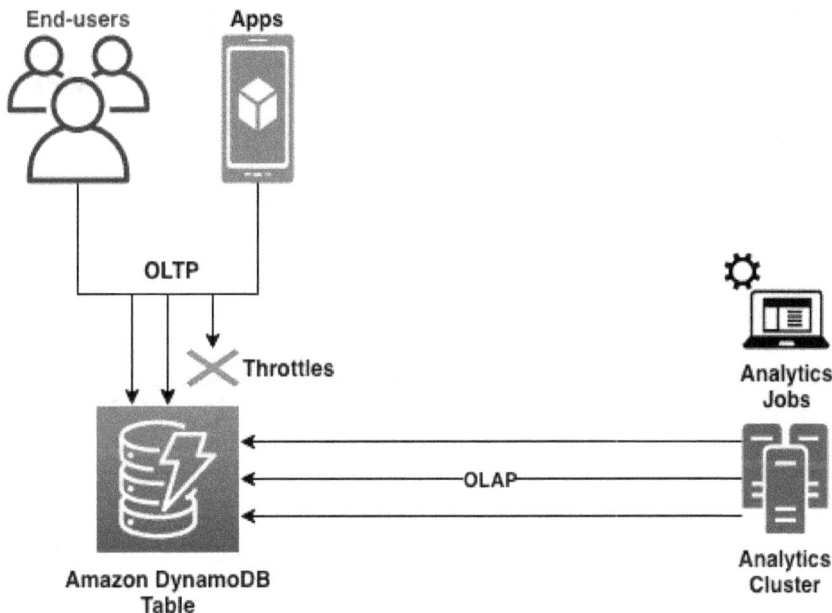

Figure 15.1 – Potential scenario of using DynamoDB for both OLAP and OLTP

Using Amazon S3 for analytics

As an alternative to scanning tables in real-time for analytics, **Amazon Simple Storage Service** (**Amazon S3**) can be integrated with your DynamoDB data since S3 is a common place for organizing your data lake. The data lake or lakehouse architecture (1) allows you to maintain datasets in different formats. You can govern access to that data using AWS LakeFormation (2), an AWS service that is purpose-built to support data management and governance with fine-grained access to rows and fields within the rows.

To leverage S3 for your DynamoDB data, you can use the native full table export functionality of DynamoDB to dump your table data into an S3 bucket. This allows you to get your DynamoDB data into S3, available to be read any number of times, and you pay a fraction of the cost to read without potentially causing interruptions to production OLTP applications accessing data from the live DynamoDB table.

The following is a figure illustrating a high-level idea of separating the OLAP access from the OLTP by leveraging a copy of the DynamoDB data in S3:

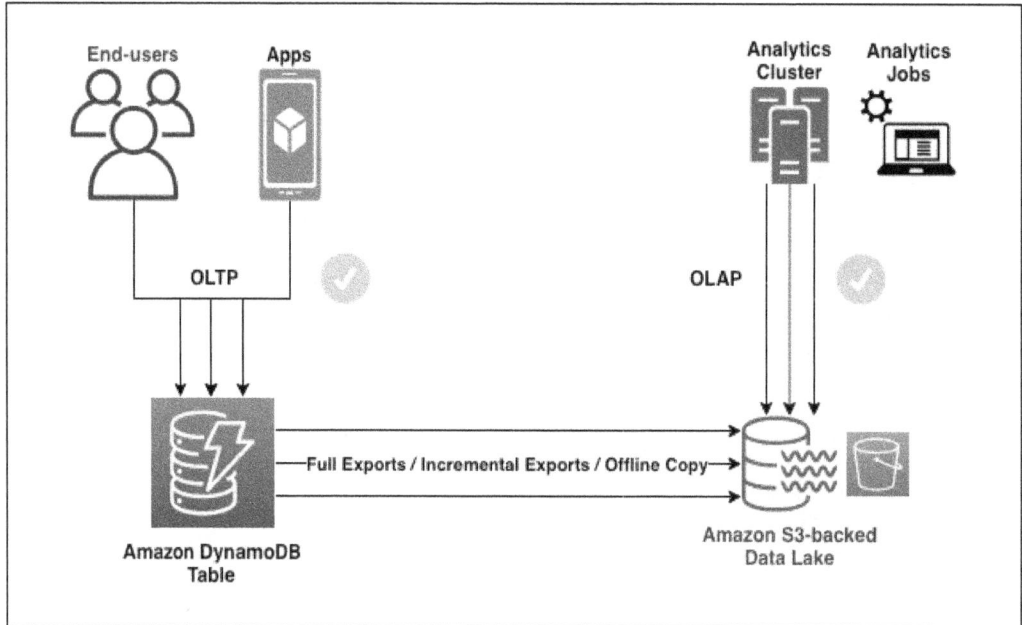

Figure 15.2 – Solution with OLAP and OLTP access separated

Complexity in analytics pipelines

When discussing complexities within the analytics world, obtaining raw data is just the tip of the iceberg. Although moving raw data from a database into an S3 bucket is a crucial step in any analytics pipeline, it is necessary to orchestrate **Extract-Transform-Load** (**ETL**) processes on this data to make it suitable and organized for your data lake in an optimal manner.

For example, while the native export formats supported by DynamoDB at the time of publishing are DynamoDB JSON and Amazon Ion, popular data lake formats include Apache Iceberg (3), Apache Parquet (4), and Apache Hudi (5), which are optimal for distributed analytics. Amazon Athena (6), a popular distributed SQL interactive query service, can be about **80% faster** when querying partitioned Parquet data and over **90% cheaper** compared to querying data in JSON format (7). Therefore, depending on the scale of your data lake and the data that needs to be processed, transforming data into optimal data lake formats is key to an effective data strategy.

Transforming data into these formats adds complexity to the analytics pipeline. Even if DynamoDB could export data directly in Parquet format, additional transformations might still be necessary to optimize it for any organization's specific analytical processing.

The following figure illustrates a high-level idea of leveraging a native DynamoDB export to S3 and running Glue ETL on the DynamoDB JSON data to convert it into a format such as Apache Parquet, which is one of the most optimal data formats for distributed analytics technologies such as Amazon Athena:

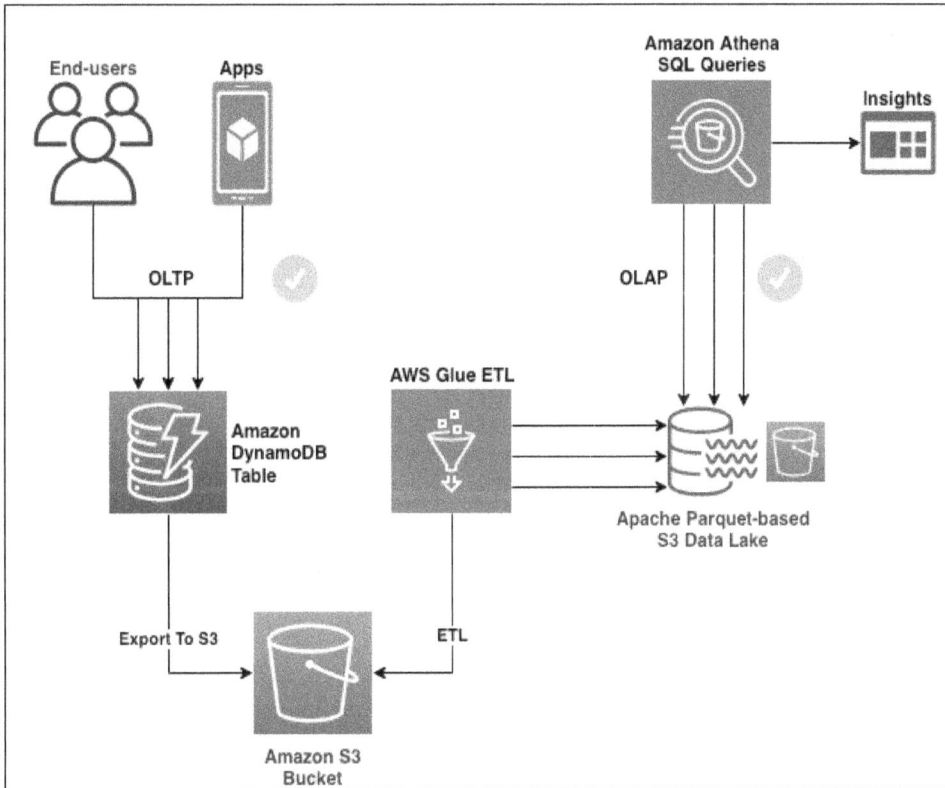

Figure 15.3 – Architecture leveraging AWS Glue to convert data into
Apache Parquet and Amazon Athena for analytics

Incremental exports and real-time processing

The complexity of analytics requirements can vary significantly. For instance, a banking application must record different transactions on your account, each affecting the overall account balance. To ensure the reliability of the accounting software, the bank also needs to record every update to the account balance and make it available for periodic verification.

If the app hosts its data on DynamoDB, a full table scan or export may not be the best way to retrieve the history of changes. However, an incremental export showing only the delta of changes can be useful in such scenarios. Thankfully, DynamoDB natively supports incremental exports to S3. The team managing this part of the banking app can schedule incremental exports every few minutes to capture data changes and run them through an ETL process, making the data easily accessible in S3.

While periodic verification of accounting can be considered an offline batch job, processing banking transactions in real time for anomaly or fraud detection requires a different approach. Triggering an incremental export every few minutes may not be sufficient. In such cases, DynamoDB Streams can be leveraged instead. A DynamoDB Stream for the accounting table creates a stream record for every mutation that occurs on the table, which can be consumed in real time and used alongside technology such as Amazon Fraud Detector (8), a service that supports building and deploying fraud detection models on AWS.

The following figure illustrates a high-level idea of this approach, where an OLTP application accesses DynamoDB, DynamoDB Streams are consumed by an AWS Lambda function, and the data is processed by the Amazon Fraud Detector service to validate transactions in real time.

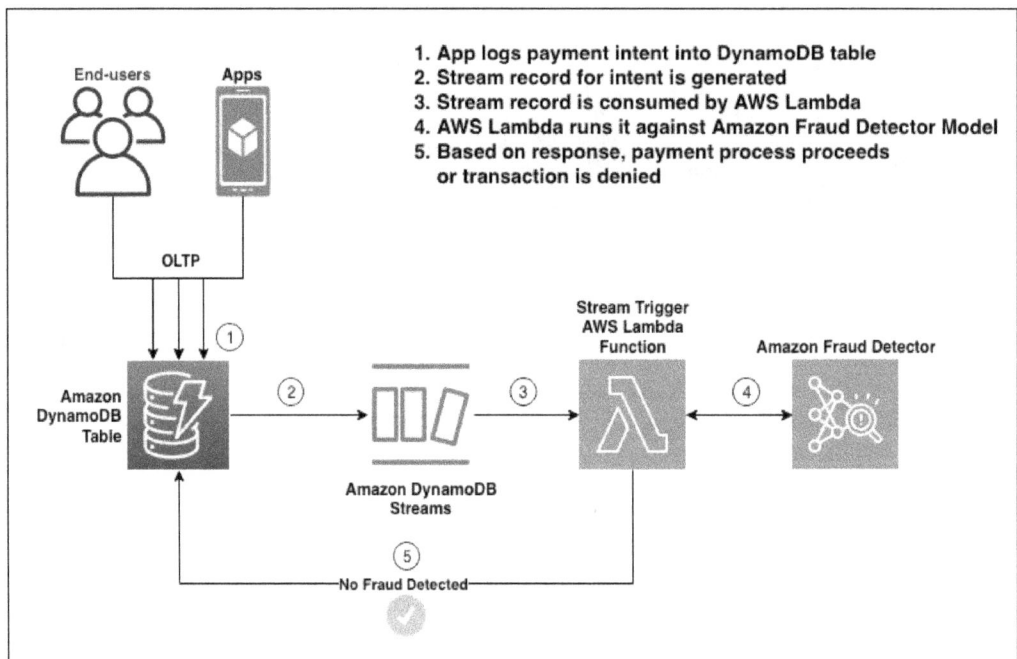

Figure 15.4 – Real-time analytical use case of fraud detection

DynamoDB Streams operate without impacting OLTP performance, offering a scalable, highly available method for real-time data processing. While Streams may be suitable for some use cases, incremental exports might be more efficient for others.

In summary, all modern organizations must have an effective data strategy, whether to support traditional number crunching for valuable business insights or to power advanced machine learning and **Generative Artificial Intelligence (GenAI)** systems. Analytics of OLTP data is a crucial component of this strategy. Adhering to the principle that analytics should not impact day-to-day business operations, DynamoDB provides several native functionalities to support analytics. The complexity of an analytics pipeline will depend on the organization's objectives and the technical expertise of the teams involved.

With this foundation, we can now explore the architectural patterns for running various kinds of analytics on DynamoDB data.

Diving into analytical patterns

In this section, we will explore different architectural patterns and their use cases for analytics on DynamoDB data. We will also discuss the advantages and considerations of each pattern, depending on specific requirements. Before diving into the patterns themselves, let us review some of the building blocks these patterns may use.

Building blocks

The following building blocks are leveraged by analytical patterns to achieve their goals. These include native DynamoDB features, other AWS services, and architectural patterns that integrate well with DynamoDB, ensuring scalability, high availability, and adherence to well-architected pillars (9).

Full table scans

While not recommended for serving OLTP requests from tables with large volumes of data, full table scans can be a straightforward way to support analytics on DynamoDB data without the need for complex workflows and analytical data pipelines. You can perform full table scans using the AWS SDKs or tools that support these scans, which could include self-written scripts or AWS Glue ETL jobs that can read through the entire table.

Full table scans will likely require rate-limiting, failure handling with backoff, and retries to ensure that OLTP access is not impacted by the bulk scan job. Full table scans are suitable if your analytics requirements involve analyzing the majority of the table data and if the analytics need to be performed occasionally, such as once a week, once a month, or in similar intervals. However, it may not be cost-efficient to perform full table scans for daily analytics jobs, as other options, such as the full table export to S3, are better suited for frequent analytics.

Full table export to S3

For scenarios where analytics jobs span multiple instances and require access to table data in several ways and at separate times, a full table export to S3 can be a suitable choice for your analytics pipeline.

When analytics need to be run on the entire dataset or a significant portion of it, leveraging DynamoDB's native feature to export the entire dataset to S3 is advantageous. For instance, if analytics tasks involve comprehensive querying and transformations across the dataset, performing a full table export can be cost-effective. Exporting the data once to S3 allows you to read it multiple times, significantly reducing costs compared to multiple full table scans.

Another benefit of full table export is that it does not consume read throughput on the live DynamoDB table, thereby avoiding impact on OLTP access from production applications during analytics operations. Although enabling **Point-in-Time Recovery (PITR)** is necessary for initiating full table exports, you can manage this efficiently by enabling PITR just before exporting and disabling it once the export is complete.

Conversely, if your analytics focus only on changes since the last job run rather than the entire dataset, utilizing incremental export to S3 may be more suitable and cost-effective.

Incremental export to S3

For scenarios where analytics are focused solely on data changes since the last job run or within a specific period, leveraging incremental export to S3 is likely the most cost-efficient approach. This method allows retrieval of old and new images of records that have changed, newly created records, or deleted records within the specified timeframe. Alternatively, you can choose to export only new or old images, which reduces costs since you are charged per GB of exported data.

Once the incremental export is stored in an S3 bucket of your choosing, you can utilize this data to execute analytics jobs and derive actionable insights. Some use cases may benefit from a hybrid approach, combining full table exports with regular incremental exports to maintain a comprehensive DynamoDB table copy in S3 (10).

While both full table and incremental exports may take several minutes to complete, for near real-time analytics requirements, integrating DynamoDB Streams or Kinesis Data Streams for DynamoDB into your analytics pipeline may be necessary.

DynamoDB Streams

DynamoDB Streams offers native support for near-real-time analytics by providing a stream record for every data mutation in a DynamoDB table. Whether it's a new insert, an update, or a delete operation, DynamoDB Streams promptly generates a stream record, typically within milliseconds of the data change.

These stream records can be consumed by various applications, such as AWS Lambda functions or self-managed applications using the Kinesis Client Library. Depending on the specific needs, the consumer application can process these change records internally or route them to other systems or data lakes.

For scenarios requiring integration with other AWS services within the Kinesis suite or for buffering and consolidating data into an S3 data lake using a low-code or no-code approach, Kinesis Data Streams for DynamoDB might be a more suitable option for your analytics pipeline.

Kinesis Data Streams for DynamoDB

Kinesis Data Streams for DynamoDB enables seamless integration of change records from DynamoDB tables with various downstream technologies. This integration extends to the following:

- **Kinesis Data Firehose**: Enables aggregation, encryption, and transformation of data before loading it into destinations such as S3, OpenSearch, Redshift, and other supported services (11)

- **Kinesis Data Analytics**: Facilitates near-real-time analysis, including generating time-series analytics, powering real-time dashboards, and creating operational metrics in real time (12)

- **Amazon Managed Service for Apache Flink**: Offers capabilities such as Kinesis Data Analytics but utilizes Apache Flink (13), a widely used open source technology renowned for real-time data analytics

These downstream integrations empower organizations to leverage DynamoDB's real-time change data effectively across a spectrum of analytical and operational use cases.

If your teams are well-versed with Spark-based ETL processing, or need more customization in your analytical processing pipelines, then AWS Glue should be leveraged.

AWS Glue ETL

AWS Glue stands out as one of the most prominently featured AWS services in this book, second only to DynamoDB itself. This recognition is well-deserved due to its versatile capabilities. AWS Glue operates in both batch and streaming modes, making it adaptable to various data processing scenarios.

You can utilize AWS Glue to conduct full table scans or export entire tables to S3, preparing the data for advanced analytics with Apache Spark. Its robust Apache Spark-based APIs empower you to execute complex analytical tasks, from generating real-time insights to supporting dashboard creation and more.

Moreover, AWS Glue offers extensive connectivity options, including connectors for S3, JDBC, and ODBC. This flexibility enables powerful integration of AWS Glue's data processing capabilities with your preferred data warehouse or **Business Intelligence** (**BI**) toolset.

Amazon Athena

Amazon Athena allows you to analyze S3 data using standard SQL. It integrates well with data exported from DynamoDB, especially when the data is transformed into optimal data lake formats such as Parquet or ORC. Athena's serverless nature and pay-per-query pricing make it an efficient option for querying large datasets stored in S3.

With these building blocks covered, we can now review specific analytics use cases and how these components can be integrated to build scalable, resilient, and feature-rich analytics pipelines.

Diving deeper into analytical use cases

The following are some common use cases where analytics may be required for DynamoDB data. Let us dive deeper into each one and explore different architectural solutions based on the specific requirements.

Powering business dashboards and periodic reports

Business dashboards are essential tools that provide actionable insights to organizational leaders, often requiring data processing to extract meaningful information. Similarly, periodic reports derived from OLTP data also require processing and transformation for insight extraction.

The choice of analytical building blocks depends largely on the required timeliness of insights. For instance, if dashboards or reports are generated on a weekly or monthly basis, a straightforward approach could involve using a full table scan. AWS Glue jobs or Amazon Athena could perform these scans, processing the entire dataset to produce dashboards or reports. This method assumes that the dashboard or report requires insights from the entire dataset (or the majority of it) and is generated infrequently.

Alternatively, for dashboards or reports needing higher-frequency updates, such as daily reports with minimal staleness, a different analytical pipeline is necessary. This pipeline might involve maintaining a fully updated copy of DynamoDB data in S3, refreshed transactionally every few hours. Achieving this requires a combination of a one-time full table export to S3 and periodic incremental exports in a transactional format such as Apache Iceberg. The conceptual pipeline is illustrated in the following figure:

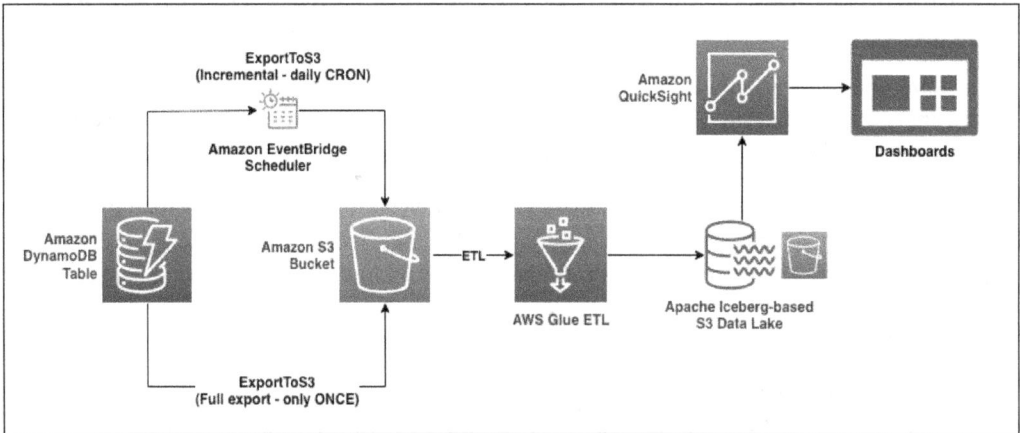

Figure 15.5 – Architecture for supporting dashboard use cases using native export features

As depicted, periodic incremental exports to S3 are triggered periodically using Amazon EventBridge Scheduler (14) and merged into the DynamoDB data copy stored in Apache Iceberg format. AWS Glue ETL can be utilized here to facilitate the merging of incremental exports with the S3 data copy.

In scenarios where DynamoDB data serves multiple analytical use cases beyond dashboards and reports—such as full-text search, geospatial queries, or text auto-completion—a zero-ETL pipeline using Amazon OpenSearch service could be a more suitable solution. This approach efficiently supports these specialized use cases without requiring extensive data transformation processes.

Powering batch jobs and downstream data sharing

In large organizations with multiple teams collaborating on a unified product or service, sharing data across teams post-processing is often essential. For instance, in an e-commerce setup, a data science team might analyze order data to generate user recommendations. After processing, they might share these recommendations with the platform team, possibly in CSV format stored in an S3 bucket, a **File Transfer Protocol (FTP)** destination, or a dedicated recommendation database.

From a DynamoDB standpoint, when batch jobs require access to the entire dataset for tasks such as dashboards and reports, the approach depends on the frequency of these jobs. For infrequent batch jobs, performing a full table scan on DynamoDB may be sufficient. Alternatively, exporting the full table to S3 could be more cost-effective, though it would require additional orchestration of the batch job workflow.

For batch jobs interested only in incremental changes since the last run, DynamoDB's incremental export to S3 feature is highly beneficial.

In most scenarios involving batch jobs or downstream data sharing, leveraging AWS Glue ETL jobs with either full or incremental exports from DynamoDB is recommended. AWS Glue ETL simplifies data format conversion and integrates seamlessly with DynamoDB, reducing manual effort significantly.

Powering machine learning systems

To effectively support machine learning pipelines using OLTP data, scalability, and cost-efficiency often requires models or technologies capable of incremental training. While occasional full-model rebuilds may be necessary, regular improvements typically rely on incremental training, which requires access to incremental changes or exports.

DynamoDB's native **incremental export to S3** feature proves invaluable in scenarios involving incremental training of machine learning models. By capturing every mutation in the table, including updates, inserts, and deletes, this feature provides comprehensive data for model training. For instance, in a banking application, if a transaction decreases an account balance twice, machine learning models can leverage each update to refine their understanding over time.

It is important to note that tasks such as anomaly detection or fraud prevention require real-time processing and fall outside the scope of typical machine learning system requirements.

Powering search and vector systems

GenAI has significantly popularized vector search across various technological domains. The substantial costs associated with training and refining **Foundation Models (FMs)** make it impractical for many organizations. Therefore, **Retrieval Augmented Generation (RAG)** (15) allows leveraging base FMs alongside external data sources to provide contextual information and maintain control over sensitive data, while ensuring high-quality output. These external sources may include unstructured data such as PDFs, images, and media, which are converted into vector embeddings and stored in databases equipped with advanced similarity search algorithms (16).

For applications requiring interactive capabilities enhanced by GenAI, such as e-commerce platforms managing product catalogs, integrating vector search with DynamoDB data is essential.

Even if your e-commerce application aims to provide a basic full-text or rich-text search functionality, allowing users to easily search for products using a search bar on your website, you would still require advanced search capabilities for DynamoDB data.

For both of these scenarios, as well as others requiring vector search or full-text search capabilities for your DynamoDB data, your analytics pipeline might benefit from leveraging a specialized search technology such as Amazon OpenSearch. Since you still require the data to reside in DynamoDB for low-latency, high-throughput access, DynamoDB and OpenSearch must remain synchronized, which would require data replication from DynamoDB to OpenSearch within the analytical pipeline. AWS offers a zero-ETL integration between DynamoDB and OpenSearch, eliminating the need for you to manage replication and synchronization manually and allowing AWS to handle these tasks.

This integration also supports real-time small transformations using DynamoDB Streams. Initially, the integration can perform a full export to S3 when it is set up for the first time, and then maintains synchronization between the two data stores in near-real time using DynamoDB Streams. OpenSearch provides capabilities for both full-text search through text-search indexes and semantic search using vector indexes.

The zero-ETL integration itself takes the form of a YAML or JSON template that you can create to specify the source DynamoDB table(s), the destination OpenSearch endpoints, and all the necessary routing and transformations for processing DynamoDB data before it is loaded into OpenSearch.

Opting to implement your own replication and synchronization between these two data stores may initially seem more cost-effective. This perception stems from the fact that AWS Lambda functions, which you might use to consume DynamoDB Streams in a self-implemented system, do not incur read costs on the DynamoDB Stream itself. In contrast, other stream consumer technologies, such as Kinesis Client Library (KCL)-based applications incur charges based on high-frequency polling of DynamoDB Streams. This approach could potentially increase the overall cost of integration compared to using AWS Lambda with DynamoDB Streams.

However, managing a self-implemented pipeline may divert attention and resources away from your core business activities. Therefore, I recommend utilizing the zero-ETL integration provided by AWS. This approach allows you to focus more on your primary business objectives while AWS manages the complexities of data replication and synchronization between DynamoDB and OpenSearch.

Powering data warehousing

Organizations often require data warehousing to structure data for complex analytics and to power BI tools. To facilitate operational reporting, historical data analysis, predictive analytics, and integration with BI tools such as Tableau, Microsoft PowerBI, or Amazon QuickSight, it is essential to periodically copy incremental changes from DynamoDB to a data warehouse. One such powerful data warehouse that supports these capabilities is Amazon Redshift.

To periodically copy incremental DynamoDB data into Redshift, you can leverage the zero-ETL integration between the two services (17). Like the zero-ETL integration between DynamoDB and OpenSearch, this integration offers a fully managed solution for periodically and incrementally copying DynamoDB data into a Redshift table. This streamlined process eliminates the need to manually build a pipeline involving periodic incremental exports of DynamoDB data, waiting for export completion, consolidating data into a Redshift-compatible format, and handling potential failures.

The zero-ETL solution, on the other hand, lands the DynamoDB data into a Redshift table in a completely managed way. You are charged only for the size of the incremental exports and the Redshift storage, with no additional costs for the zero-ETL integration itself. To enable incremental exports, PITR must be enabled on the DynamoDB table. This is often already enabled on production tables to allow restoration to any point within the last 35 days or to protect against data corruption or accidental deletions.

Since DynamoDB is schemaless and Redshift is not, the zero-ETL integration aims to efficiently ingest the highly flexible DynamoDB data into the strictly structured Redshift table. The target table in Redshift follows this schema:

- **Partition key**: A column with the same name as the partition key attribute, containing the partition key value
- **Sort key**: A column with the same name as the sort key attribute, containing the sort key value, if the DynamoDB table has a sort key
- **Value**: A single column that holds the entire DynamoDB item in DynamoDB JSON format, including the partition key and sort key

The following figure illustrates the schema of the target table in Redshift created as part of a zero-ETL integration setup between a DynamoDB table and a Redshift serverless cluster:

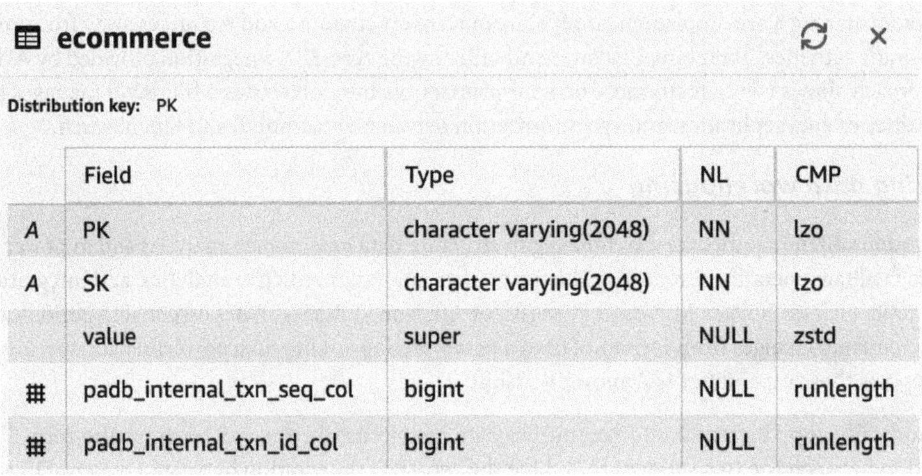

ecommerce

Distribution key: PK

	Field	Type	NL	CMP
A	PK	character varying(2048)	NN	lzo
A	SK	character varying(2048)	NN	lzo
	value	super	NULL	zstd
#	padb_internal_txn_seq_col	bigint	NULL	runlength
#	padb_internal_txn_id_col	bigint	NULL	runlength

Figure 15.6 – Zero-ETL integration with Redshift: target table schema

From the preceding figure, `PK` and `SK` are the partition key and sort key attribute names copied by the zero-ETL integration from my DynamoDB table, with `value` being the column with the whole DynamoDB JSON item. In addition to these columns, there are a couple of internal columns, `padb_internal_txn_seq_col` and `padb_internal_txn_id_col` that support the zero-ETL functioning.

For a sample e-commerce dataset in DynamoDB containing products and orders, the target table in Redshift would look like the following figure:

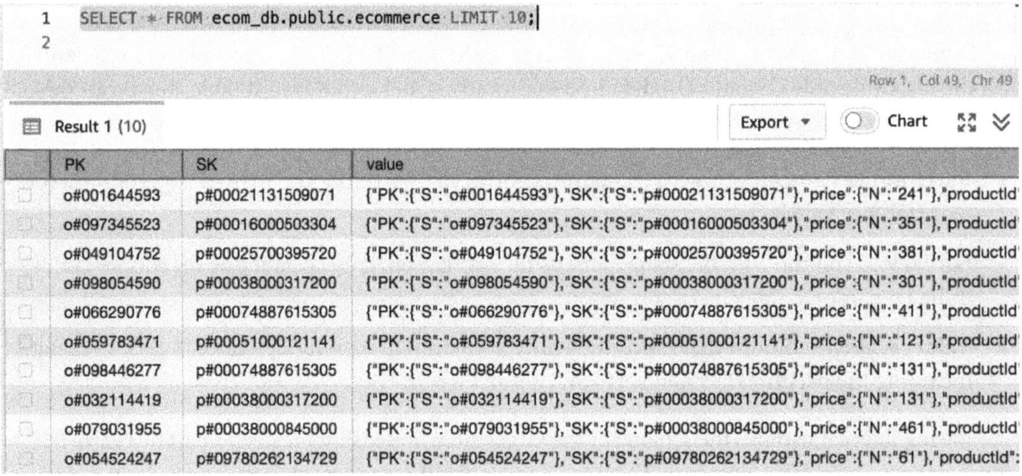

```
SELECT * FROM ecom_db.public.ecommerce LIMIT 10;
```

Row 1, Col 49, Chr 49

Result 1 (10) Export Chart

PK	SK	value
o#001644593	p#00021131509071	{"PK":{"S":"o#001644593"},"SK":{"S":"p#00021131509071"},"price":{"N":"241"},"productId'
o#097345523	p#00016000503304	{"PK":{"S":"o#097345523"},"SK":{"S":"p#00016000503304"},"price":{"N":"351"},"productId'
o#049104752	p#00025700395720	{"PK":{"S":"o#049104752"},"SK":{"S":"p#00025700395720"},"price":{"N":"381"},"productId'
o#098054590	p#00038000317200	{"PK":{"S":"o#098054590"},"SK":{"S":"p#00038000317200"},"price":{"N":"301"},"productId'
o#066290776	p#00074887615305	{"PK":{"S":"o#066290776"},"SK":{"S":"p#00074887615305"},"price":{"N":"411"},"productId'
o#059783471	p#00051000121141	{"PK":{"S":"o#059783471"},"SK":{"S":"p#00051000121141"},"price":{"N":"121"},"productId'
o#098446277	p#00074887615305	{"PK":{"S":"o#098446277"},"SK":{"S":"p#00074887615305"},"price":{"N":"131"},"productId'
o#032114419	p#00038000317200	{"PK":{"S":"o#032114419"},"SK":{"S":"p#00038000317200"},"price":{"N":"131"},"productId'
o#079031955	p#00038000845000	{"PK":{"S":"o#079031955"},"SK":{"S":"p#00038000845000"},"price":{"N":"461"},"productId'
o#054524247	p#09780262134729	{"PK":{"S":"o#054524247"},"SK":{"S":"p#09780262134729"},"price":{"N":"61"},"productId":

Figure 15.7 – Zero-ETL integration with Redshift: target table preview

It is evident that most analytical pipelines would not terminate with the Redshift table in the form illustrated in the preceding figure. Typically, data warehouses would have tables structured quite differently from how the zero-ETL integration ingests data. Given that DynamoDB is schemaless and that the dataset may contain items with diverse attributes, achieving a highly structured format in Redshift can be challenging.

If your DynamoDB table data adheres to a somewhat strict schema, you might need to set up a data transformation task as a post-step to this zero-ETL ingestion into Redshift. This task would take data from the zero-ETL generated Redshift table, transform it into the desired structure, and then store the data in a different, analytics-ready Redshift table. Such a scenario would treat the zero-ETL integration Redshift table as a staging table. This approach remains advantageous because you do not have to manage a full-fledged pipeline that utilizes various AWS services to orchestrate the entire zero-ETL process.

The native integration also provides several Amazon CloudWatch metrics, such as data lag between the DynamoDB table and the Redshift table and bytes transferred between the two.

With that, we have covered the common analytical requirements modern applications have and how to go about designing them with the transactional data backed by DynamoDB. While analytics may deal mostly with reading data from DynamoDB, bulk processing in terms of updating DynamoDB table data in bulk is often a need to accommodate evolution in applications. Let us learn in-depth about these bulk processing jobs and how to think about them for DynamoDB next.

Bulk data processing

Although it is not strictly analytics, bulk data processing is crucial for managing large datasets effectively. This could involve truncating outdated data to optimize storage costs or supporting schema evolution in NoSQL databases. As application access patterns change or new features are added, datasets may need updates with new attributes, modifications to existing attributes, or removal of deprecated ones. Additionally, bulk processing can be essential for adding new attributes to create new secondary indexes and support schema evolution.

To explore when bulk processing is needed for DynamoDB tables, consider these example use cases.

Use case 1 – creating a new Global Secondary Index (GSI)

Consider a scenario where your application, two years post-launch, would benefit from a new GSI to support a new feature. Suppose that the sort key of this GSI needs to be a composite attribute, generated by concatenating two existing attributes such as status and timestamp. If this composite `Status#Timestamp` attribute is not already present in your dataset, you will need to add it to all relevant items using a bulk processing job. Importantly, this must be done without downtime, as the DynamoDB table may still be in production and serving live traffic.

Use case 2 – backfilling time to live attributes

Another common bulk processing requirement in DynamoDB involves backfilling tables with a **Time to Live (TTL)** attribute. As a reminder, DynamoDB TTL helps manage data by automatically deleting items that have outlived their relevance. To use TTL, each item must have an attribute specifying the epoch time for when it should be deleted. This feature is particularly useful for data that loses relevance after a certain period of inactivity or a predefined timeframe.

Often, customers realize the need for TTL after their table has grown significantly, sometimes to hundreds of gigabytes or even terabytes. In such cases, they need to backfill a TTL attribute for every item in the table to enable automatic data expiry by DynamoDB.

New GSI creation and backfilling TTL are common scenarios requiring bulk processing for DynamoDB tables. These tasks often involve performing a full table scan and potentially updating every item in the scan result.

Next, let us explore these bulk processing jobs in more detail to understand how they help manage and evolve DynamoDB data effectively.

Options for bulk data processing

Reviewing the example use cases in the preceding section, bulk processing jobs may be necessary to update data in DynamoDB tables. These scenarios arise from experiences with DynamoDB users, such as adding new attributes to terabytes of data or backfilling the TTL attribute after determining that much of the data is no longer needed. To efficiently manage these bulk processing tasks on DynamoDB tables, we will explore various approaches and strategies that are commonly employed.

Performing parallel scans with custom applications or tools

When you need to add new GSI sort keys or backfill TTL attributes across all items in a DynamoDB table, you will likely need to do a full table scan. However, not all methods that can scan tables can also update and write data back to DynamoDB. For example, Amazon Athena can scan tables thoroughly but cannot modify or write data back to DynamoDB.

One straightforward solution is to create your own application or tool for performing full table scans and updating DynamoDB data afterward. This tool can run on various platforms such as Amazon Elastic Compute Cloud (Amazon EC2) instances or containers. To ensure smooth operation without disrupting live customer traffic, it is crucial to implement rate-limiting using DynamoDB's `ReturnConsumedCapacity` feature for both the `Scan` (18) and the `UpdateItem` (19) APIs. This helps control how much read and write throughput your application uses per second, preventing it from overwhelming the table's capacity during bulk processing. It is also important to handle errors effectively, with strategies such as back-off and retries such as those used in AWS SDKs.

Alternatively, if you prefer a more integrated approach where rate-limiting and error handling can be implemented with relative ease, AWS Step Functions and AWS Lambda provide a robust solution for orchestrating bulk processing tasks.

Step Functions Workflow with Lambda

AWS Step Functions (20) offers a managed approach to create serverless workflows that integrate various AWS services to execute tasks in a defined sequence. These workflows, defined as state machines, can include Lambda functions, Amazon **Simple Notification Service** (**SNS**) notifications, ECS tasks, AWS Glue job initiations, and more.

For our bulk processing requirements, AWS Step Functions can orchestrate a state machine consisting of two Lambda functions and utilize Step Functions' native Map feature. The workflow is illustrated in the following figure:

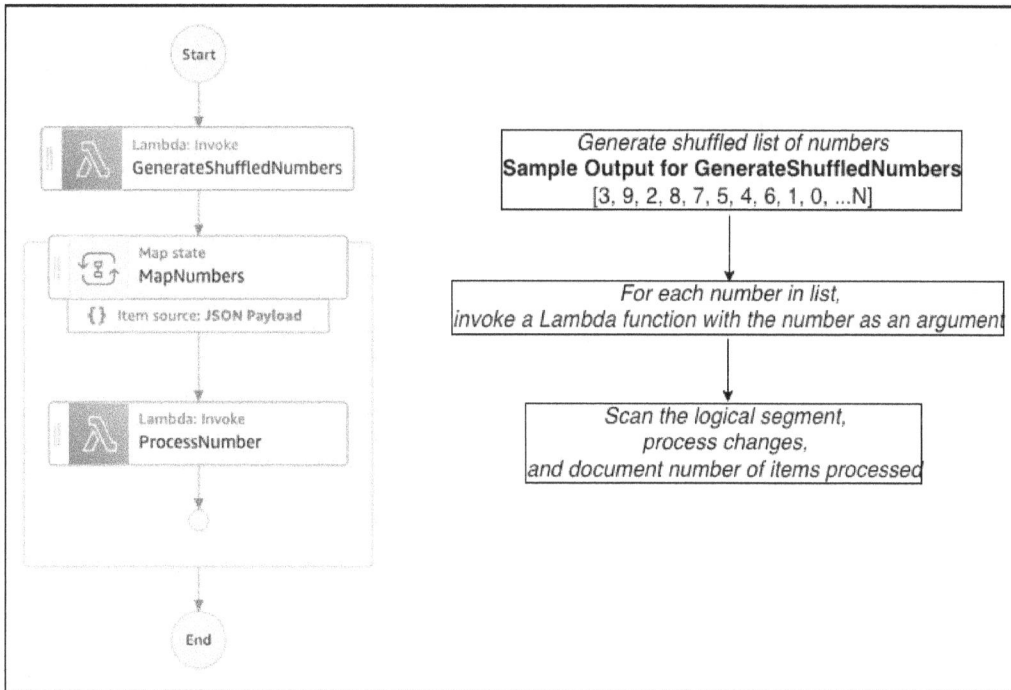

Figure 15.8 – AWS Step Functions state machine diagram

In the workflow illustrated in the preceding figure, a single Lambda function initiates the process by generating a list of shuffled numbers. These numbers represent segments for a parallel scan across DynamoDB. It is important to note that DynamoDB's Scan API supports both sequential and multi-threaded parallel scans. The latter allows multiple threads to scan different segments of the DynamoDB table simultaneously, significantly enhancing scan efficiency.

The `GenerateShuffledNumbers` Lambda function, depicted in the following code snippet, prepares the list of shuffled numbers for the Step Functions state machine:

```python
import random
import time

def lambda_handler(event, context):
    numbers = list(range(0, 100))
    random.shuffle(numbers)
    run_id = time.time()
    return {
        "numbers": numbers,
        "runId": run_id
    }
```

Subsequently, Step Functions utilizes its native map functionality to invoke multiple instances of the second Lambda function, `ProcessNumber`. Each instance handles a specific segment allocated by the previous Lambda function. The role of each `ProcessNumber` Lambda invocation is to conduct a scan within its designated segment, update data as necessary, and record the count of updated items in a single DynamoDB item. This aggregated count serves as a reporting statistic comprising all processed items across the DynamoDB table. The following is the `ProcessNumber` Lambda function code:

```python
import json
import boto3
from botocore.config import Config
import time

DDB_CLIENT = boto3.client('dynamodb', region_name='eu-west-1',
    config=Config(max_pool_connections=150, retries = {
        'max_attempts': 20,
        'mode': 'standard'
    }))

def lambda_handler(event, context):
    segment_number = event["number"]
    run_id = event["runId"]
    # name of table to scan
    table_name = '<mytable>'
    # total segments to be same as length of list
    # returned by GenerateShuffledNumbers
    total_segments = 100

    return ("Ending Invocation with count:" + str(
        parallel_process(
```

```
            run_id,
            table_name,
            segment_number,
            total_segments)))


def parallel_process(run_id: str, table_name: str,
    segment_number: int, total_segments: int = 30) -> None:
    count=0
    kwargs = {
        'TableName': table_name,
        'TotalSegments': total_segments,
        'Segment': segment_number,
        'Select': 'COUNT'
    }
    while True:
        scan_response = DDB_CLIENT.scan(**kwargs)
        # bulk processing logic to go here
        # for item in scan_response:
        #     do something
        count += scan_response.get('Count', 0)
        last_evaluated_key = scan_response.get(
            'LastEvaluatedKey', None)
        kwargs.update(
            {'ExclusiveStartKey': last_evaluated_key})

        if last_evaluated_key is None:

            # log total number of items scanned
            # can use transact_write_items() for idempotency
            DDB_CLIENT.update_item(
                # name of table to log item counts to
                TableName='<result table>',
                Key= {
                    'uuid': {'S' : table_name},
                    'sort_id': {'N': str(run_id)}
                },
                UpdateExpression="SET item_count = \
                    if_not_exists(item_count, :zero_var) + :new_var",
                ExpressionAttributeValues={
                    ':zero_var': {'N': '0'},
                    ':new_var' : {'N': str(count)}})
            return count
```

While the preceding `ProcessNumber` Lambda function does not include rate-limiting, you can manage some degree of rate control by adjusting Lambda's reserved concurrency settings (21). Alternatively, similar bulk processing capabilities can be achieved using Amazon **Simple Queue Service** (**SQS**) in conjunction with AWS Lambda, which we will explore next.

Distributed map-reduce using SQS and AWS Lambda

If you are familiar with distributed computing frameworks that implement map-reduce, this approach shares similarities. It utilizes an SQS **First-in-First-out** (**FIFO**) queue, which guarantees exactly-once processing semantics (22), alongside a pair of Lambda functions. The following is a figure that illustrates this pattern:

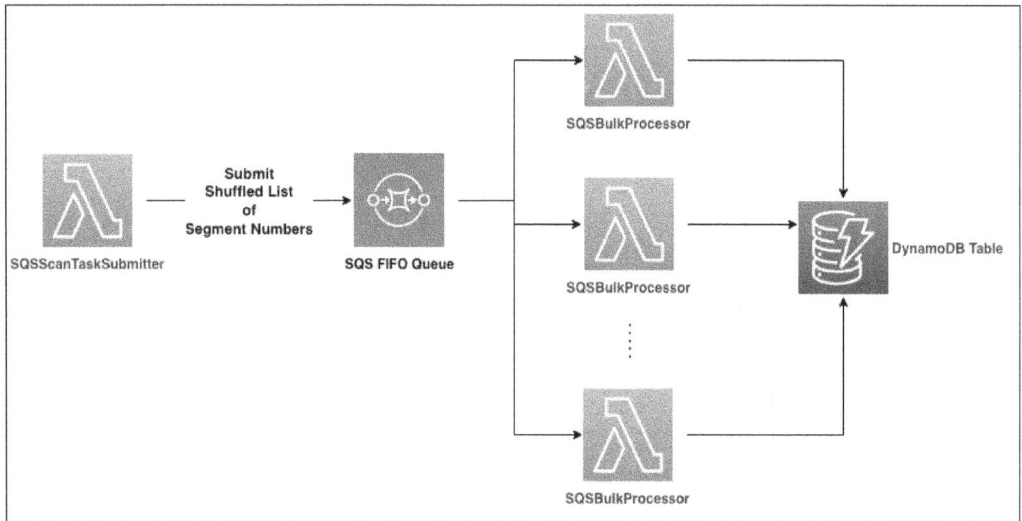

Figure 15.9 – Architecture of distributed map-reduce using SQS and AWS Lambda

Similar to the approach involving AWS Step Functions, this pattern begins with an `SQSScanTask-Submitter` Lambda function that initiates by shuffling a list of scan segments and placing them sequentially into a FIFO SQS queue. The queue's consumer is another Lambda function triggered for each queue record, which corresponds to a scan segment number. This Lambda function executes a scan of the designated segment, processes the data, and logs the number of processed items into a single DynamoDB table item for reporting purposes.

The processing logic within the second Lambda function is similar to that of the `ProcessNumber` Lambda function from the previous pattern. The following figure illustrates the logic of the initial `SQSScanTaskSubmitter` Lambda function that drops the shuffled list of scan segments into a FIFO SQS queue:

```
import json
import boto3
```

```python
import random
import time
import uuid

sqs_client = boto3.client("sqs", region_name="eu-west-1")
TOTAL_SEGMENTS = 1000

# name of table to parallel scan and bulk process
TABLE_NAME="<mytable>"

# SQS can parallelize processing using MessageGroups
# generating 200 groups for 1000 segments
# such that 200 Lambdas can process simultaneously
# and each group has 5 segments to process sequentially
group_shuff = [str(uuid.uuid4()) for x in range(0, 200)]

def lambda_handler(event, context):
    run_id = time.time()
    for segment in range(0, TOTAL_SEGMENTS):
        send_message({
            "TableName": TABLE_NAME,
            "TotalSegments": TOTAL_SEGMENTS,
            "Segment": segment,
            "RunId": run_id
        })
    return {
        'statusCode': 200,
        'body': json.dumps('All Tasks Submitted')
    }

def send_message(message):
    response = sqs_client.send_message(
        QueueUrl="https://sqs.eu-west-1.amazonaws.com/XXXX/table_
scanner.fifo",
        MessageBody=json.dumps(message),
        MessageGroupId=random.choice(group_shuff)
    )
    print(response)
```

Lambda's concurrency can be managed through reserved concurrency settings, although it is recommended to implement rate limiting using the `ReturnConsumedCapacity` response and an effective algorithm such as the token bucket algorithm (23) to ensure optimal rate limiting.

Both the Step Functions and SQS patterns offer flexibility and customization but require users to implement Lambda functions with rate limiting, failure handling, and bug-free processing logic. While these patterns eliminate the need for managing infrastructure, like most native AWS services, they do place considerable implementation responsibility on the user. If managing such responsibility is not desirable, an alternative approach involves leveraging AWS Glue and its native DynamoDB connector, which automates much of the heavy lifting. Let us explore this option next.

Glue ETL for DynamoDB

AWS Glue seamlessly integrates with DynamoDB for data access, using a connector supported by the Glue team that interfaces DynamoDB APIs through native Spark or Glue APIs. The Glue DynamoDB connector also incorporates native support for rate-limiting and failure handling. When reading from a DynamoDB table using Glue's native APIs, it performs a rate-limited, distributed parallel scan. Similarly, for writing to DynamoDB, Glue executes rate-limited, fault-tolerant `BatchWriteItem` API calls in the background for bulk writes. The Glue docs on the DynamoDB connector are a must-read (24).

However, it is important to consider that the Glue connector interfaces only with the `Scan` and `BatchWriteItem` APIs of DynamoDB. Operations such as `GetItem`, `Query`, or `UpdateItem` are not natively supported by the connector. Consequently, all write operations use the `BatchWriteItem` API, and read operations utilize the `Scan` API, which performs a full table scan. The following code provides examples of read and write operations using the Glue DynamoDB connector:

```
# reading from DynamoDB table using parallel Scan
dyf = glue_context.create_dynamic_frame.from_options(
    connection_type="dynamodb",
    connection_options={
        "dynamodb.input.tableName": test_source,
        "dynamodb.throughput.read.percent": "1.0"
    }
)
# writing into DynamoDB table using BatchWriteItem
glue_context.write_dynamic_frame_from_options(
    frame=dyf,
    connection_type="dynamodb",
    connection_options={
        "dynamodb.output.tableName": test_sink,
        "dynamodb.throughput.write.percent": "1.0"
    }
)
```

Due to limitations in the DynamoDB APIs it interfaces with, the Glue connector does not support direct updates to existing data. When performing read and write operations within the same Glue ETL job, there is a risk of overwriting items in the target DynamoDB table that share the same key schema. This characteristic restricts the use of Glue ETL standalone for tasks such as updating data with new attributes or backfilling TTL values on live production tables actively serving CRUD operations.

It is possible to implement Glue jobs to issue `UpdateItem` API calls by utilizing Python boto3 methods parallelized through Spark, rather than relying on the native Glue DynamoDB connector for writing data.

Typically, using the `boto3` library in a distributed Spark environment such as AWS Glue is cautioned against. This is because, by default, boto3 code execution may happen on a single Glue worker, undermining the benefits of parallelism offered by a distributed Spark environment. However, Spark functions can be combined with boto3-based methods to parallelize execution like Spark's native methods can.

The following is a snippet demonstrating how a boto3-based function can be executed in a distributed Spark environment to issue `UpdateItem` API calls. The complete example is available in the GitHub repository (25) associated with this chapter:

```
# read from S3 export
df = glueContext.create_dynamic_frame.from_options(
    connection_type="s3",
    connection_options={
        "compressionType": "gzip",
        "paths": ['s3://path/to/exported/data/']
    },
    format="json",
    transformation_ctx="df"
).toDF()


# issue UpdateItem API calls for each item
def process_item_update(boto3_table, item):
    # issue update_item calls here
    #
    # update_response = table.update_item(
    #       Key={'pk': 'foo', 'sk': 'bar'},
    #       ConditionExpression=Attr(NewAttribute).not_exists(),
    #       UpdateExpression='SET NewAttribute = :new_attr_val;,
    #       ExpressionAttributeValues={
    #           'NewAttribute': 'NewAttributeValue'},
    #       ReturnConsumedCapacity='INDEXES'
    #       )
    # handle throttles and condition check failure exceptions
    #
    # return update_response

def execution_for_each_spark_partition(partitionData):
    # initialize boto3 clients on each Glue executor
    ddbclient = boto3.resource('dynamodb',
```

```
        region_name='eu-west-1', config=config)
    table = ddbclient.Table(TABLE_NAME)

    for item in partitionData:
        response = process_item_update(boto3_table, item)
        # use response and output of ReturnConsumedCapacity to
        # rate-limit here

# convert to Spark RDD and parallelize
df1.rdd.foreachPartition(execution_for_each_spark_partition)
```

From the preceding example, it is clear that when using the `UpdateItem` API directly with boto3 in a distributed Spark environment such as AWS Glue, the responsibility for rate-limiting and handling failures falls on the developer. The following is an example snippet demonstrating how to implement rate-limiting and failure handling in a Glue ETL script:

```
# get job infrastructure details to parallelize processing efficiently
TASKS_PER_EXECUTOR = int(spark.sparkContext.getConf().get(
    "spark.executor.cores"))
NUM_EXECUTORS = int(spark.sparkContext.getConf().get(
    "spark.executor.instances"))
print("#### Tasks per executor: %d | Num executors: %d"
      % (TASKS_PER_EXECUTOR, NUM_EXECUTORS))

WRITE_THROUGHPUT_PERCENT = "1.0"
SPLITS_STR = str(NUM_EXECUTORS * TASKS_PER_EXECUTOR)

# use infrastructure details and WRITE_THROUGHPUT_PERCENT
# to compute max WCU per spark partition
# MAX_WCU_PER_TASK obtained will be used to rate-limit on
# individual spark partition/Glue worker level
if table.billing_mode_summary is not None and \
    table.billing_mode_summary[
        'BillingMode'] == 'PAY_PER_REQUEST':
    TOTAL_WCU_TARGET = float(WRITE_THROUGHPUT_PERCENT) * 40000
else:
    TOTAL_WCU_TARGET = float(WRITE_THROUGHPUT_PERCENT) * \
        table.provisioned_throughput['WriteCapacityUnits']

MAX_WCU_PER_TASK = float(TOTAL_WCU_TARGET)/(
    TASKS_PER_EXECUTOR * NUM_EXECUTORS)

print("#### Max WCU per task: %f | Tasks per executor: %d | \
    Num executors: %d | Splits: %s" % (MAX_WCU_PER_TASK,
```

```
            TASKS_PER_EXECUTOR, NUM_EXECUTORS, SPLITS_STR))

# issue UpdateItem API calls for each item
def process_item_update(boto3_table, item):
    update_response = None
    try:
        # bulk processing logic here
        # do something
        #
    # handle ProvisionedThroughputExceededException and
    # ConditionCheckFailedExceptions
    # if ConditionCheckFailedException, return None
    except ClientError as e:
        if e.response['Error']['Code'] == \
            'ConditionalCheckFailedException':
            # Condition failed, do nothing
            pass
        elif e.response['Error']['Code'] == \
            'ProvisionedThroughputExceededException':
            # Request throttled, retries exceeded, will try again
            update_response = process_item_update(
                boto3_table, item)

    return update_response

# This gets executed on each Glue worker in parallel.
# On each worker, this will be further parallelized across
# TASKS_PER_EXECUTOR (default=4) threads
def execution_for_each_spark_partition(partitionData):
    global MAX_WCU_PER_TASK

    # initialize boto3 client
    # table = ddbclient.table(TABLE_NAME)

    rate = MAX_WCU_PER_TASK # unit: WCU
    per  = 1 # unit: seconds
    allowance = rate # unit: WCU
    last_check = get_time() # unit: seconds

    task_item_count = 0 # counter to log progress

    for item in partitionData:
        current = get_time()
```

```
        time_passed = current - last_check;
        last_check = current
        allowance += time_passed * (rate / per)
        if allowance > rate:
            allowance = rate

        if allowance < 1.0: # no tokens remaining
            delta_time = float(get_time() + 1) - time.time()
            if delta_time > 0.0:
                time.sleep(delta_time)

        response = process_item_update(table, item)
        task_item_count += 1
        # Using consumed capacity per request to rate limit
        if response is not None:
            total_consumed_capacity = response[
                'ConsumedCapacity']
            tokens_to_deduct = total_consumed_capacity['Table'][
                'CapacityUnits']
            # since LSI WCU is also consumed from table,
            # we deduct additional tokens for rate limiting
            if 'LocalSecondaryIndexes' in \
                total_consumed_capacity:

                lsi_key_name = list(total_consumed_capacity[
                    'LocalSecondaryIndexes'].keys())
                lsi_name = lsi_key_name[0]
                tokens_to_deduct += total_consumed_capacity[
                    'LocalSecondaryIndexes'
                ][lsi_name]['CapacityUnits']
            allowance -= tokens_to_deduct
            tokens_to_deduct = 0
        else:
            # response is None,
            # due to ConditionCheckFailedException
            allowance -= 1
    print("#### Items processed for current task: %d"
        % task_item_count)
```

Implementing a Glue ETL script in the manner shown in the previous snippet allows you to safely introduce new composite sort key attributes or backfill TTL attributes into your production DynamoDB tables.

To execute bulk data processing using any of the discussed patterns or tools, it is essential to update your application beforehand. This update should incorporate the new composite attributes or TTL configurations for any new data that will be written into DynamoDB. By doing so, you ensure that the entire dataset, including new writes received by the live DynamoDB table during and after the bulk processing job, is consistently updated.

This section covered various bulk processing patterns, their advantages, and considerations when using them, particularly in live production tables actively handling CRUD operations. Next, we will explore the optimizations that are critical for running analytics or bulk processing jobs on DynamoDB tables. These optimizations include best practices for DynamoDB itself and other AWS services that integrate with these patterns.

Optimizing usage of DynamoDB and other AWS services for analytics

Optimizing the use of DynamoDB and other AWS services for analytics involves several best practices, particularly when integrating them with DynamoDB for bulk processing and analytics tasks. Let us review some of these practices now.

Rate-limiting read and write throughput consumption

When performing analytics or bulk processing tasks directly on DynamoDB, it is crucial to implement effective rate-limiting mechanisms. These measures prevent these operations from impacting the performance of live OLTP traffic handled by the same DynamoDB tables.

While rate-limiting based on **Transactions Per Second** (**TPS**) or **Requests Per Second** (**RPS**) can be effective, using **Read Capacity Units** (**RCU**) and **Write Capacity Units** (**WCU**) for rate-limiting is more accurate. The ReturnConsumedCapacity property, supported by all dataplane APIs in DynamoDB, can be combined with a token bucket algorithm to implement precise rate-limiting.

This approach ensures that the rate limiting is aligned with the actual capacity consumed by your operations, providing more granular control over resource usage and helping avoid throttling on your DynamoDB tables.

When using AWS Glue ETL's native DynamoDB connector, you can control read and write throughput consumption with properties such as dynamodb.throughput.read.percent and dynamodb. throughput.write.percent. These properties range from 0 to 1.5 and are based on the configured read and write throughputs of the DynamoDB table in provisioned capacity mode.

For on-demand mode, a 100% or 1.0 ratio equals 40,000 **Read Request Units** (**RRU**) or 40,000 **Write Request Units** (**WRU**). In provisioned mode, if the read throughput is set to 1.0, the Glue connector will first issue a DescribeTable API call to obtain the provisioned capacity for reads and then rate-limit to consume 100% of that capacity.

For on-demand tables, a 0.5 ratio for read throughput means the Glue connector will rate-limit to consume about 20,000 RRUs per second. The Glue connector calculates these ratios at the start of the read or write operation, so changes in provisioned capacity during the activity will not affect the rate-limiting of the Glue job. You can find more details on configuring parallelism with the DynamoDB connector in the Glue docs (24).

It is recommended to configure these ratios with a buffer to accommodate any growth in application traffic when running analytical or bulk jobs.

Leveraging exponential back-off and retries

Just like rate-limiting, implementing effective failure handling is crucial for analytics or bulk processing jobs. AWS SDKs offer simple ways to manage back-offs and retries, which can be leveraged for this purpose. Given that analytics or bulk processing should be secondary to the OLTP live traffic served by the application, you can configure higher and less aggressive retries with longer back-off periods for these jobs.

Choose a retry strategy that ensures that OLTP live traffic remains prioritized over bulk processing tasks. The AWS docs (26) provide guidance on retry behaviors supported by AWS SDKs. Additionally, the Amazon Builder's Library (27) offers valuable insights on timeouts, retries, and back-offs, which are highly recommended readings.

Preventing rolling hot keys during bulk processing

For bulk processing jobs that involve a full table scan followed by executing updates, such as adding new attributes, backfilling TTL values, or removing deprecated attributes, it is essential to shuffle the scanned table data across a non-key attribute or a random string to prevent rolling hot keys before executing the writes or updates.

When performing a full table scan, DynamoDB returns data in the natural order of how it is stored across partitions. All data from the same underlying DynamoDB table partition is returned to a single worker or the entire application. If the bulk processing job issues create, update, or delete operations on this scanned data, multiple writes will target the same partition simultaneously. This can lead to request throttling since these writes may exceed the per-partition limit of 1,000 WCU per second.

This Scan behavior is why, in earlier bulk processing patterns involving SQS and AWS Step Functions, we shuffled the list of scan segments. By doing so, during the write phase, the Lambda functions interact with different logical and physical parts of the dataset non-sequentially.

If the scanned data is not shuffled across a non-key attribute or a random string, the writes in the bulk processing job might exhibit low throughput consumption (approximately 1,000-1,400 WCU on average) compared to your configured write ratio. If the overall bulk processing job is rate-limited to under 1,000 WCU per second, shuffling the data is unnecessary as the job is unlikely to exceed the per-partition limits.

Leveraging serverless concurrency control options

When using AWS Lambda for analytical or bulk processing jobs, you can exercise some control over the overall RPS and DynamoDB throughput consumption with reserved concurrency. Reserved concurrency allows you to set the maximum number of simultaneous Lambda function instances. New instances are only created if the current number of running instances is below this limit, helping to control the parallelism of reads and writes in the bulk processing job. If the traffic consumed by the job exceeds your estimates, you can reduce the reserved concurrency limit to decrease the number of parallel Lambda instances and manage read or write requests more effectively.

In addition to Lambda concurrency controls, you can also limit downstream processing by using features within Lambda Event Source Mapping between SQS and Lambda, such as `BatchSize`, `BatchWindow`, `ParallelizationFactor`, and others.

Cost-optimizing analytics and predictive bulk processing

Analytics or bulk processing jobs are generally predictable, allowing you to control when they start and roughly how long they will run. To cost-optimize these jobs that scan or update DynamoDB table data, ensure that the DynamoDB table is configured in provisioned capacity mode. Provisioned capacity mode is recommended for predictable workloads, such as full table scans for analytics or bulk processing jobs.

You can switch to on-demand mode once every 24 hours. If your table is in on-demand mode serving unpredictable traffic during the day, orchestrate a workflow to convert the table to provisioned capacity mode just before the nightly batch jobs run. Once the jobs are complete, switch the table back to on-demand mode to handle unpredictable traffic the next morning. Even if the bulk jobs run once a week or month, the cost profiles of on-demand compared to provisioned mode make it worthwhile to use provisioned mode and orchestrate workflows for switching capacity modes.

Remember, there are no performance differences between the two capacity modes, so there is no performance benefit of one over the other.

These are some of the key considerations for optimizing AWS service usage for analytics and bulk processing use cases. Let us wrap this chapter up with a summary.

> **Important note**
> Consider this a reminder to delete any resources you may have created while going through this chapter in your AWS account.

Summary

This chapter explored the need for, and complexity of, analytics within modern data systems. Businesses increasingly rely on data-driven decisions, requiring efficient analysis and processing of large data volumes. The complexity arose from diverse data sources and varied formats, as well as the requirement to process data in near-real time without disrupting live applications. Effective analytical patterns addressed these challenges, ensuring scalability, reliability, and performance.

We dove into various analytical patterns, discussing methods and tools such as AWS Glue, AWS Step Functions, and SQS for seamless integration with DynamoDB. Highlighting AWS Glue's native DynamoDB connector, which supports rate-limiting and failure handling, the section explained how to leverage it for robust data operations. Additionally, it examined patterns involving Lambda functions, SQS, and Step Functions for managing workflows and facilitating efficient bulk data processing.

In the section on bulk data processing, we outlined strategies for handling full table scans, preventing rolling hot keys, and updating items with new attributes. Emphasis was placed on shuffling data to avoid request throttling and maintain optimal throughput. Cost optimization techniques, such as switching between provisioned and on-demand capacity modes to manage predictable workloads, were also discussed. Lastly, the chapter provided best practices for optimizing DynamoDB and other AWS services usage, including rate-limiting, leveraging exponential back-off and retries, and using reserved concurrency with AWS Lambda to balance analytical processing with live application demands.

In the next and final chapter of this book, we will explore migrations to DynamoDB. We will begin by evaluating migration opportunities and gathering essential data from existing database estates to inform our decisions. Then, we will dive into migration strategies tailored to different use cases and constraints.

References

1. AWS blog – lake house architecture: https://aws.amazon.com/blogs/big-data/harness-the-power-of-your-data-with-aws-analytics/

2. AWS LakeFormation: https://aws.amazon.com/lake-formation/

3. Apache Iceberg: https://iceberg.apache.org/

4. Apache Parquet: https://parquet.apache.org/

5. Apache Hudi: https://hudi.apache.org/

6. Amazon Athena: https://aws.amazon.com/athena/

7. AWS blog – *Top 10 Performance Tuning Tips for Amazon Athena*: https://aws.amazon.com/blogs/big-data/top-10-performance-tuning-tips-for-amazon-athena/

8. Amazon fraud detector: https://aws.amazon.com/fraud-detector/

9. AWS docs – *AWS Well-Architected Framework*: https://docs.aws.amazon.com/wellarchitected/latest/framework/welcome.html

10. AWS docs – export to S3: https://docs.aws.amazon.com/amazondynamodb/latest/developerguide/S3DataExport.HowItWorks.html

11. AWS docs – Kinesis Firehose destinations: https://docs.aws.amazon.com/firehose/latest/dev/create-destination.html

12. AWS docs – Kinesis analytics: https://docs.aws.amazon.com/kinesisanalytics/latest/dev/what-is.html

13. AWS docs – Amazon Managed Service for Apache Flink: https://docs.aws.amazon.com/managed-flink/latest/java/what-is.html

14. AWS docs – Amazon EventBridge Scheduler: https://docs.aws.amazon.com/scheduler/latest/UserGuide/what-is-scheduler.html

15. What is RAG: https://aws.amazon.com/what-is/retrieval-augmented-generation/

16. The role of vector databases in GenAI: https://aws.amazon.com/blogs/database/the-role-of-vector-datastores-in-generative-ai-applications/

17. Zero-ETL between DynamoDB and Redshift: https://aws.amazon.com/about-aws/whats-new/2023/11/amazon-dynamodb-zero-etl-integration-redshift/

18. AWS docs – scan API reference: https://docs.aws.amazon.com/amazondynamodb/latest/APIReference/API_Scan.html#DDB-Scan-request-ReturnConsumedCapacity

19. AWS docs – UpdateItem API reference: https://docs.aws.amazon.com/amazondynamodb/latest/APIReference/API_UpdateItem.html#DDB-UpdateItem-request-ReturnConsumedCapacity

20. AWS docs – AWS Step Functions: https://docs.aws.amazon.com/step-functions/latest/dg/welcome.html

21. AWS docs – AWS Lambda reserved concurrency: https://docs.aws.amazon.com/lambda/latest/dg/configuration-concurrency.html

22. AWS docs – SQS FIFO: https://docs.aws.amazon.com/AWSSimpleQueueService/latest/SQSDeveloperGuide/FIFO-queues-exactly-once-processing.html

23. Wikipedia – *Token Bucket Algorithm*: https://en.wikipedia.org/wiki/Token_bucket

24. AWS docs – Glue DynamoDB connections: https://docs.aws.amazon.com/glue/latest/dg/aws-glue-programming-etl-connect-dynamodb-home.html

25. GitHub – book artifacts: `https://github.com/PacktPublishing/The-Definitive-Guide-to-Amazon-DynamoDB/tree/main`

26. AWS docs – SDK retry behavior: `https://docs.aws.amazon.com/sdkref/latest/guide/feature-retry-behavior.html`

27. Amazon Builder's Library – *Timeouts, retries, and backoff with jitter*: `https://aws.amazon.com/builders-library/timeouts-retries-and-backoff-with-jitter/`

16
Migrations

In the dynamic tech landscape, applications are not static. They grow and adapt to changing business needs, user demands, and technological advancements. Over time, applications undergo numerous modifications driven by factors such as scalability needs, business strategy shifts, or non-technical considerations. This chapter focuses on one critical aspect of application evolution: migrating the database layer, specifically to Amazon DynamoDB.

An application's journey is like navigating a complex expedition. Initially designed for specific users and needs, it might see its user base expand, new features introduced, and infrastructure load surge. This scalability demand leads to a reevaluation of the application's architecture.

Scaling an application involves more than adding servers or increasing computing power; it's a holistic process affecting various layers, including the database. For example, scaling a traditional relational database may involve complex sharding and partitioning or offloading read-heavy operations to caching mechanisms. Often, these strategies can lead to operational complexity, increased maintenance, and performance bottlenecks.

This is where **migration** comes in. Evolving the database layer involves deciding whether to adapt the existing database or transition to a system better aligned with the application's growth. DynamoDB offers **seamless scalability**, **high availability**, and **low latency**, offering an efficient path for handling large data volumes and traffic spikes while minimizing operational challenges.

This chapter guides you through approaching and executing a migration to DynamoDB. Whether facing database performance issues, a growing user base, or seeking a modern and flexible database solution, this chapter serves as your guide.

You will learn to identify signs indicating when migration is needed, such as sluggish query response times or difficulties with a growing dataset. Recognizing these signs early helps make informed decisions about your application's future.

We will explore various migration approaches for different scenarios, offering practical advice and strategies. Whether considering a phased migration to minimize disruptions or one with acceptable downtime to harness DynamoDB's potential, you'll find valuable insights.

Throughout this chapter, real-world examples and best practices will paint a picture of the migration landscape. We'll explore common challenges faced during migration, such as data transformation and ensuring data consistency, and provide strategies to mitigate these obstacles effectively.

By the end of this chapter, you will understand how to qualify DynamoDB as a migration target, approach the migration process in various scenarios, and key considerations for migrating applications to DynamoDB.

In this chapter, we are going to cover the following main topics:

- Qualification for DynamoDB

- Homogenous and heterogenous migrations

- High-level migration strategies

- Key considerations for migrations

Qualification for DynamoDB

Before deciding if DynamoDB is right for your application, recognize common signs that signal the need for a database migration. Your current database setup can often provide insights indicating it's time for a change. These indicators help assess the migration opportunity, regardless of the target database.

Let us explore the qualification process for these indicators next.

Qualifying the migration itself

To assess your current database setup, focus on key data points, such as performance metrics. These metrics reveal your database's performance during peak and low usage times. Consider how performance issues impact your business operations:

- Evaluate data related to performance, scalability, storage growth, and functionality. Determine what works well and what causes problems for your application and team.

- Factor in scalability needs as your application grows. Use these insights to shortlist potential databases for migration that meet your application's requirements.

Next, you will want to narrow down your shortlist by selecting databases that align with your business objectives. Prioritize databases that can potentially minimize future migrations, scale seamlessly, optimize utility, and satisfy both your business and finance teams.

For substantial migration projects, it is advisable to conduct a **proof of concept** (**POC**) with a subset of the shortlisted databases to confirm their feasibility and compatibility with your application. After completing the qualification process, you can confidently select the most suitable database for your application.

To thoroughly evaluate a migration opportunity, begin by collecting data about your current database solution. Next, let us dive into the process of gathering this essential data.

Gathering insights

As we have already covered, the **performance metrics** of your database would be a critical source of information to gain insights from. But performance metrics are not the only data source you must tap into to learn about your existing database setup and qualify a migration opportunity. It is important to consider several factors such as performance, functionality, scaling requirements, traffic patterns, and cost, to really make an informed decision about whether you must plan a migration of your existing database.

For performance insights, use tools relevant to your current database solution. If your database is hosted on a cloud provider such as AWS, platform-specific tools are available. For example, with **Amazon Relational Database Service (Amazon RDS)**, you can access CPU and memory utilization metrics through Amazon CloudWatch. If you manage your databases or host them on-premises, use open source or third-party tools for similar information.

Again, depending on the technology, it could be an inbuilt or external tool. For example, it could be a mere `pg_stat_statements` extension for a Postgres database (1) regardless of where it is hosted, or an AWR (automatic workload repository) report from an Oracle database (2). In the case of NoSQL databases, they may have a set of tools or profilers that could be used to obtain the desired insights. For example, MongoDB databases support different tools and resources (3) for performance monitoring depending on where they were deployed.

Now, let us explore the performance-related metrics in more detail.

Performance metrics

The performance metrics to consider for a database depend on the nature of the solution. For serverless, fully managed databases, you may not need to monitor resource utilization metrics. However, if your database involves managing instances, servers, and CPU/memory allocations, these metrics become crucial for understanding performance.

Regardless of the metrics available, what do the resource utilization metrics for your database solution tell you? The following are some of the questions you must be able to answer by looking at different performance metrics of your database solution. This is not an exhaustive list:

- Do the metrics indicate a resource bottleneck of any kind? This could be CPU, memory, disk read/write, or network.
- Are there long wait times on the CPU you have allocated for your database?
- Do your storage volumes show high disk read and write latencies?
- Are you exhausting any thread pools you may or may not have control over?

Performance is not only resource utilization. Your customer's observed end-to-end application latency is a critical metric you must learn from. Some of the questions your application-level performance metrics should allow you to answer are as follows:

- Are your application's read and write latencies impacted due to the database during peak hours of usage?

- How often are you able to meet your own goals of application availability? How much availability impact is caused by database issues?

- Does your application often see query wait times?

Based on the answers to the preceding questions, you can evaluate how well your existing database solution meets your needs. If the responses indicate significant issues that can't be resolved through configuration tweaks, it's a strong indicator that migration may be necessary.

Next, let us explore how expected traffic patterns can influence your assessment of a potential migration target.

Traffic patterns and scaling requirements

The traffic patterns of your application are crucial in selecting the right migration target. Your database choice should align with these patterns to avoid wasted resources and unnecessary costs. It must efficiently handle both low-traffic periods and peak traffic times.

Ideally, your database should offer a pay-as-you-go approach, scaling seamlessly with your application's traffic. This ensures that expenses increase proportionally with traffic growth, providing cost efficiency and scalability.

When qualifying a database target for your application, consider the following questions about your application's current traffic patterns. Keep in mind that this list is not exhaustive, and your specific needs may vary. Nevertheless, these questions provide a solid starting point for most scenarios:

- Do you think the database setup would be able to handle twice your expected growth of traffic and data storage for the current year and next?

- Would your database consume weeks of developer time for scaling to support your next big customer event?

- Is there a considerably large gap between your average and peak usage of the database? If yes, do you always pay for peak usage of your database?

- Do you have spiky, unpredictable traffic where you must always have to provide lots of infrastructure on the database level to support sudden surges in traffic?

- Did you have to scale your database instance vertically recently, and do you foresee vertically scaling the database again in the short- to mid-term future?

If answers to most of the preceding questions tend to your existing database not being most suited for the application, that is a data point in the migration direction.

Next, let us review some other aspects of databases that could not be categorized as one of the preceding, and are more database technology dependent.

Other technical indicators

The significance of the indicators discussed can vary based on your current database setup and your application's specific requirements. Your application may depend on features specific to your database technology, some of which may not be performing optimally. Your database choice should align with your application's needs rather than forcing your application to adapt to database limitations. Prioritize a database that suits your application's unique requirements and functionality.

If your database uses read replicas to enhance read scaling, be aware that this introduces additional overhead on the primary instance involved in replication. This overhead is typically justified compared to having the primary instance handle all traffic. When using replicas for scaling, consider whether your application experiences peak lags in replication. As your application grows, replica lag may become more pronounced. Anticipate whether increasing replica lag will cause issues and define what level of lag is unacceptable based on customer experience and internal standards. Having a rough estimate of acceptable lag can help you avoid similar challenges in the future.

Next, let us learn how analyzing the costs of operating your existing database could help you qualify for a migration opportunity.

Costs

Database costs include not only server expenses but also the investments in skills and human resources needed to maintain the database's performance and operation. Assess your application's traffic growth rate and compare it to the growth rate of your **total cost of ownership** (**TCO**) for the database solution.

This analysis will help determine whether your current database solution remains feasible for the organization or whether alternative options offer a significantly lower TCO, making them better suited for your application.

The following are some questions you must answer about the costs of your existing database solution:

- Do you need that shiny functionality that your database charges you for, whether you use it or not? Do you need the durability and availability SLAs that your existing database provides or are you okay to compromise on either aspect for a large cost benefit?

- Do you need the three-way replication of data across data centers for high durability, or are you okay with a single data center hosting a particular AdTech-related dataset that you need temporarily anyway?

- Do you need super-fast read and write performance on a dataset that is infrequently accessed? Are you okay with using colder storage for your dataset that costs a fraction of what your existing database is costing?

After collecting all the insights about your current database solution, you will be well-equipped to make an informed decision regarding the need for a database migration. Your final choice will not be solely guided by functional and technical factors; non-technical aspects such as organizational changes, mergers, acquisitions, and strategic partnerships may also play a role. However, from a technical perspective, the considerations mentioned previously encompass most of what you should evaluate when qualifying for a database migration.

Now that we have understood what insights to form the basis of qualifying your migration, let us learn about when to qualify a migration to DynamoDB in particular.

Qualify a migration to DynamoDB

Throughout this book, we have explored many DynamoDB features and applications. Now, let us analyze when DynamoDB is suitable for migration and when it might not be the best solution for your use case. The following are key aspects of DynamoDB to consider when evaluating it as a migration target.

Single digit performance

Does your application require lightning-fast, single-digit millisecond performance for both reads and writes? DynamoDB excels at this, especially for small items or rows under 10 KB. However, if alternatives can meet your speed requirements and offer a lower TCO, they might be worth considering.

Well-defined access

Understanding your application's access patterns is crucial in determining whether DynamoDB is a suitable migration target. If you can confidently identify 60-70% of these patterns, DynamoDB may be the right choice. Here, we assume you have a production application and a clear understanding of your top access patterns.

Example – financial services (FinServ) use case

In a FinServ scenario, users may need to access transaction history and specific transactions for actions or flagging. The primary access pattern could be "listing transactions by accountID," with additional patterns such as the following:

- Fetching authentication methods for accountID
- Retrieving stakes in mutual fund folios
- Retrieving buy or sell transactions

These patterns dominate the OLTP landscape, while other patterns related to daily reporting or machine learning model training for fraud detection make up the remaining 20-30%. Such use cases align well with DynamoDB.

Conversely, if your use case heavily involves **semantic searching** or extensive **geospatial lookups**, it is advisable to consider other purpose-built databases. While DynamoDB can be complemented with a secondary data store for these scenarios, relying on it for access patterns it does not natively support might result in unnecessary complexity.

Additionally, if your application requires multiple optional filters, evaluate whether these patterns can be separated into distinct access patterns. If they cannot, DynamoDB might not be ideal. For example, if you need users to apply various optional filters on a dashboard, managing this in DynamoDB could be cumbersome. You would have to create and maintain numerous combinations of these filters with different global secondary indexes, increasing complexity and cost.

Scalability

DynamoDB excels in scalability. Unlike your current database solution, which might face read/write bottlenecks and require scaling methods such as read replicas or sharding that need ongoing maintenance, DynamoDB offers automatic scaling and seamless sharding. For scenarios requiring exceptionally high throughput per key, you can implement application-level sharding patterns. These patterns are well documented and involve minimal effort once the initial design is complete.

Serverless benefits

DynamoDB's serverless design offers significant advantages by avoiding issues such as CPU and memory bottlenecks. It eliminates concerns about storage disk read/write latencies and removes the need to maintain continuously running servers. Instead, you can use a pay-as-you-go approach with on-demand capacity mode. This flexibility is especially beneficial for applications with significant traffic fluctuations between average and peak loads.

Costs

Evaluating costs is crucial when considering DynamoDB for migration. Often, DynamoDB offers a lower cost profile compared to other databases. However, unpredictable write-heavy workloads with frequent large-item updates may not see the same cost advantages, making write costs for such use cases a potential concern. I've seen cases where write costs were high for write-focused workloads involving large data items.

DynamoDB offers native features, such as capacity reservations, and commercial programs, such as the AWS **Migration Acceleration Program (MAP)** (4), to mitigate some of the migration cost concerns. Predictable write-heavy workloads with gradual traffic peaks can still benefit from DynamoDB. Storage and read costs are generally lower, and its fully managed nature reduces the need for a large DBA team, positively impacting TCO.

This information should help you decide if migrating to DynamoDB is necessary. Let's explore the migration process if you choose to proceed.

Homogeneous and heterogeneous migrations

Once you have determined that DynamoDB is a suitable migration target (with final confirmation possibly coming after a POC), the next step is to outline the migration process. Generally, database migrations are classified as either **homogeneous** or **heterogeneous**, depending on the source and destination technologies.

Homogeneous migrations involve moving between databases of the same technology. For example, migrating from a self-managed on-premises PostgreSQL database to a fully managed Amazon RDS for PostgreSQL is a homogeneous migration. Although the database technologies are the same, other aspects like management and scalability may differ.

Heterogeneous migrations, in contrast, involve moving between databases of different technologies. For example, modernizing from Amazon RDS for SQL Server to DynamoDB is a heterogeneous migration. This type of migration requires addressing differences in data models, query languages, and other fundamental aspects of the databases.

Next, let's explore these migration categories in more detail.

Homogeneous migrations

In the context of DynamoDB, homogeneous migrations involve transferring data between DynamoDB tables. This can be done within the same region or account, or across different regions and accounts. Although these migrations may not align perfectly with the broader migration strategies discussed earlier in this chapter, they are a common practice and worth addressing.

For homogeneous migrations within DynamoDB, the primary focus is on copying data between tables, ensuring consistency and accuracy throughout the process.

The following is an illustration of a homogeneous migration of data from a DynamoDB table in one account to a DynamoDB table in another account. The migration job here could be any of the tools and approaches you use to execute the migration.

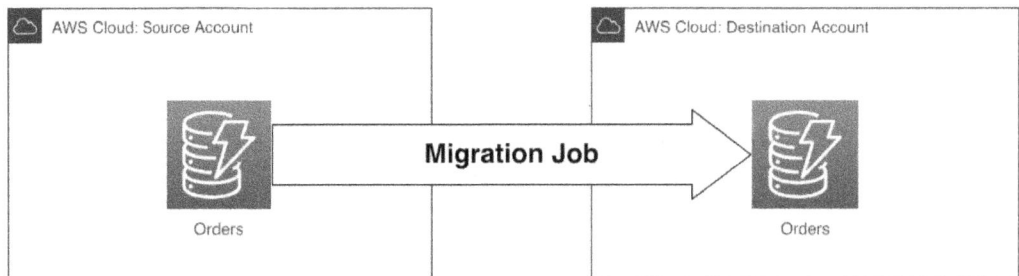

Figure 16.1 – Example of cross-account homogeneous migration of DynamoDB data

Usually, in most cases of homogeneous migrations involving DynamoDB, you may not find it necessary to alter the data model for the target database. Nonetheless, there are instances where migrating data from one DynamoDB table to another with a distinct schema can still be considered a homogeneous migration. This type of migration is often driven by two primary motivations: targeted optimization of specific access patterns within the application or the natural evolution of the data model in response to evolving business requirements.

Next, let us understand homogeneous migrations in the context of DynamoDB with the help of some examples.

Examples of homogeneous migrations

Consider an organization using DynamoDB and AWS, following best practices for security by maintaining separate AWS accounts for different teams or microservices. Organizational restructuring might group all metering services under a new team, retaining another core microservice in its original team. This would require planning and executing a migration to consolidate DynamoDB tables associated with metering into a single AWS account while keeping the core service in its original account. This situation will require a homogeneous migration of DynamoDB data between accounts, which may involve downtime.

Another scenario I frequently encounter as a solutions architect for DynamoDB is customers aiming to relocate their DynamoDB tables to different regions. This move is often driven by the availability of a new AWS region closer to the user base. Occasionally, large-scale migrations between DynamoDB tables in different regions can also lead to cost optimization, as AWS regions may have slightly different pricing. At scale, these slight differences can result in significant savings, provided that data privacy regulations permit moving regions.

With these examples of homogeneous migrations in mind, let us now dive into heterogeneous migrations in more detail.

Heterogenous migrations

If you are considering a migration to DynamoDB, you are likely dealing with a heterogeneous migration. These migrations are often more complex, usually requiring data transformations before moving data into DynamoDB. These transformations adapt the data to fit DynamoDB's NoSQL design patterns for better efficiency and scalability.

For example, consider migrating from a MySQL database to a DynamoDB table. Such migration may involve various tools and technologies to handle data model changes. Whether your source database is on-premises, as shown in the figure, or in the cloud, the migration is still considered heterogeneous.

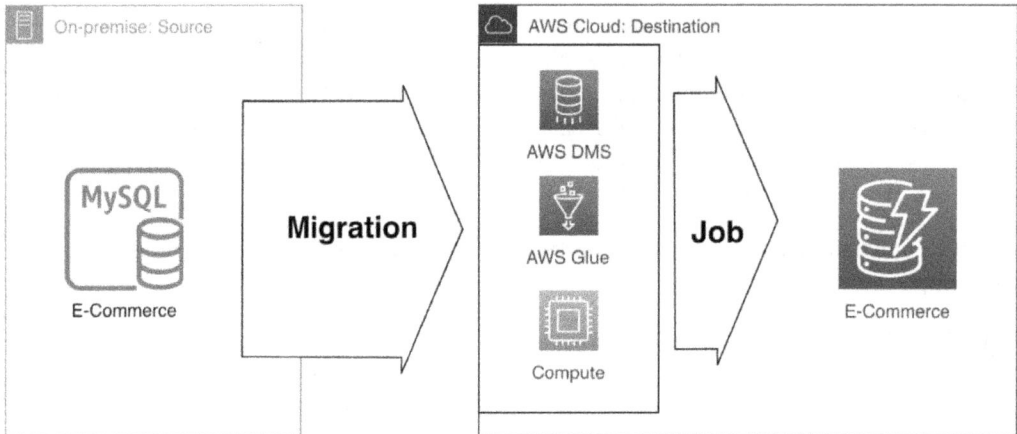

Figure 16.2 – Example of heterogeneous migration from MySQL to DynamoDB

For target data models involving multiple entities, understanding DynamoDB design patterns is crucial. Before setting up your migration pipeline, it is essential to build your target data model. Unlike traditional RDBMS, NoSQL databases such as DynamoDB require a well-defined data model before planning your migration. To gain a better understanding of these design patterns, refer to *Chapters 4 to 8* of this book, which focus on data modeling.

With this background, let us dive into heterogeneous migrations through various examples.

Examples of heterogeneous migrations

Consider a team working to modernize a large event ticketing system by migrating from a SQL Server database to DynamoDB, a classic example of heterogeneous migration. The original SQL Server setup utilized a highly normalized schema with data distributed across numerous tables, that required compute-intensive runtime joins that caused performance issues, even on the largest server available. These complex join queries performed poorly during high-traffic periods, such as when tickets for popular events went live. Additionally, concurrency problems arose due to the locking mechanisms inherent in traditional databases.

To address these challenges and consider their defined access patterns and cost factors, the team decided to migrate to DynamoDB. This involved transforming their RDBMS data model into one suited for DynamoDB, designed for scalability and high concurrency. The most crucial step was de-normalizing entities into pre-built items, which required more writes but ensured app access never hit scaling bottlenecks. Additionally, they utilized various design patterns, such as DynamoDB Streams and AWS Lambda for asynchronous tasks, such as maintaining counts, and Transaction APIs for atomic operations across multiple entities. To handle high concurrency, they used conditional write requests with optimistic concurrency control.

Here is another example – consider a news website team migrating their data from MongoDB to DynamoDB. The site provided users with news articles relevant to their location or neighborhood. Although MongoDB was already a NoSQL database, it used database-specific features for analytical tasks and stored entire articles as single JSON documents. This design, while functional for queries based on news article IDs and neighborhood IDs, led to performance issues due to daily spikes from analytical jobs on the same host.

For the migration to DynamoDB, the team needed to revamp their data model to enhance the performance of the summary news feed view and separate analytical tasks from OLTP performance. They applied vertical partitioning to optimize the summary view and utilized DynamoDB's native export features to handle analytical processing without affecting the database's primary operations.

I hope the preceding examples clarify both homogeneous and heterogeneous migrations with DynamoDB. With this understanding, let us move on to discussing high-level migration strategies.

High-level migration strategies

With DynamoDB as your chosen target and a clear understanding of your migration type – homogeneous or heterogeneous – you should now have an initial version of your target data model. This model should address most, if not all, of your application's access patterns and may include secondary indexes, streaming options, and a separate analytical/reporting pipeline. It could involve simple key-value access or more complex design patterns to manage relationships between multiple entities, potentially leading to a single-table design.

In this section, the focus shifts to creating and executing a migration plan to transition your application from its current database to DynamoDB. We will start by exploring online and offline migration strategies.

Online and offline migrations

Online migrations occur while the application remains active, often termed **zero-downtime migrations**. These require careful coordination to transfer data from the source database to DynamoDB, followed by a cutover where the application switches from the old database to DynamoDB. This approach is critical since data continues to be updated during the migration, and the target DynamoDB database must handle these real-time changes to ensure a smooth end-user experience.

The following diagram outlines a standard online migration process from system V1 to the new system V2:

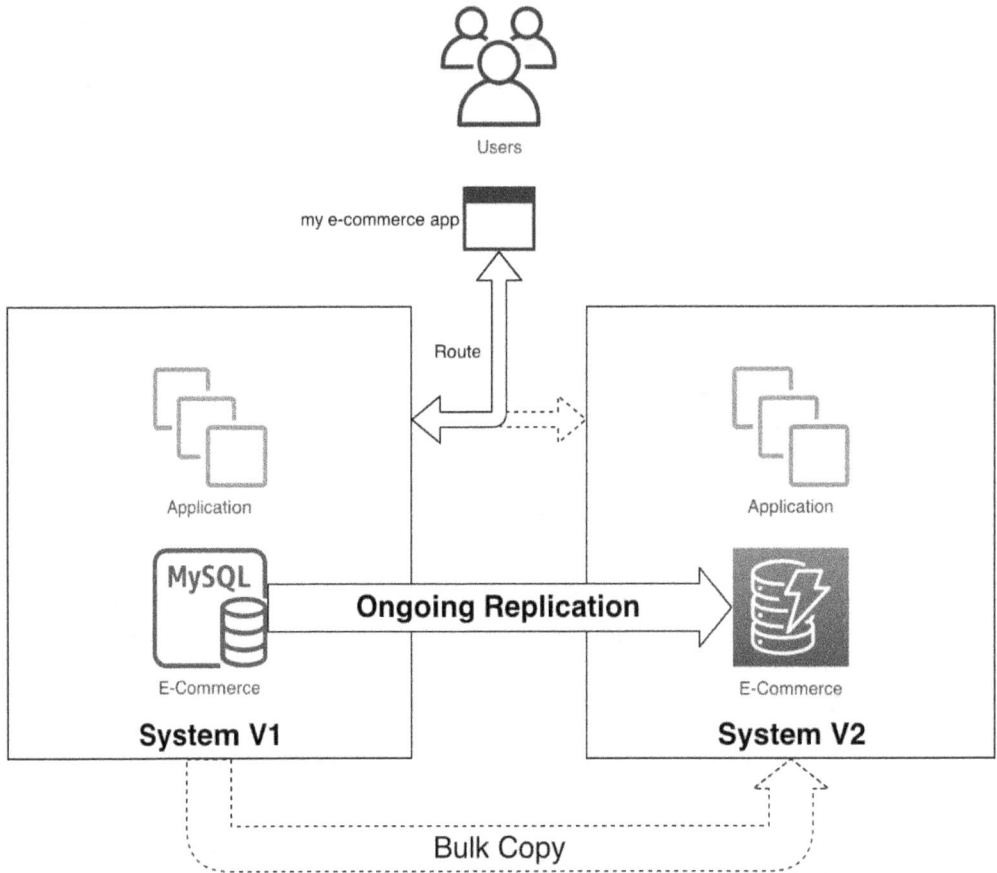

Figure 16.3 – Example of online migration while users continue to interact with the application seamlessly

During an online migration, the application stays live, managing ongoing read and write traffic. The bulk data transfer begins and completes, supplemented by continuous replication to keep the source and DynamoDB in sync. The final step involves switching users from the old system (system V1) to the new one (system V2).

These migrations can last from weeks to months due to their complexity. Although they may temporarily raise costs, these usually normalize after the migration is complete. Online migrations are often required for critical workloads.

In contrast, offline migrations occur during scheduled application downtime, such as planned maintenance windows. These migrations are simpler than online migrations, with fewer moving parts, leading to a lower TCO. Since the data is static during the migration, there is no need to continuously capture and update the target database as in online migrations.

The following diagram depicts a standard offline migration scenario, where users are notified in advance of a scheduled maintenance window:

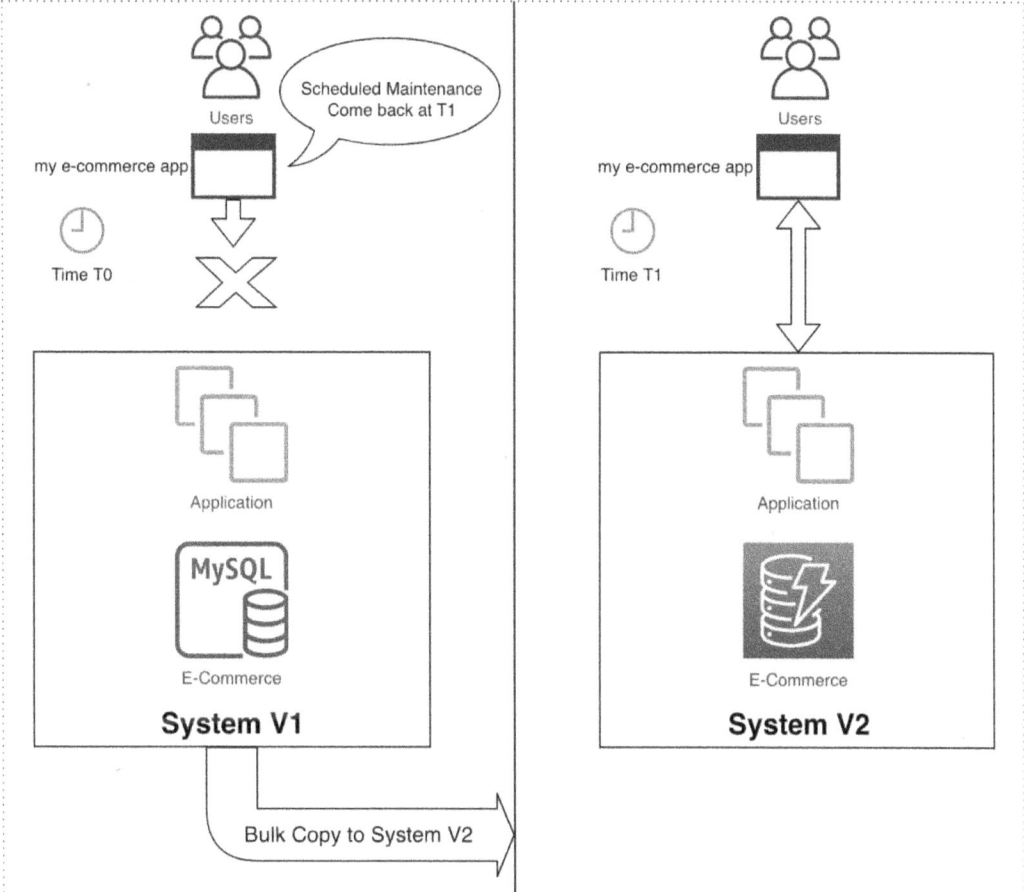

Figure 16.4 – Example of offline migration with downtime in the form of scheduled maintenance

During the designated downtime, the migration – including bulk copying and data transformation – proceeds smoothly, provided it completes within the maintenance window. Once this period ends, users can confidently start interacting with the new system V2 without disruption.

However, many live applications may not have the option for an offline migration.

Regardless of the migration type, having a **rollback strategy** is essential in case issues arise. Understanding the differences between online and offline migrations provides the foundation for the next section, where we will explore the tools and engineering patterns needed for a successful migration.

Common tools and building blocks for migration

Every DynamoDB customer previously used a different database technology and had to choose among various tools and services for migrating their data. Based on the experiences of numerous customers, this section outlines some key tools and best practices to help with your migration plan. While not exhaustive, this list includes foundational tools and practices likely to be valuable for your migration project. Consider these as essential building blocks for creating a comprehensive migration strategy.

Bulk copy / initial data transfer

A key step in migrating data to DynamoDB is the initial data transfer. This is essential for both online and offline migrations. In offline migrations, this step involves a full copy of your data. For online migrations, it is part of a phased migration plan. For heterogeneous migrations, you may need to transform the data during this process. In contrast, for homogeneous migrations within the same AWS account or across regions, data model transformation might not be necessary.

AWS Glue / Apache Spark-based tools

To perform initial data transfer, tools such as Apache Spark and Apache Hive are commonly used. These tools utilize distributed computing frameworks to parallelize the workload, helping estimate processing times and handle data transformations. In the AWS cloud, AWS Glue is a managed, serverless service that offers Apache Spark-based computing and supports various databases, including DynamoDB. AWS Glue connects to a range of databases such as SQL Server, Oracle, MySQL, Postgres, MongoDB, Amazon DocumentDB, Amazon Redshift, Snowflake, and others that support JDBC (Java Database Connectivity) connections. For more details on AWS Glue's connection options, refer to the AWS docs (5).

AWS Database Migration Service (DMS)

AWS DMS is another managed service by AWS that helps with end-to-end data migration. Beyond handling the initial data transfer, AWS DMS supports ongoing replication of changes from your source database to DynamoDB, which is crucial for online migrations. It also allows for basic data model transformations using a simple JSON-like transformation configuration format.

However, for heterogeneous migrations to DynamoDB, especially when transitioning from an RDBMS to a single-table design in DynamoDB, DMS might not be the best fit due to DynamoDB's schemaless and denormalized nature. DMS is more suitable for straightforward data model transformations and supports various open source and third-party databases as sources, though DynamoDB itself cannot be a source. For a detailed list of supported source databases, refer to the AWS docs (6).

Using native backup/restore features

For homogeneous migrations, DynamoDB offers built-in features for backing up, restoring, and copying backups across AWS accounts or regions. These features are effective when no data model transformation is needed. For example, if you are moving DynamoDB data to a different AWS account without altering the data model, these native tools are both efficient and straightforward.

If your migration involves data transformations, you will need to use tools that support DynamoDB as both a source and a target. To copy native backups across AWS accounts, both the source and target accounts must be part of the same AWS organization. If they are not, you can still use the Export to Amazon Simple Storage Service (Amazon S3) and Import from S3 features, but this approach might require additional steps for orchestration.

While it is usually best to rely on existing tools for the initial data transfer, there are cases where custom scripts or tools might be necessary if your requirements are not met by available options. Since database migrations can be relatively rare, creating and maintaining your own tool can often be excessive.

Change data capture for ongoing replication

For online migrations, maintaining synchronization between the source and target databases is crucial until you fully transition to DynamoDB. This requires ongoing replication in addition to the initial data transfer.

Modern databases typically offer change logs that record every mutation, which are essential for keeping both databases in sync during the migration. To handle this, you will likely need to use an open source or third-party tool to consume and transform these change logs before loading them into DynamoDB, as each database technology has its specific tools for change log management.

AWS DMS can replicate change data capture logs from various RDBMS technologies but may not support all source databases (7). If your source database does not provide change logs, dual writes to both the source and target databases might be necessary to ensure synchronization.

Dual writes for ongoing replication

One commonly employed pattern for maintaining synchronization between the source and target databases during migration to DynamoDB involves implementing dual writes at the application level. In this approach, you may still require an initial data copy as part of your online migration, but ongoing replication is managed by updating the application to make changes to both the source and target DynamoDB databases. Any creations, updates, or deletions of data in the source database must also be mirrored in the DynamoDB data. This approach enables you to gradually phase out write operations to the source database as you progress through the migration. Moreover, you likely needed to develop a data access layer for interacting with DynamoDB, making dual writes a dual-purpose solution.

The following figure illustrates an intermediary phase of the dual writes approach.

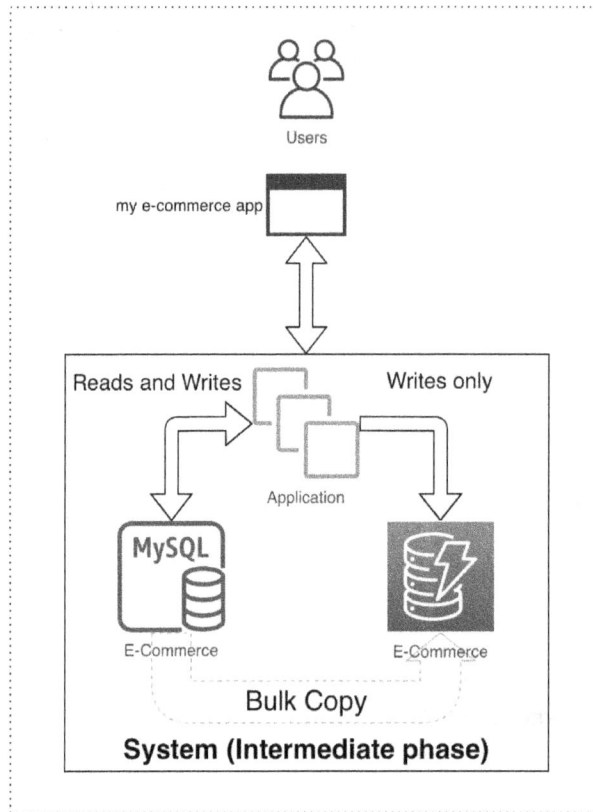

Figure 16.5 – Intermediary phase of dual writing to both old and new data stores by the application

While users maintain uninterrupted access to the application, modifications are made to enable the application to write to the new data store alongside its existing interactions with the old data store. Reads continue to be made to the old data store until validations are made to ensure consistency among the two data stores.

Key emphasis must be placed on handling updates and deletes in the dual writes approach. If the target DynamoDB table is not in sync with the source database, it may not have the required item to execute updates or deletes routed to the new system. Several approaches can be implemented:

- **Use an Amazon Simple Queue Service (SQS) FIFO (First-In-First-Out) queue or a streaming technology**: This can queue all updates and delete requests, processing them only after the initial bulk load is complete.

- **Issue additional read requests to the source database**: This fetches records necessary for updates and uses Tombstone records (8) for deletes.

These approaches can be complex compared to blocking updates and deletes during a maintenance window. If you proceed with dual writes without downtime, even for updates and deletes, you must carefully arrange the order of events and leverage DynamoDB's features for making conditional writes.

Typically, you can synchronize dual writes with the initial data transfer, either by conducting the initial data transfer before enabling dual writes or by having both steps perform conditional writes to DynamoDB, with the condition being that the intended change is more recent than the existing data in DynamoDB. For example, if the initial data transfer sets the last update time of an item to T1, and the application performing dual writes intends to write data to DynamoDB at time T2, you must ensure that T2 is later than T1. This can be achieved through conditional writes in DynamoDB (9).

Network-level routing

Network-level routing allows a smooth transition from the old system to the new system that would be using DynamoDB. After ensuring that the source and target data are synchronized, you can gradually shift read traffic from the old system to the new system using network-level routing. This approach involves controlling the percentage of traffic directed to each system, effectively balancing the load, and reducing the risk of disruption.

This method aligns with **blue/green deployment** strategies, where you can manage traffic between the two systems, though it might involve additional components for complete implementation.

The following figure illustrates an architecture that allows network-level routing for user requests to the system.

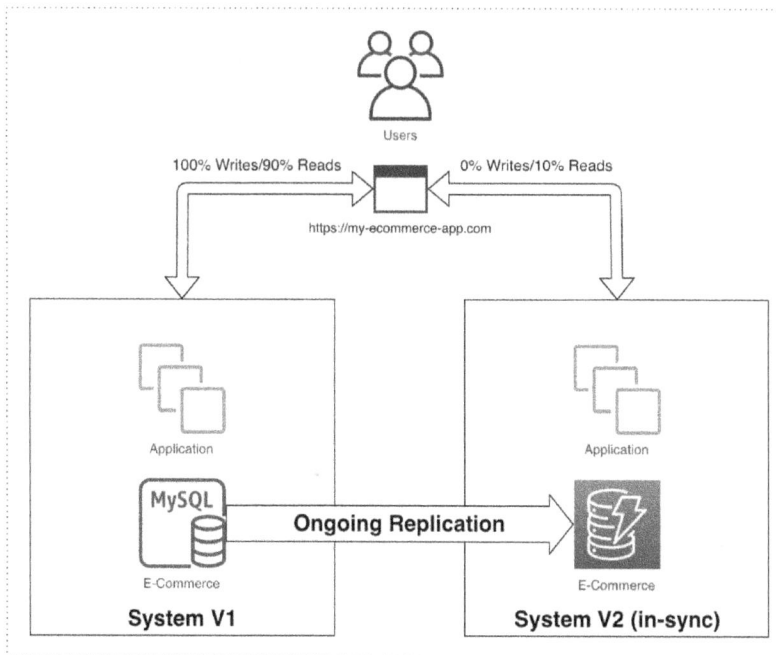

Figure 16.6 – Sample architecture that migrates read traffic using network-level routing

Following the scenario described in the preceding figure, reads are incrementally routed to the new system once both the old and new data stores achieve high consistency. Writes, on the other hand, are committed to the old system and propagated to the new system to facilitate potential rollbacks. This phased approach aims to transition reads entirely to the new system before moving writes, thereby completing the migration.

Network-level routing allows for the temporary coexistence of multiple application versions, enabling a gradual phase-out of the older version. This method provides an opportunity to monitor how the new system performs under live traffic and to verify that it meets expectations. Should any issues arise during this transition, you can roll back to the old system, which remains synchronized and available.

Application-level versioning

In addition to other phased approaches for shifting read traffic to the new system, in some cases, you might consider migrating users from the old experience to the new one, like an A/B testing style.

The following figure illustrates an approach using application-level routing of user requests to their intended data store.

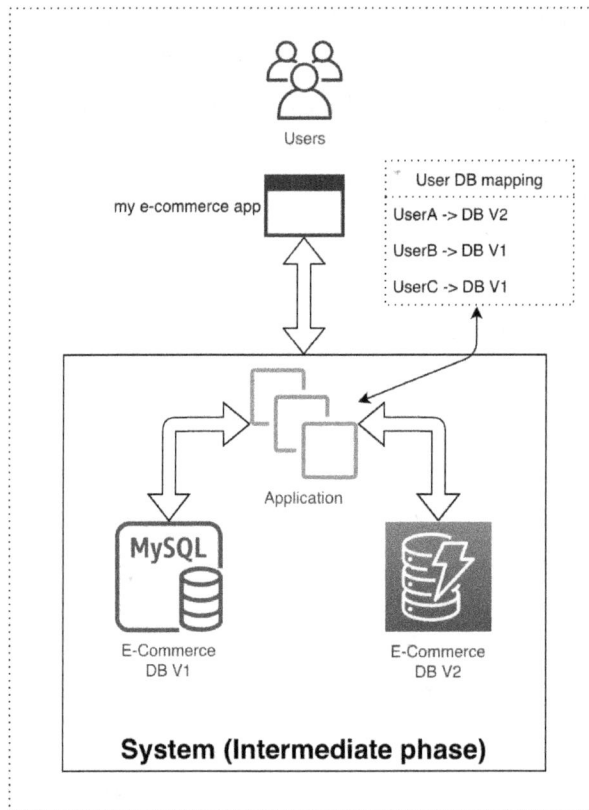

Figure 16.7 – Sample architecture highlighting application-level routing of user requests

Here is how it works: you designate a certain percentage of users to always be served by the new system using a mechanism, such as a feature flag. This could also be in the form of a user to database mapping table which is looked up first by the application to learn which database to hit to serve the request.

With this approach, you can adjust the number of users who receive reads from the new system. It may be necessary to continue dual writes to both the source and destination databases, as this allows you to revert users in case the need arises. However, you can gradually phase out the old system by using a feature flag in your user profile data, which you can remove entirely once the old system is retired and the migration is complete.

Data validation

Data validation is a crucial component of migration, ensuring that the data remains accurate after the transfer. This process typically involves comparing data between the old and new systems, which can be complex, especially during an online migration where writes are occurring on both systems.

Due to race conditions, there may be temporary inconsistencies between the systems. For example, data validation services might retrieve user data from the old system and then from the new system, potentially leading to discrepancies if changes occur between these fetches. Therefore, data validation should not be a one-time offline task but rather a continuous process, combining real-time checks with offline validation batch jobs throughout the migration.

It is common to encounter a small percentage of data validation failures due to these transient inconsistencies, but these should remain consistent over time as the systems stabilize. Allocating adequate resources to data validation is essential, as it ensures the accuracy of your data and provides confidence throughout the migration process.

With these foundational elements in mind, we can now explore migration strategies tailored to your specific needs.

Migration plans

The following outlines some of the most common migration scenarios along with their high-level migration approaches. While there may be various methods to execute these migrations, the following provides commonly followed paths. This information should serve as a foundation for constructing your own migration plan or, at the very least, a framework to get you started.

Heterogeneous migration with zero downtime (online migration)

Following are the high-level steps involved in performing a heterogeneous migration without any downtime on the application:

1. **Implement dual writes**: Begin by implementing dual writes from the application to both the old and new databases. Ensure that the application writes to both databases simultaneously.

2. **Bulk copy**: Perform a bulk copy of data from the old database to the new database. This bulk copy should use conditional puts on DynamoDB, as DynamoDB may already have received newer data written by the application during the bulk copy process. Once the bulk copy is completed, both the old and new databases should gradually converge.

3. **Invest in design transition**: Since this migration is heterogeneous, considerable time would have been dedicated to defining the target schema. Both the application and the bulk copy process must be adapted to write data into DynamoDB using the new schema.

4. **Shadow reads**: The next phase involves updating the application to perform shadow reads from DynamoDB to validate that the data matches the old system. During this phase, the old system remains authoritative for both writes and reads. The application runs validations to compare results from both databases. It's essential to run offline validation batch jobs continuously to compare data and monitor the rate of convergence. A consistent convergence rate is a key indicator of successful validation.

5. **Validations**: Successful validation over a sustained period provides confidence to proceed with the migration. The next step involves increasing the ratio of customers served authoritatively from the new database. This can be achieved through application-level routing or network-level routing, as previously described. Continue to validate results from both the old and new databases and maintain dual writes.

6. **Cut over**: After running 100% of reads authoritatively from the new system and gaining confidence in the data validation mechanism, you can finalize the migration. At this point, you can turn off writes to the old system and proceed to decommission it. This step is irreversible, so it should only be executed after thorough confidence in the data validation process.

For the bulk copy phase, AWS Glue ETL (extract, transform, and load) and AWS DMS are both viable options. AWS DMS provides no-code capabilities to connect various source databases to DynamoDB, but it may have limitations regarding version compatibility and database engines. In contrast, AWS Glue ETL allows for custom Spark code to handle data transformation and benefits from a broad open source community for support.

During the dual write phase, the focus shifts to modifying the application's data access layer to handle writes to both systems. Gradually redirecting read traffic from the old system to the new one is advisable, as a sudden switch can lead to potential issues. This phased approach helps ensure a smoother transition and minimizes the risk of operational disruptions.

Heterogeneous migration with downtime (offline migration)

In cases where downtime is permissible for the migration, the process involves fewer complexities compared to online migrations, and dual writes may not be necessary. The following high-level steps outline how to conduct such a migration:

1. **Bulk copy**: Initiate a bulk copy job to transfer data from the old database to the new DynamoDB database. As downtime is allowed, this process can be managed efficiently.

2. **Scheduled downtime**: Plan for scheduled maintenance windows that typically last a few hours. During these windows, the application should continue to support read access, but write access can be temporarily suspended on the old system.

3. **Data copy and transformation**: As soon as the maintenance window begins and writes are blocked on the old system, execute bulk copy into the new DynamoDB database. If data transformation is required, this is the stage to perform it.

4. **Data validation**: Run data validation jobs during or after the bulk copy to ensure that the data in both databases converges. Since data is not changing during the migration (unlike in online migrations), data validation jobs must demonstrate 100% convergence before proceeding.

5. **Network switch**: When data validation is successful, make a network switch to direct users to the new system, which is ready to accept both read and write traffic.

6. **Migration conclusion**: Conclude the migration by decommissioning the old system. During the network switch, if any issues or regressions are observed with the new system, it may be possible to revert to the old system.

7. **Synthetic canaries**: To ensure the new system functions as expected, consider building and running synthetic canaries for the system. These canaries should exercise all functionalities of the application continuously.

In terms of tools, Glue ETL is recommended for performing the bulk copy due to its serverless, fully managed, and naturally distributed characteristics. Timing the bulk copy job to complete as quickly as possible may require some dry runs. Fortunately, whether you use high write units for a short period or spread the same write units over a longer period, DynamoDB costs remain consistent. The same Glue ETL job can be used to run validation checks after completing the bulk copy.

Homogeneous migration with zero downtime (online migration)

In the case of a DynamoDB-to-DynamoDB migration with data model transformation requirements, you can approach it similarly to a heterogeneous migration or explore alternative methods. To achieve a homogeneous migration without any downtime, consider the following approach:

1. **Bulk copy and online replication**: Initiate a bulk copy of data from the source DynamoDB table to the target DynamoDB table. Once the bulk copy is complete, use DynamoDB Streams to replicate changes from the source table to the target table. Since both the source and target databases are DynamoDB tables, a consumer of the source table's DynamoDB stream can effectively replicate changes to keep the tables in sync.

2. **Application cutover**: The final phase of this approach involves redirecting the application to interact with the new target DynamoDB table instead of the old one. Execute this cutover during a period of relatively low traffic.

3. **Tear down**: After ensuring that the stream of changes has been completely drained, the old table and its stream can be decommissioned.

To perform the bulk copy without any transformations, you can leverage appropriate native features based on the relationship between the source and target DynamoDB tables:

- **Same AWS account**: If both tables belong to the same AWS account, consider using a simple on-demand backup of the source table followed by a restore operation to create the target table. Native on-demand restores also support cross-region restores.

- **Different AWS accounts (same AWS organization)**: If the two tables are part of different AWS accounts within the same AWS organization, you can perform an on-demand backup of the source table and copy it to the target table's AWS account using the AWS Backup service (10).

- **Different AWS accounts (no common AWS organization)**: If the two accounts are not part of the same AWS organization, utilize the native Export to S3 feature in the source account and the Import from S3 feature in the target account to execute the bulk copy.

If your migration involves data transformations, you may not be able to rely on the native on-demand backup/restore features. In this case, you can use the native Export to S3 feature in the source account and complement it with a Glue ETL job. The ETL job can read the output of the export, perform the necessary data transformations, and write the transformed data to the target DynamoDB table.

In all these approaches utilizing native features, ongoing replication is still necessary through DynamoDB Streams. Compared to the dual write method for online migrations, using Streams is typically faster and more cost-efficient in terms of TCO.

Homogeneous migration with downtime (offline migration)

Performing an offline homogeneous migration is often relatively straightforward compared to other migration types. This type of migration typically involves a bulk copy or relies on native no-code features. If data transformations are part of the migration, you may need to implement a Glue ETL job to handle these transformations, utilizing the output of the native Export to S3 feature.

For an offline homogeneous migration without data transformations, the choice between native features depends on the relationship between the source and target DynamoDB tables:

- **Same AWS account**: If both the source and target tables belong to the same AWS account, consider using the native on-demand backup and restore features. This approach is generally sufficient and does not require writing custom code.

- **Different AWS accounts (same AWS organization)**: If the source and target tables are in different AWS accounts but within the same AWS organization, you can use the AWS Backup service to facilitate cross-account data copying.

- **Different AWS accounts (no common AWS organization)**: If the source and target tables are in different AWS accounts without a common AWS organization, you can employ the native Export to S3 feature in the source account and complement it with the native Import from S3 feature in the target account for data transfer.

The choice of approach depends on your specific migration requirements and the relationship between the source and target tables.

I hope that the migration plans provide valuable guidance for planning and executing your database migration to DynamoDB. As previously mentioned, you can adapt the specific tools and technologies to align with your team's skills and technical capabilities. The technologies mentioned in this section are chosen with best practices and commonly used AWS services in mind.

Next, let us explore some key considerations for database migrations to DynamoDB.

Key considerations for DynamoDB migrations

This section outlines some of the key aspects of migrations to DynamoDB that could not be categorized as part of any other section in this chapter. These aspects are not necessarily technical, more toward common gotchas you might want to know and learn upfront, saving you time and effort.

Common concerns against migrating to DynamoDB

Working with customers looking to migrate to DynamoDB for more than six years has helped in coming up with a list of the greatest hits of concerns or questions that might come up while evaluating a migration to DynamoDB. The following are some of those concerns and some guidance to address them for your stakeholders:

- **DynamoDB is more restrictive compared to our RDBMS database**: True, DynamoDB prioritizes consistent performance over flexibility. Each operation in DynamoDB is designed to be scalable, performant, and deterministic. For instance, a Query API retrieves a maximum of 1 MB of data per request, ensuring a predictable experience. If absolute flexibility is a top priority for your application and scalability and predictability are less critical, DynamoDB might not be the best fit for your workload.

- **DynamoDB results in vendor lock-in; it's challenging to migrate out of AWS if you use DynamoDB**: Not necessarily. When considering the costs, scanning about 1 TiB of data in DynamoDB costs approximately $38. You can then export this data and import it into any other system of your choice. However, this cost does not account for potential application-level changes required during any database migration.

- **DynamoDB is unsuitable for relational data**: If you have not explored the earlier data modeling chapters of this book, it is essential to understand that DynamoDB, like NoSQL databases in general, can maintain relationships between various entities in the data model. The difference lies in the representation of data, with DynamoDB favoring a denormalized, pre-built approach compared to the highly normalized model in RDBMS databases, which rely on runtime-intensive joins to assemble data on demand.

- **DynamoDB is primarily for use cases with high operation volume**: Not necessarily. While DynamoDB is frequently associated with large-scale applications, it is relevant to applications of any size. DynamoDB tables are used to store a variety of data, from configurations and user profiles to system metadata and user or game state. The advantages of a serverless, scalable, pay-as-you-go database apply regardless of the scale.

- **DynamoDB is not suitable for reporting use cases**: DynamoDB is primarily designed for high-performance OLTP access by applications. However, reporting requirements are common in various applications, and DynamoDB provides native support for exporting data to Amazon S3, which can serve as your data lake. This export can be a complete data dump or incremental, making it suitable for complex analytical reporting use cases.

- **DynamoDB lacks support for a specific feature that our current database natively supports**: Different technologies may have varying approaches to support certain use cases. It is essential to evaluate whether the specific feature your current database supports is crucial for your workload. If so, consider whether DynamoDB offers straightforward alternatives to achieve similar outcomes. If not, it is important to recognize that DynamoDB may not be the right fit for your specific requirements.

Ensuring two-way doors throughout the migration

While this concept is well known in software engineering, it is crucial to continuously evaluate each step of your migration plan: *Is this a step where a robust rollback plan is necessary in case unexpected issues arise?* If the answer is yes, ensure that a solid rollback plan is integrated into your strategy. This includes taking backups and maintaining an archive of data as needed. Adopting this mindset instills greater confidence in your migration plan. This principle is not specific to DynamoDB; it applies to any database migration.

Leverage AWS for assistance

While the availability of assistance varies based on factors like your relationship with AWS, they often provide both technical and commercial support to significantly aid your migration efforts. This support may or may not include migration credits, but it does accelerate the migration process. It can encompass best practices, guidance, hands-on assistance, or even credits and discounts, particularly if transitioning from an on-premises environment or another cloud provider. The **AWS MAP** (4) is one such initiative that has facilitated many migrations to AWS. As part of MAP, AWS can also connect you with local partners and teams, which can be valuable for planning and executing your migration. Engaging with this support early in your project can help avoid potential issues and optimize resource use.

> **Important note**
> Consider this as a reminder to delete any resources you may have created while going through this chapter in your AWS account.

Summary

In this chapter, we delved into the intricacies of database migrations, with a particular focus on migrating to DynamoDB. We began by exploring the process of qualifying a migration opportunity within your organization. We emphasized the importance of data-driven decision-making, highlighting the key data points required to evaluate the need for a database migration. These metrics encompass not only database-level statistics but also the overall end-user experience. The mantra here is *Don't try to fix what isn't broken.*

Next, we shifted our attention to the process of determining whether DynamoDB is a suitable destination for your database migration. We examined DynamoDB's strengths and assessed how well they align with the application's needs and objectives. It's crucial not to compromise your application's goals merely to fit DynamoDB. Therefore, conducting thorough due diligence is essential when considering DynamoDB as your data's new home.

With DynamoDB as our target database, we explored various migration scenarios, including online and offline migrations, and heterogeneous and homogeneous migrations. Each combination comes with its unique considerations. We discussed the building blocks such as bulk copy, ongoing replication, dual writes, and native features that can be leveraged to formulate your migration plan. By the end of this chapter, we have a clear understanding of how different migration plans should be structured based on specific scenarios and requirements.

This marks the end of this chapter (not quite – look ahead for the cheat sheets), and the book as well! Hopefully, the learnings and guidance from this book help you excel in your role.

Cheat sheets

Here are cheat sheets that you can use to grasp the various migration plans and share them with your team. These succinct references will aid in understanding the different migration strategies:

Phases of Dual Write Approach				
Phase	Authoritative Read/Write System			
	Old System Read	Old System Write	DynamoDB Read	DynamoDB Write
Update application to dual write	Yes	Yes		Yes
Bulk copy	Yes	Yes		Yes
Switch authoritative reads		Yes	Yes	Yes
Complete migration			Yes	Yes

Table 16.1 – Phases of dual write approach cheat sheet

Homogeneous Migrations				
	Bulk Copy			Ongoing Replication
Migration Scenario	**Export to S3** + **Import from S3**	**Export to S3** + **AWS Glue**	**AWS Backup**	**DynamoDB Streams** + **AWS Lambda**
Online Transformations		Yes		Yes
Online No Transformations	Option		Option	Yes
Offline Transformations		Yes		
Offline No Transformations	Option		Option	

Table 16.2 – Homogeneous migrations high-level plan cheat sheet

We also reviewed the different native data movement features that can help with homogeneous migrations. To learn more about the native DynamoDB features around data movement like backup, restore, and others, see *Chapter 11*.

Heterogeneous Migrations				
Migration Scenario	**Dual Writes**	**AWS Database Migration Service (DMS)**	**Change Data Capture (CDC)**	**AWS Glue**
Online Transformations	Replication	Bulk Copy		
Online Transformations		Bulk Copy + Replication		

Online Transformations		Bulk Copy	Replication	
Online Transformations			Replication	Bulk Copy
Offline Transformations		Bulk Copy		
Offline Transformations				Bulk Copy

Table 16.3 – Heterogeneous migrations high-level plan cheat sheet

References

1. Postgres pg_stat_statements: `https://www.postgresql.org/docs/current/pgstatstatements.html`

2. AWR reports: `https://docs.oracle.com/en-us/iaas/performance-hub/doc/download-awr-report.html`

3. MongoDB Performance: `https://www.mongodb.com/docs/manual/administration/analyzing-mongodb-performance/`

4. AWS Migration Acceleration Program: `https://aws.amazon.com/migration-acceleration-program/`

5. Connection types and options for ETL in AWS Glue for Spark: `https://docs.aws.amazon.com/glue/latest/dg/aws-glue-programming-etl-connect.html`

6. AWS Database Migration Service: `https://docs.aws.amazon.com/dms/latest/userguide/CHAP_Source.html`

7. Creating tasks for ongoing replication using AWS DMS: `https://docs.aws.amazon.com/dms/latest/userguide/CHAP_Task.CDC.html`

8. Tombstone (data store): `https://en.wikipedia.org/wiki/Tombstone_(data_store)`

9. Conditional writes: `https://docs.aws.amazon.com/amazondynamodb/latest/developerguide/WorkingWithItems.html#WorkingWithItems.ConditionalUpdate`

10. AWS Backup for DynamoDB: `https://docs.aws.amazon.com/amazondynamodb/latest/developerguide/backuprestore_HowItWorksAWS.html`

11. GitHub Repo: `https://github.com/PacktPublishing/Amazon-DynamoDB---The-Definitive-Guide`

Index

‹packt›

packtpub.com

Subscribe to our online digital library for full access to over 7,000 books and videos, as well as industry leading tools to help you plan your personal development and advance your career. For more information, please visit our website.

Why subscribe?

- Spend less time learning and more time coding with practical eBooks and Videos from over 4,000 industry professionals

- Improve your learning with Skill Plans built especially for you

- Get a free eBook or video every month

- Fully searchable for easy access to vital information

- Copy and paste, print, and bookmark content

Did you know that Packt offers eBook versions of every book published, with PDF and ePub files available? You can upgrade to the eBook version at packtpub.com and as a print book customer, you are entitled to a discount on the eBook copy. Get in touch with us at customercare@packtpub.com for more details.

At www.packtpub.com, you can also read a collection of free technical articles, sign up for a range of free newsletters, and receive exclusive discounts and offers on Packt books and eBooks.

Other Books You May Enjoy

If you enjoyed this book, you may be interested in these other books by Packt:

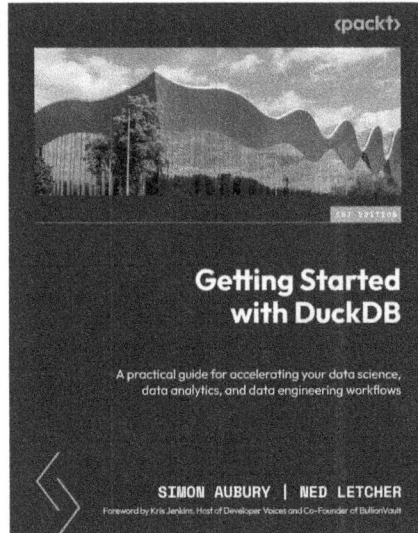

Getting Started with DuckDB

Simon Aubury, Ned Letcher

ISBN: 978-1-80324-100-5

- Understand the properties and applications of a columnar in-process database
- Use SQL to load, transform, and query a range of data formats
- Discover DuckDB's rich extensions and learn how to apply them
- Use nested data types to model semi-structured data and extract and model JSON data
- Integrate DuckDB into your Python and R analytical workflows
- Effectively leverage DuckDB's convenient SQL enhancements
- Explore the wider ecosystem and pathways for building DuckDB-powered data applications

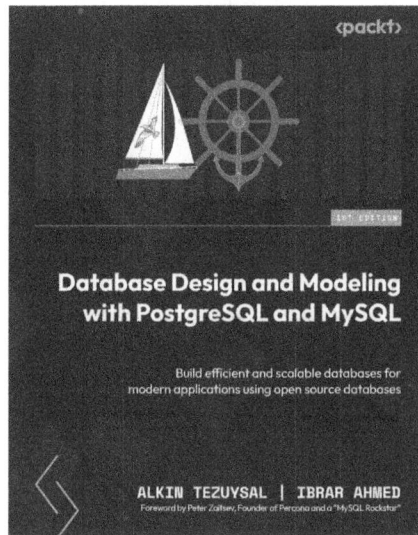

Database Design and Modeling with PostgreSQL and MySQL

Alkin Tezuysal, Ibrar Ahmed

ISBN: 978-1-80323-347-5

- Design a schema, create ERDs, and apply normalization techniques
- Gain knowledge of installing, configuring, and managing MySQL and PostgreSQL
- Explore topics such as denormalization, index optimization, transaction management, and concurrency control
- Scale databases with sharding, replication, and load balancing, as well as implement backup and recovery strategies
- Integrate databases with web apps, use SQL, and implement best practices
- Explore emerging trends, including NoSQL databases and cloud databases, while understanding the impact of AI and ML

Packt is searching for authors like you

If you're interested in becoming an author for Packt, please visit `authors.packtpub.com` and apply today. We have worked with thousands of developers and tech professionals, just like you, to help them share their insight with the global tech community. You can make a general application, apply for a specific hot topic that we are recruiting an author for, or submit your own idea.

Share Your Thoughts

Now you've finished *Amazon DynamoDB – The Definitive Guide*, we'd love to hear your thoughts! Scan the QR code below to go straight to the Amazon review page for this book and share your feedback or leave a review on the site that you purchased it from.

`https://packt.link/r/1-803-24689-8`

Your review is important to us and the tech community and will help us make sure we're delivering excellent quality content.

Download a free PDF copy of this book

Thanks for purchasing this book!

Do you like to read on the go but are unable to carry your print books everywhere?

Is your eBook purchase not compatible with the device of your choice?

Don't worry, now with every Packt book you get a DRM-free PDF version of that book at no cost.

Read anywhere, any place, on any device. Search, copy, and paste code from your favorite technical books directly into your application.

The perks don't stop there, you can get exclusive access to discounts, newsletters, and great free content in your inbox daily

Follow these simple steps to get the benefits:

1. Scan the QR code or visit the link below

https://packt.link/free-ebook/978-1-80324-689-5

2. Submit your proof of purchase
3. That's it! We'll send your free PDF and other benefits to your email directly

www.ingramcontent.com/pod-product-compliance
Lightning Source LLC
Chambersburg PA
CBHW081040220326
41598CB00038B/6939